Advances in Diagnostic and Therapeutic Ultrasound Imaging

Artech House Series
Bioinformatics & Biomedical Imaging

Series Editors
Stephen T. C. Wong, The Methodist Hospital Research Institute
Guang-Zhong Yang, Imperial College

Advances in Diagnostic and Therapeutic Ultrasound Imaging, Jasjit S. Suri, Chirinjeev Kathuria, Ruey-Feng Chang, Filippo Molinari, and Aaron Fenster, editors

Biological Database Modeling, Jake Chen and Amandeep S. Sidhu, editors

Biomedical Informatics in Translational Research, Hai Hu, Michael Liebman, and Richard Mural

Genome Sequencing Technology and Algorithms, Sun Kim, Haixu Tang, and Elaine R. Mardis, editors

Life Science Automation Fundamentals and Applications, Mingjun Zhang, Bradley Nelson, and Robin Felder, editors

Microscopic Image Analysis for Life Science Applications, Jens Rittscher, Stephen T. C. Wong, and Raghu Machiraju, editors

Next Generation Artificial Vision Systems: Reverse Engineering the Human Visual System, Maria Petrou and Anil Bharath, editors

Systems Bioinformatics: An Engineering Case-Based Approach, Gil Alterovitz and Marco F. Ramoni, editors

Advances in Diagnostic and Therapeutic Ultrasound Imaging

Jasjit S. Suri
Chirinjeev Kathuria
Ruey-Feng Chang
Filippo Molinari
Aaron Fenster

Editors

ARTECH HOUSE

BOSTON | LONDON
artechhouse.com

Library of Congress Cataloging-in-Publication Data
A catalog record for this book is available from the U.S. Library of Congress.

British Library Cataloguing in Publication Data
A catalogue record for this book is available from the British Library.

ISBN-13: 978-1-59693-144-2

Cover design by Igor Valdman

© 2008 ARTECH HOUSE, INC.
685 Canton Street
Norwood, MA 02062

DISCLAIMER OF WARRANTY

10 9 8 7 6 5 4 3 2 1

Contents

CHAPTER 2

Despeckle Filtering in Ultrasound Imaging of the Carotid Artery   37

Preface

The use of ultrasound imaging in medicine has undergone an enormous increase in the last 10 years, due to strong technological improvements and to the availability of more accurate and effective strategies for the representation of the ultrasound informative content. During the last decade, the 3-D visualization technique was introduced into clinical practice; new contrast agents were developed to improve the ultrasound diagnosis of heart, liver, brain, and vessel diseases; and highly focused ultrasound beams were used to perform tissue ablation.

This book covers the most recent advances in clinical and therapeutic ultrasound imaging in medicine. It consists of six parts, each dedicated to important clinical applications that represent the state of the art in the fields of ultrasound-based medical diagnosis or therapy.

Part I covers the recent advances in ultrasound instrumentation. Chapter 1 reviews the development of 3-D ultrasound devices and visualization strategies adopted in the analysis of complex anatomical structures and in monitoring and guiding interventional procedures. Chapter 2 deals with the most innovative filtering techniques to cope with speckle noise.

Part II is devoted to vascular applications. Chapter 3 reviews the 3-D application to carotid artery analysis. Chapter 4 proposes the quality evaluation of carotid ultrasound images. Chapter 5 describes the most modern techniques to perform carotid wall segmentation.

The classification of breast lesions is the topic covered by Part III. Chapter 6 shows the 3-D classification of breast lesions. Chapter 7 deals with strain images and elasticity measure. Chapter 8 focuses on the breast lesion classification by means of a support vector machine.

Ultrasound imaging has played a fundamental role in cardiology since the early development of the first duplex devices. In Part IV, Chapter 9 shows the new diagnostic possibilities offered by the 4-D ultrasound devices. Chapter 10 illustrates innovative contributions in the assessment of systolic dysfunction and heart failure. Chapter 11 introduces innovative diagnostic methodologies based on phonocardiography.

In Part V, Chapter 12 provides a thorough description of ultrasound imaging for prostate cancer detection and shows how recent techniques are capable of creating extremely accurate 3-D maps.

Part VI is devoted to ultrasound therapy. Chapter 13 illustrates the techniques to perform treatment planning by ultrasound imaging and simulation. Chapter 14 deals with kidney and liver applications of focused ultrasounds. Chapter 15 describes the recent advances in ultrasound thermal ablation.

On the technical side, this book covers the aspects of 2-D and 3-D image segmentation by deformable geometric and parametric models, of feature analysis and classification by using intelligent techniques, and of denoising. Registration, realignment, and normalization are specific techniques presented in the chapter relative to therapy.

This book is the product of a relatively large international team of ultrasound experts in several fields, ranging from bioengineering to medicine, from clinical practice to technological development, and from advanced diagnosis to the further frontiers of therapy.

Jasjit S. Suri
Chirinjeev Kathuria
Ruey-Feng Chang
Filippo Molinari
Aaron Fenster
Editors
June 2008

PART I

Recent Advances in Ultrasound Instrumentation

CHAPTER 1

3-D Ultrasound Imaging

Aaron Fenster, Zhouping Wei, and Donal Downey

1.1 Introduction

Since the discovery of x-rays, two-dimensional (2-D) radiographic imaging has been the most widely used approach for imaging anatomic structures, diagnosing disease, and guiding interventions. Unfortunately, 2-D x-ray imaging cannot provide physicians with a complete set of information for imaging the human body. The introduction of computed tomography (CT) in the early 1970s revolutionized diagnostic radiology by providing physicians with images of real three-dimensional (3-D) anatomic structures, reconstructed from contiguous tomographic 2-D images of the human body. However, the development of multislice CT imaging, cone-beam CT imaging, and 3-D magnetic resonance imaging (MRI) has provided a major impetus to the field of 3-D visualization, which has, in turn, led to the development of a wide variety of applications in diagnostic medicine.

3-D visualization techniques of the interior of the human body have increased the accuracy of diagnoses, the complexity of interventions, and the safety of interventions to patients. In the past three decades, developments of 3-D CT, MRI, and computed rotational angiography have accelerated and improved the use of 3-D visualization in many interventional procedures. Because new minimally invasive 3-D image-guided techniques provide important benefits to patients (i.e., less pain, less trauma, faster recovery), health care providers (i.e., faster and safer procedures, better outcomes), taxpayers (i.e., shorter, less expensive hospital stays), employers (i.e., less absenteeism), and hospitals (i.e., improved outcomes with shorter lengths of stay), they are rapidly replacing traditional practices.

Although 3-D CT, MRI, and x-ray techniques are important imaging modalities, 3-D ultrasound (US) has been shown to play a major role in medical diagnostics and minimally invasive image-guided interventions [1, 2]. Three-dimensional US imaging has provided high-definition images of complex anatomic structures and pathology to diagnose disease and to monitor and guide interventional procedures. Researchers have begun to incorporate 3-D visualizations into ultrasound instrumentation as well as integrate 3-D ultrasound into biopsy and therapy procedures [3–5] for daily use.

This chapter reviews the development of 3-D US and describes our application of 3-D US for image-guided prostate brachytherapy.

1.2   Disadvantages of 2-D Ultrasound

Two-dimensional US imaging is an inexpensive, compact, and highly flexible imaging modality that allows users to manipulate a transducer in order to view various anatomic structures of the human body; however, 2-D US suffers from the following disadvantages, which 3-D ultrasound imaging attempts to correct:

- 2-D US imaging requires that users mentally integrate many images to form an impression of the anatomy and pathology in 3-D.
- 2-D US imaging is controlled manually; therefore, it is difficult to relocate anatomic positions and orientations when imaging a patient. Monitoring the progression and regression of pathology in response to therapy requires a physician to reposition the transducer at a particular location.
- 2-D US imaging does not permit the viewing of planes parallel to the skin. Diagnostic and interventional procedures sometimes require an arbitrary selection of the image plane.
- 2-D US imaging for measurements of organ or lesion volume is variable and at times inaccurate. Diagnostic procedures and therapy/surgery planning often require accurate volume delineation and measurements.

During the past two decades, many researchers have attempted to develop efficient 3-D US imaging techniques [6–12]. However, due to the enormous computational requirements of acquiring, reconstructing, and viewing 3-D US information on low-cost systems (in real time or near real time), progress was slow. In the past decade, advances in low-cost computer technology and visualization techniques have made 3-D US imaging a viable technology. Several research laboratories have demonstrated the feasibility and utility of 3-D US imaging; recently, several companies have begun to provide 3-D visualization features on their US scanners.

The remainder of this chapter summarizes the various approaches used in the development of 3-D US imaging systems. Particular emphasis is placed on the geometric accuracy of viewing and measuring anatomic structures in 3-D as well as the use of this technology in interventional applications.

1.3   3-D Ultrasound Scanning Techniques

Although 2-D arrays are available for real-time 3-D imaging, most low-cost 3-D US imaging systems use conventional one-dimensional (1-D) US transducers to acquire a series of 2-D US images. These images are reconstructed to form a 3-D image. A wide variety of approaches have been developed to determine the position and orientation of the 2-D images within the 3-D image volume. The production of a 3-D US image volume, without any distortions, requires that three factors be optimized:

- The scanning technique must be either rapid (e.g., real time) or gated to avoid image artifacts due to involuntary, respiratory, or cardiac motion.

- The locations and orientations of the acquired 2-D US images must be accurately known to avoid geometric distortions in the 3-D US image, which could lead to errors of measurement and guidance.
- The scanning apparatus must be simple and convenient to use, such that the scanning process can be easily added to an examination or interventional procedure.

During the past two decades, four 3-D US imaging approaches have been developed and used: mechanical scanning; free-hand scanning with position sensing; free-hand scanning without position sensing; and 2-D array scanning for dynamic 3-D ultrasound, or four-dimensional (4-D) US. These four approaches are discussed in the next four sections. Other approaches are also discussed in later sections.

1.4 Mechanical Scanning

Mechanical scanners use a motorized mechanism to translate, tilt, or rotate a conventional 2-D US transducer while a series of 2-D US images is rapidly acquired by a computer. Because the scanning geometry is predefined and precisely controlled, the relative position and orientation of each 2-D US image is known accurately.

Using novel computational algorithms, the US system either reconstructs the set of acquired 2-D US images in real time or stores the images in its memory; then, using predefined geometric parameters that describe the orientation and position of the set of acquired 2-D US images within the 3-D image volume, the US system (or an external computer) reconstructs and displays the 3-D US image. The angular or spatial interval between each 2-D image is adjustable so that each interval can be optimized to minimize the scanning time while adequately sampling the volume [13].

Various types of scanning mechanisms have been developed in order to translate or rotate the 2-D US transducer in the required manner. From integrated 3-D US probes that house the scanning mechanism within the transducer housing, to external fixtures that mechanically hold the housing of a conventional 2-D US probe, mechanical scanners vary in size.

Integrated 3-D US probes are typically larger and easier to use than 3-D US systems that use external fixtures with conventional 2-D US probes. However, integrated 3-D probes require a special US machine that can interface with them. External mechanical 3-D scanning fixtures are generally bulkier than integrated probes; however, external fixtures can be adapted to hold any conventional US transducer, obviating the need to purchase a special 3-D US machine. Improvements in image quality (e.g., image compounding) and flow information (e.g., Doppler imaging) offered by any conventional US machine can, therefore, be achieved in three dimensions.

3-D mechanical scanning offers these three advantages: short imaging times, high-quality 3-D images, and fast reconstruction times. However, its bulkiness and weight sometimes make it inconvenient to use. As shown in Figure 1.1, three basic types of mechanical scanners have been developed and used: linear scanners, tilt scanners, and rotational scanners.

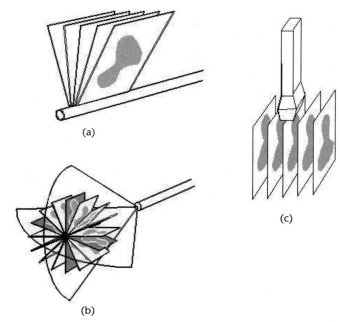

Figure 1.1   Schematic diagram of 3-D ultrasound acquisition methods. (a) Side-firing transrectal ultrasound transducer being mechanically rotated for a 3-D US scan. The acquired images have equal angular spacing. (b) Rotational scanning mechanism used to acquire 3-D US images. The acquired images have equal angular spacing. (c) Linear mechanical scanning mechanism. The acquired images have equal spacing.

1.4.1   Linear Mechanical 3-D Scanners

Linear scanners use a motorized drive mechanism to translate the transducer across the skin of the patient. As the transducer is moved, 2-D images are acquired at regular spatial intervals so that they are parallel and uniformly spaced. The translating speed and sampling interval can be varied to match the sampling rate of the frame rate for the US machine in order to match the sampling interval (half) of the elevational resolution of the transducer [13].

The predefined geometry of the acquired 2-D US images allows a 3-D image to be reconstructed while a set of 2-D US images is acquired. Because the 3-D US image is produced from a series of conventional 2-D US images, its resolution will not be isotropic. In the direction parallel to the acquired 2-D US images, the resolution of the restructured 3-D image will be equal to the original 2-D US images; however, in the direction perpendicular to the acquired 2-D US images, the resolution of the restructured 3-D image will be equal to the elevational resolution of the transducer. Because the resolution of the 3-D US image is poorest in the 3-D scanning direction, a transducer with reasonable elevational resolution should be used for optimal results [14].

Linear scanning has been successfully implemented in many vascular B-mode and Doppler imaging applications, particularly for carotid arteries [15–24] and tumor vascularization [20, 25–27]. Figure 1.2 shows two examples of linearly scanned 3-D US images made with an external fixture.

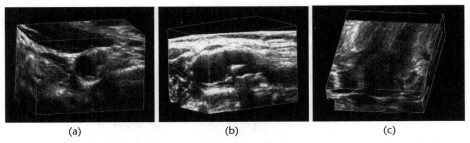

(a)   (b)   (c)

Figure 1.2   An example of a 3-D ultrasound image of the carotid arteries obtained with the mechanical 3-D scanning approach. The 3-D US image has been sliced to reveal the details of the atherosclerotic plaque in the carotid arteries. (a) Transverse view, (b) longitudinal view, and (c) view in which the 3-D US image has been sliced parallel to the skin to produce a view not possible with conventional US techniques.

1.4.2   Tilt 3-D Mechanical Scanners

The tilt scanner approach uses a motorized drive mechanism to tilt a conventional US transducer about an axis parallel to the face of the transducer while a fan of images radial to the tilting axis is acquired (at 0.5° to 1.0° intervals). An integrated 3-D US probe or an external fixture produces the tilting motion. In each case, the housing of the probe remains fixed on the skin of the patient while the US transducer is angulated. Using a predefined but adjustable angular separation between the acquired images to yield the desired quality for the image, users can adjust the range of the angular tilt to sweep a large region of interest [26, 28–32]. Special 3-D US systems with integrated 3-D US probes well suited for abdominal and obstetrical imaging have been successfully demonstrated by several companies [33–37].

The predefined geometry of the acquired 2-D US images allows for real-time 3-D image reconstruction; however, the resolution will not be isotropic. The resolution will degrade as the distance from the axis of rotation is increased. Because the geometry of the acquired 2-D images is fan-like, the distance between the acquired US image plane and the axis of the transducer increases (with increasing depth). This increase results in a decrease in the spatial sampling and spatial resolution of the reconstructed 3-D image. As the distance from the axis increases, the spreading of the beam in the elevational direction and in the acquired image plane produces a degradation of resolution [38].

1.4.3   Endocavity Rotational 3-D Scanners

The rotational scanning approach uses an external fixture to rotate an endocavity probe, for example, a transrectal ultrasound (TRUS) probe, about its long axis. A set of acquired 2-D images is produced parallel to the axis of a side-firing linear transducer array to form a fan of images radial to the rotational axis. For 3-D TRUS imaging of the prostate, the probe is typically rotated from 80° to 110° [39, 40]. As illustrated in Figure 1.3, endocavity scanning has been successfully used to image the prostate [15, 16, 20, 39, 41] and guide 3-D ultrasound cryosurgery [5, 42–46].

Several modifications to this scanning approach have been developed. While a set of 2-D images is acquired, a motorized mechanism is used to rotate an end-firing

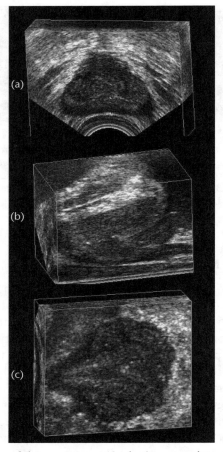

Figure 1.3 A 3-D US image of the prostate acquired using an endocavity rotational 3-D scanning approach (rotation of a TRUS transducer). The transducer was rotated around its long axis while 3-D US images were acquired and reconstructed. The 3-D US image is displayed using the cube view approach and has been sliced to reveal: (a) a transverse view, (b) a sagittal view, and (c) a coronal view, not possible using conventional 2-D US techniques.

endocavity transducer array about a fixed axis that perpendicularly bisects the transducer array by at least 180°. This scanning geometry acquires images by sweeping out a conical volume about the rotational axis of the transducer. However, the resolution of the 3-D image will not be isotropic. Because the spatial sampling is highest near the axis of the transducer and lowest away from the axis of the transducer, the resolution of the 3-D image will degrade as the distance from the rotational axis of the transducer is increased. Similarly, the axial and elevational resolution will decrease as the distance from the transducer is increased. The combination of these effects will cause the 3-D image resolution to vary—highest near the transducer and the rotational axis, and lowest away from the transducer and rotational axis.

The 3-D rotational scanning is sensitive to the motion of the operator and to the motion of the patient. Because the acquired 2-D images intersect along the rotational axis of the transducer, any motion during the scan will cause a mismatch in the acquired planes, resulting in the production of artifacts in the 3-D image. Artifacts will also occur if the axis of rotation is not accurately known; however, proper calibrations can remove this source of potential error. Although handheld 3-D rotational scanning of the prostate and uterus can produce excellent 3-D images (see Figure 1.3), for optimal results in long procedures, such as prostate brachytherapy and cryotherapy, the transducer and its assembly should be mounted onto a fixture. Rotational scanning techniques have successfully been used for transrectal imaging of the prostate and endovaginal imaging of the ovaries [20, 39–41].

1.5   Free-Hand Scanning with Position Sensing

To overcome problems of size and weight, researchers have developed devices that do not require an internal or external motorized fixture. To measure its position and orientation, a sensor is mounted on the transducer [47]. As its position and orientation are recorded, the transducer is manipulated over the anatomy of the body to reconstruct the 3-D image. Because the relative locations for each of the 2-D images are not predefined, an operator must move the transducer over the anatomy at an appropriate speed to ensure that the spatial sampling is proper and to ensure that the set of 2-D images does not have any significant gaps. Researchers have developed several free-hand scanning approaches: tracked 3-D US with articulated arms, free-hand 3-D US with acoustic sensing, free-hand 3-D US with magnetic field sensing, and image-based sensing (speckle decorrelation).

1.5.1   Tracked 3-D US with Articulated Arm Sensing

Position and orientation sensing is achieved by mounting the ultrasound transducer onto a mechanical arm system that has multiple joints. Potentiometers, which are located at the joints of the movable arms, provide the information necessary to calculate the relative position and orientation of the acquired 2-D images. This arrangement allows the operator to manipulate the transducer while the computer records the set of 2-D images and the relative angles of all of the arms. Reasonable 3-D image quality is achieved by keeping the number of movable joints small and the lengths of each arm short so that the arms do not flex. This approach is often used in external beam radiotherapy (before treatment fraction) to localize the prostate with a transabdominal transducer [48].

1.5.2   Free-Hand 3-D US with Acoustic Tracking

Developed in the 1980s, acoustic tracking was one of the first methods used to produce 3-D images [49–55], however, it has not been used in recent years. This method uses an array of three sound-emitting devices mounted on a transducer (e.g., spark gaps) and an array of fixed microphones mounted above the patient to receive sound pulses continuously from the emitters while the transducer is manipulated. The position and orientation of the transducer for each of the acquired 2-D images are calculated using the speed of sound (in air), the measured times of flight from each emitter to each microphone, and the fixed locations of the microphones.

1.5.3   Free-Hand 3-D Scanning with Magnetic Field Sensing

Free-hand 3-D scanning with magnetic field sensors has been used successfully in many diagnostic applications, including echocardiography, obstetrics, and vascular imaging [16, 56–68]. As a result, this imaging technique has become the most successful and the most popular free-hand 3-D scanning approach. A transmitter placed near the patient is used to produce a spatially varying magnetic field, and a small receiver containing three orthogonal coils (with six degrees of freedom) is mounted on the probe and used to sense the strength of the magnetic field. By measuring the strength of the three components of the local magnetic field, the position and orientation of the transducer are calculated as each 2-D image is acquired.

The magnetic field sensors are small and unobtrusive devices that allow the transducer to be tracked with fewer constraints than any other free-hand 3-D scanning approach. However, electromagnetic interference can compromise the accuracy of the tracking, and geometric distortions can occur in the final 3-D image if ferrous (or highly conductive) metals are located nearby. Modern magnetic field sensors produce excellent images and are now less susceptible to sources of error; however, the position of the sensors relative to the ultrasound image must be calibrated accurately and precisely. Numerous calibration techniques have been developed to minimize these potential sources of error [69–76].

1.5.4   3-D US Tracked by Speckle Decorrelation

Mechanical scanners, acoustic tracking scanners, and magnetic field sensing scanners all require sensors to track the position and orientation of an ultrasound transducer; however, researchers have developed a tracking technique that uses speckle decorrelation to determine the relative positions of the acquired US images [77]. When a source of coherent energy interacts with a scatterer, the pattern of the reflected spatial energy will vary, due to interference, and appear speckled in an image. Ultrasound images are characterized by image speckle, which can be used to create images of moving blood [78], images of moving tissue scatterers, and images of stationary tissue scatterers. If the tissue scatterers remain stationary, then two sequential images generated by the scatterers will be correlated and the speckle pattern will be identical. However, if one of the images is moved with respect to the first, then the degree of decorrelation will be proportional to the distance moved; the exact relationship will depend on the beam width in the direction of the motion [79, 80].

Speckle decorrelation techniques can be complicated by the fact that a pair of adjacent 2-D US images may not be parallel to each other. Thus, the rotational (or tilt) angle of one image with respect to the second image must be determined by subdividing each of the images into smaller cross-correlated images. As a result, a pattern of decorrelation values, which is used to determine a pattern of distance vectors, between the two images is generated. These vectors are then analyzed to determine the relative position and orientation of the two 2-D US images.

1.6   Free-Hand 3-D US Scanning Without Position Sensing

The 3-D US images can be produced without position sensing. By assuming a prede-fined scanning geometry, users manipulate the transducer over the patient while a set of 2-D US images is acquired and reconstructed to form a 3-D image. Because no geometrical information is recorded during the motion of the transducer, the opera-tor must move the transducer at a constant linear or angular velocity so that each of the 2-D images is obtained at regular intervals [15]. However, because this approach does not guarantee that the 3-D US image is geometrically accurate, it cannot be used for measurements.

1.7   2-D Array Scanning for Dynamic 3-D Ultrasound (4-D US)

Mechanic and free-hand scanning techniques produce a set of 2-D US images by using a conventional 1-D array transducer. Because a set of 2-D US images is used to reconstruct the 3-D US image, the transducer must be moved so that the set of acquired 2-D US images covers the region of interest. However, a 2-D array allows the transducer to remain stationary; an electronic scanner is then used to sweep the ultrasound beam over the entire volume [81–88], allowing for the acquisition of a set of 3-D images to occur in real time (i.e., 4-D US imaging).

In this approach, a 2-D phased array of transducer elements is used to transmit a broadly diverging beam of ultrasound away from the array, sweeping out a vol-ume shaped like a truncated pyramid. The returned echoes detected by the 2-D array are processed to display a set of multiple planes in real time. Using various image rendering techniques, users can interactively control and manipulate these planes to explore the entire volume. This approach is successfully used in echocardiology [37, 89–91].

1.8   3-D Ultrasound Visualization

Many algorithms have been developed to help physicians and researchers visualize and manipulate 3-D medical images interactively. Because US images suffer from image speckle and poor tissue–tissue contrast, the display of a 3-D US image plays a dominant role in the ability of a physician to obtain valuable information. Although many 3-D US display techniques have been developed and used, two of the most fre-quently used techniques are multiplanar reformatting (MPR) and volume rendering (VR) [92–94].

1.8.1   Multiplanar Reformatting

The MPR technique reformats a 3-D US image to display a set of 2-D US planes. To view the anatomy in three dimensions, users interact with a utility to extract a desired plane from the 3-D image. The US planes are then displayed with 3-D cues.

Three MPR approaches are commonly used to display 3-D US images. Figure 1.4 illustrates the *crossed-planes approach*. Single or multiple planes are presented

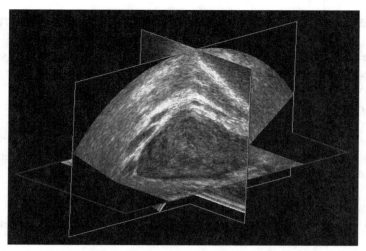

Figure 1.4 The 3-D US image of the prostate shown in Figure 1.3, but displayed using the crossed-planes approach, in which the extracted planes are intersecting with each other.

in a view that shows their correct relative orientations. As seen in Figure 1.4, these planes typically intersect each other. Users can then select and move each plane (parallel or obliquely) to any other plane to reveal the desired views. Figure 1.5 illustrates the *cube-view approach*. The extracted set of 2-D US images is texture mapped onto the faces of a polyhedron. Users can select any face of the polyhedron and move it (parallel or obliquely) to any other plane. The appropriate set of 2-D US images is then extracted in real time and texture mapped onto the new face. The appearance of a "solid" polyhedron provides users with 3-D image-based cues, which relates the manipulated plane to the other planes [7, 95–97]. In a third approach, three orthogonally *extracted planes* are displayed together with 3-D cues, such as lines on each extracted plane to designate its intersection with the other planes. These lines can be moved in order to extract and display the desired planes [5, 93, 98].

1.8.2 Volume Rendering Techniques

VR techniques are frequently used to view 3-D CT and MRI images that have been projected onto a 2-D plane. This approach is extensively used to view 3-D US fetal images [36, 90, 91] and 4-D US cardiac images [35, 37, 99]. VR software uses ray-casting techniques to project a 2-D array of lines (rays) through a 3-D image [100–105]. The volume elements (voxels) intersecting each ray are weighted, summed, and colored in a number of ways to produce various effects. Three VR techniques are commonly used to view 3-D and 4-D US images: maximum intensity projection, translucency rendering, and surface enhancement.

Because they project all of the information onto a 2-D plane, many VR techniques are not well suited for viewing the details of soft tissues in 3-D B-mode ultrasound. Instead, VR techniques are best suited for viewing anatomic surfaces that are distinguishable in 3-D B-mode US images, including limbs and fetal face surround by amniotic fluid [93, 98], tissue–blood interfaces such as endocardial surfaces and

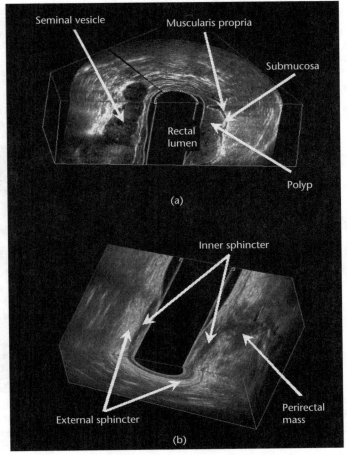

Figure 1.5   The 3-D US rectal images obtained by the mechanical rotational mechanism, which rotates a side-firing transrectal transducer around its long axis by 180°. The 3-D US images are viewed using the MPR formatting approach. (a) A 3-D US image showing a stage T1 rectal polyp invading the submucosa. (b) A 3-D US image showing histological confirmed recurrent prostatic cancer.

inner vascular surfaces (see Figure 1.6), and structures where B-mode clutter has been removed from power or color Doppler 3-D images [15].

1.9   3-D Ultrasound-Guided Prostate Therapy

Continuous improvements in image quality and miniaturization have made 3-D US an indispensable tool for diagnosing and assessing many diseases. However, therapy planning, surgery planning, and intraoperative guidance have developed slowly because of the poor organ contrast and narrow presentations of conventional 2-D US images. Hence, physicians have relied on CT and MRI, which produce images of higher definition, to interpret anatomy for planning therapy and guiding surgery.

The availability of 3-D US imaging and real-time visualization techniques has stimulated the use of 3-D US to applications in interventional techniques.

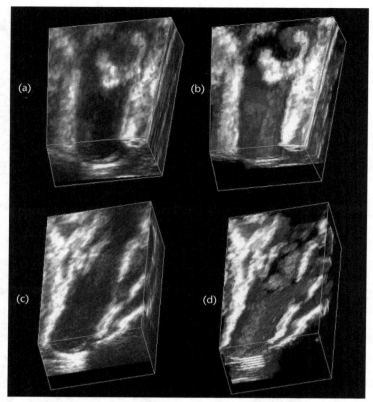

Figure 1.6   Two 3-D ultrasound images of the carotid arteries showing both the MPR and VR approaches. (a) MPR view of a 3-D US image sliced parallel to the skin in the longitudinal direction. (b) The same image as in part (a) but volume rendered, showing the surface of the carotid artery and the details of the plaque. (c, d) MPR and VR views of another 3-D US image of a carotid artery.

1.9.1   Early-Stage Prostate Cancer Management

The wide availability of the prostate-specific antigen (PSA) test and the increased awareness of the public about prostate cancer have dramatically increased the number of men who have been diagnosed with early-stage prostate cancer [106]. For men with early-stage prostate cancer, radical prostatectomy (the surgical removal of the prostate) is often not a viable option due the morbidity (incontinence and impotence) of the procedure. External beam radiotherapy can be effective; however, it requires long treatment times and results in the irradiation of normal tissue [107]. Technological and pharmaceutical advances have led to the development of new treatment options, such as brachytherapy, cryosurgery, hyperthermia, interstitial laser photocoagulation (ILP), and photodynamic therapy (PDT). Of these new treatments, brachytherapy is considered not only the most advanced but also the most important treatment option for early-stage prostate cancer [108–110].

The following section describes a 3-D US-guided prostate brachytherapy application that has been developed in our laboratory.

1.9.2   US-Guided Prostate Brachytherapy

Low-dose-rate (LDR) brachytherapy is presently the most common form of prostate brachytherapy. By placing approximately 80 radioactive seeds (e.g., ^{125}I) into the prostate, the organ can be subjected to high doses of radiation while sparing the surrounding tissue [108–110]. The delivery of a high conformal dose to the prostate requires that the radioactive sources be properly planned and accurately positioned within the prostate [111]. Thus, efficient and accurate 3-D imaging of the prostrate are required to plan and guide the procedure.

The first phase of the procedure involves preimplantation dose planning (preplanning). CT or US images are used to delineate the margins of the prostate and the structures sensitive to radiation. Using the information obtained from CT or US images, as well as the radiation distribution around the seeds, physicians can determined the distribution of the radioactive seeds in the prostate [112]. The preplan is then used to implant the seeds while the patient is positioned in approximately the same position as in the preplan. The implantation of the radioactive seeds is typically carried out using TRUS guidance. After implantation, CT or fluoroscopy is used to determine the actual positions of the seeds, which may be different from the planned positions due to implantation errors or seed migration. The actual seed positions are then used to calculate a postimplantation dose plan. If significant deviations of dose are detected, then additional seeds may be implanted.

Although CT and fluoroscopy are still used to determine the positions of the seeds, 3-D US is beginning to establish itself as an effective modality for planning and guiding brachytherapy procedures. Rapid scanning and viewing of the prostate using 3-D US will allow a complete intraoperative procedure to be performed [38]. 3-D US will permit preimplant dose planning and seed implantation during the same session; therefore, problems of repositioning, prostate motion, prostate size/contour changes, and image registration between imaging modalities can be avoided. 3-D US images will also permit immediate postimplant verification to detect and correct implantation errors immediately. We have been developing an intraoperative 3-D US brachytherapy system that uses software to contour the prostate, segment and track the needles, and detect the seeds for postplanning dosimetry of the prostate.

We have reported on our development of a 3-D US imaging technique for the diagnosis of prostate cancer [39, 113], prostate cryosurgery [4, 5], and prostate brachytherapy [45, 46, 114, 115]. The following section briefly describes our 3-D US-guided prostate brachytherapy approach.

1.9.3   3-D TRUS-Guided Brachytherapy System

The introduction of our 3-D TRUS imaging system has overcome several limitations of the 2-D TRUS guided brachytherapy system by providing: (1) an interactive view of the prostate, (2) a more efficient segmentation of the prostate boundary, and (3) a process to monitor the insertion of the needle during implantation [2, 5, 39, 115–118]. These developments, coupled with several efficient 3-D reconstruction and visualization software tools, have allowed for the development of an intraoperative system capable of dynamic reoptimization and intraoperative postplanning verification.

The following sections describe the integration of robotic aids and implementation of 3-D TRUS in the development of an intraoperative system for prostate brachytherapy.

Our approach uses a one-hole needle guide, which is attached to the arm of a robot so that the trajectory of the needle can be changed in response to the movements of the robot. Once the robot has positioned the needle, the physician inserts the needle into the patient's prostate. The position of the guidance hole can be moved so that the needle can target any point identified in the 3-D TRUS image [45]. In addition, we have developed a near real-time method for not only automatically segmenting the needle and prostate but also for tracking brachytherapy needles during oblique insertion. Tracking allows us to follow the needle as it exits the 2-D TRUS image plane.

A schematic diagram and a photograph of our robot and 3-D TRUS-guided system are shown in Figures 1.7 and 1.8, respectively. The prototype system consists of an A465 industrial robotic system (Thermo-CRS, Burlington, Ontario, Canada) with six degrees of freedom and a 3-D TRUS imaging system developed in our laboratory [45]. The one-hole needle guide is attached to the arm of the robot and a software module transforms the coordinate system of the robot to the coordinate system of the 3-D TRUS. This transformation allows the hole of the needle guide and, hence, the trajectory of the needle to be displayed and tracked with the 3-D TRUS images [45].

The following sections describe our 3-D TRUS acquisition system and its role in prostate segmentation, dosimetry, calibrations, and evaluations of errors using calibration, needle tracking, and seed implantation.

1.9.4 3-D TRUS Imaging System

Our results were obtained using a B-K Medical 2102 Hawk ultrasound machine (B-K, Denmark) with a side-firing 7.5-MHz transducer; however, our 3-D TRUS system can be coupled to any US machine using a side-firing transducer. The side-firing linear array US transducer is mounted on a rotational mover [116]. To generate a 3-D image volume in the shape of a fan, the motorized mover rotates the transducer approximately 100° around its long axis [39]. As the transducer is rotated, the

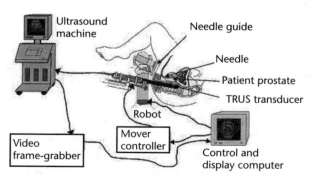

Figure 1.7 Schematic diagram of the robot-aided and 3-D US-guided prostate brachytherapy system. The diagram shows that the TRUS transducer is attached to a rotational 3-D scanning mechanism and is inserted into the rectum below the prostate. The 3-D US scanning mechanism is mounted on a robotic device used to guide the brachytherapy needles into the prostate.

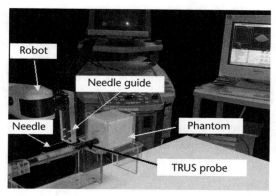

Figure 1.8 A photograph of the robot-aided and 3-D US-guided prostate brachytherapy system, showing the TRUS transducer inserted into an agar phantom used to test the system. (*From:* [45]. © 2004 American Association of Physicists in Medicine. Reprinted with permission.)

2-D US images acquired from the US machine are digitized by a frame grabber at 0.7° intervals at 30 Hz for approximately 9 seconds. The acquired 2-D US images are reconstructed in real time to form a 3-D US image. 3-D visualization software is used to view the 3-D US image [116]. Figure 1.3 earlier in this chapter shows an example of several views of a 3-D TRUS image acquired in this manner.

1.9.5 Prostate Segmentation

Manually outlining prostate margins is a time-consuming and tedious process. Thus, a semi- or fully automated prostate segmentation technique that is accurate, reproducible, and fast is required. Because 3-D US images suffer from shadowing, speckle, and poor contrast, procedures for fully automated segmentation can result in unacceptable errors. Our approach has been to develop a semiautomated prostate segmentation algorithm that allows users to correct errors [119–122]. Because the details of our prostate segmentation technique have been described elsewhere [119–122], this section merely summarizes the details of our technique.

Our approach is to automatically segment the boundary of the prostate into sequential slices of 2-D images. These boundaries are then merged to form the complete shape of the prostate. Our 2-D segmentation technique uses model-based initialization and a discrete dynamic contour (DDC) model to refine the initial boundary of the prostate [123]. Once the prostate has been initially sliced and segmented, the boundary is propagated automatically to the next slice and segmented until the entire prostate has been completed. This process requires three phases: (1) The DDC is initialized by an approximate outline of the object, (2) the initial outline is deformed automatically to fit the desired organ boundary, and (3) the boundary is propagated repeatedly and deformed until the complete prostate has been segmented.

1.9.5.1   Initialization

The initialization phase requires the operator to select four or more points on the boundary of the prostate in a 2-D image slice of the 3-D image. Two points on the midline of the prostate define an axis of symmetry; two points on both sides of the base of the prostate define the width [Figure 1.9(a)]. Additional points may be used if the prostate deviates significantly from an elliptical shape. Using the initialization points, the size, approximate shape, and initial boundary of the prostate can be computed using a precomputed model equation [119] or using a spline technique [see Figure 1.9(a)].

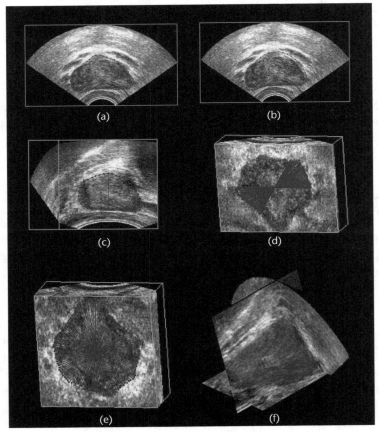

Figure 1.9   Steps showing the 3-D prostate segmentation procedure. (a) The user identifies four points on the prostate boundary in a 2-D section. These points are used to calculate an initial boundary for the segmentation. (b) The initial prostate boundary in the 2-D section is segmented by refining the initial boundary using the DDC approach. (c) The 3-D segmentation is completed by repeatedly propagating the refined boundary in one slice to the next and refining it. (d) View of the prostate sliced perpendicular to the rotational slicing axis showing the boundaries after propagation and refining through about 45°. (e) View of the prostate sliced perpendicular to the rotational slicing axis showing the boundaries after the 3-D segmentation has been completed. (f) The boundaries have been surfaced and displayed with the crossed-planes approach. (*From:* [45]. © 2004 American Association of Physicists in Medicine. Reprinted with permission.)

1.9.5.2   Refinement

In the boundary refinement phase, the initial contour is decomposed into vertices that are separated by approximately 10 to 20 pixels. As seen in (1.1), which is based on simple dynamics, this initial boundary is used in the operation of the DDC. A weighted combination of the internal force ($\grave{}f_i^{\text{int}}$), the image force ($\grave{}f_i^{img}$), and the damping force ($\grave{}f_i^{d}$) is applied to each ith vertex of the DDC, resulting in the total force ($\grave{}f_i^{tot}$). This boundary will evolve until the total force acting on each vertex is zero.

$$\grave{}f_i^{tot} = w_i^{\text{int}}\grave{}f_i^{\text{int}} + w_i^{img}\grave{}f_i^{img} + \grave{}f_i^{d} \tag{1.1}$$

The relative weights for the image force and internal force are w_i^{img} and w_i^{int}, respectively. These forces are vector quantities having both magnitude and direction [124].

The external force at each vertex is proportional to the local image gradient, which causes the vertex to seek regions of high image gradient. These regions are typically located at the boundary of the imaged prostate. The internal force mimics an elastic force that is proportional to the edge length. This force determines the rigidity of the boundary, minimizes the local curvature at each vertex, and keeps the boundary smooth in the presence of image noise. Adjusting the weighting of the internal and external forces can optimize the segmentation of the prostate. We have typically used $w_i^{img} = 1.0$ and $w_i^{int} = 0.3$ for each vertex. Figure 1.9(b) shows the results of a refined boundary superimposed onto a prostate image that demonstrates good segmentation.

1.9.5.3   Propagation

The segmented 2-D boundary of the prostate is extended to form a 3-D image by propagating a contour onto an adjacent image slice. The deformation process is then repeated by radially slicing the prostate at a constant angle (e.g., 3°) that intersects along an axis that is close to the center of the prostate [122] as shown in Figures 1.9 and 1.10.

The accuracy of the segmentation algorithm was tested by comparing its results with the results obtained using manual planimetry. Using the volume of the prostate obtained by manual planimetry as a reference, the errors in the semiautomated approach ranged from an underestimate of 3.5% to an overestimate of 4.1%. The mean error was found to be −1.7% with a standard deviation of 3.1% [122]. Segmentations of the prostate require approximately 5 seconds when implemented on a 1-GHz PC.

1.9.6   Dosimetry

We use AAPM TG-43 formalism, which uses predetermined dosimetry data in dose rate evaluation [125]. After delineating the organs, users select the type of radioactive source and enter its calibration data. The possible area of insertion is outlined, and the preplan is produced using approximately 20 needles, which can be oriented in oblique trajectories to avoid any interference with the pubic arch.

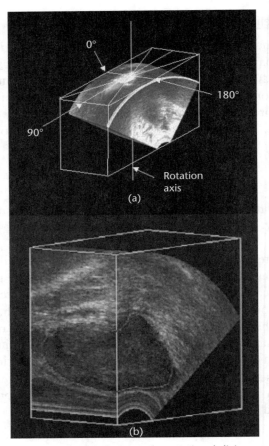

Figure 1.10 (a) A 3-D US prostate image showing the rotational slicing geometry used to segment the prostate in 3-D. (b) A 3-D US image showing the segmented 3-D boundary of the prostate. (*From:* [122]. © 2003 American Association of Physicists in Medicine. Reprinted with permission.)

The isodose curves are displayed on the 3-D TRUS image in real time as well as on a view that is surface rendered, which shows the needles and seeds. Because each needle can be individually activated or deactivated, the modified isodose curves can be observed instantly. Users can evaluate the plan using dose–volume histograms for each organ and make the necessary modifications. Once the needle has been retracted, the actual seed locations are determined or are assumed from the location of the needle, and the new isodose curves are displayed. Once the needle has been retracted, users have the option of modifying the plan according to the real seed locations. Figure 1.11 shows an example of the use of the preplanning software for planning an oblique trajectory of a needle.

1.9.7 Calibration of the Coordinate Systems

The goal of the calibration procedure is to determine the transformation between the coordinate system of the 3-D TRUS image and the coordinate system of the robot. This procedure involves the following two steps: (1) map the coordinate sys-

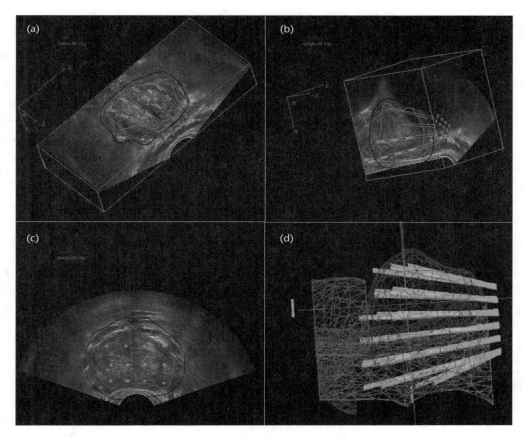

Figure 1.11 The 3-D images from a typical prostate brachytherapy dose plan with oblique needle trajectories for use with 3-D TRUS guidance and robotic aids. Our 3-D visualization approach allows display of a texture-mapped 3-D cube view of the prostate, extracted planes, and graphical overlays of surfaces and contours. (a) Coronal view with delineated organs, needles, seeds and isodose curves, (b) sagittal view, (c) transverse view, and (d) surface rendered view showing the organs and needles with seeds.

tem of the transducer to the 3-D TRUS image (image calibration), and (2) map the coordinate system of the robot to the transducer (robot calibration). The transformation between any two coordinate systems is found by solving the orthogonal Procrustes problem as follows: Given two 3-D sets of points, $K = [k_j]$ and $L = [l_j]$ for $j = 1, 2, ..., N$, and a rigid-body transformation, $\mathbf{F}: l_j = \mathbf{F}(k_j) = \mathbf{R}k_j + \mathbf{T}$, where $\mathbf{R}$ is a 3×3 rotation matrix and $\mathbf{T}$ is a 3×1 translation vector, minimize the cost function as shown:

$$C = \frac{1}{N} \sum_{j=1}^{N} \left\| l_j - \left(R k_j + T \right) \right\|^2 \tag{1.2}$$

A unique solution exists if, and only if, the sets of points K and L contain at least four noncoplanar points [126].

To calibrate the image, we used a phantom that contained three sets of nylon strings with diameters of 1 mm orthogonally positioned in a Plexiglas box. Each set of strings was immersed in an agar gel and placed 1 cm apart. The coordinates of the intersections of the strings were determined in the coordinate system of the US transducer from the design of the phantom. The coordinates of the points in the 3-D TRUS image coordinate system were determined by imaging the intersections of the strings in 3-D. Using the coordinates of the intersections created by the orthogonally positioned strings, we determined the transformation that mapped the coordinate system of the transducer to the 3-D TRUS image by solving (1.2) [45].

To calibrate the robot, we used two orthogonal plates mounted on a device to hold the transducer and drilled five hemispherical divots on each plate to produce homologous points in the coordinate system of the transducer and in the coordinate system of the robot. The coordinates of the divot centers in the coordinate system of the transducer were known from the design of the plate, and the coordinates of the divot centers in the coordinate system of the robot were determined by moving the robot so that it sequentially touched the divots with a stylus tip, which was attached to the arm of the robot. Using the coordinates of the divots in the two coordinate systems, we determined the transformation that mapped the coordinate system of the transducer to the coordinate system of the robot by solving (1.2).

1.9.8 Needle Segmentation

Our approach to robot-assisted brachytherapy allows for the implantation of the needle to be nonparallel, meaning the conventional template does not require a set of rectilinear holes. The needle may, therefore, be inserted in an oblique trajectory, which will result in the image of the needle exiting the real-time 2-D US image. Thus, it is necessary to track the tip of the needle as it is being inserted into the prostate in order to ensure proper placement of the needle and to avoid implanting the seeds at incorrect locations.

Our approach to tracking uses near real-time 3-D US imaging to segment the needle in three dimensions; display the oblique sagittal, coronal, and transverse views; and highlight the trajectory of the needle [46, 117, 118, 127]. Because the needle may be angled approximately 20° from the orientation of the 2-D US plane, and a series of 2-D images may be acquired at 30 images per second, a new 3-D image may be formed in less than 1 second. From these 3-D images, a needle may be automatically segmented [46, 117, 118], and the three planes needed to visualize the insertion of the needle may be displayed immediately. Figure 1.12 shows the results of tracking the insertion of a needle into a prostate.

1.10 Evaluation of 3-D TRUS-Guided Brachytherapy System

Using test phantoms, we evaluated the performance of our robot-aided system and 3-D TRUS-guided brachytherapy system. The following sections discuss the methods and results of calibration as well as the accuracy of needle placement, needle angulation, targeting, tracking, and seed implantation for the 3-D TRUS-guided brachytherapy system.

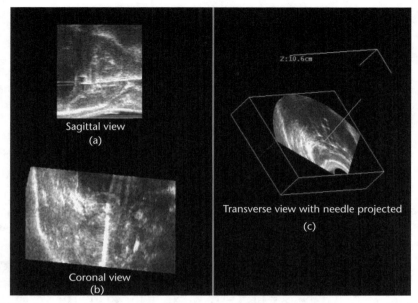

Figure 1.12 Example of the views displayed during oblique needle tracking in a patient's prostate. (a) Oblique sagittal view showing the oblique plane with the needle, and (b) oblique coronal view showing the plane with the needle (this view cannot be obtained with conventional US system). (c) The 3-D TRUS image was sliced in a transverse direction showing the segmented needle trajectory.

1.10.1 Methods of Calibration

Evaluations of the calibrations of the image and robot were performed by analyzing the accuracy of the point-based, rigid-body registration and calculating the fiducial localization error (FLE), fiducial registration error (FRE), and target registration error (TRE) [128].

1.10.1.1 Fiducial Localization Error

The FLE is a statistical measure of the error in locating the fiducial points used in the registration process [129]. We have assumed that the mean value of the error in locating the fiducial points is zero and have calculated the root-mean-square (rms) distance between the exact and calculated fiducial positions using [130]:

$$\text{FLE}^2 = \sigma_x^2 + \sigma_y^2 + \sigma_z^2 \tag{1.3}$$

The variances of the errors in locating the fiducial points along the three orthogonal axes are given by σ_x^2, σ_y^2, and σ_z^2. Each of the variances is calculated using (1.4):

$$\sigma_{ij}^2 = \frac{1}{n-1} \sum_{k=1}^{n} \left(x_{ijk} - \overline{x}_{ij} \right)^2 \tag{1.4a}$$

$$\overline{x}_{ij} = \frac{1}{10}\sum_{k=1}^{10} x_{ijk} \qquad (1.4b)$$

where $j = 1, 2, 3$ are the three orthogonal axes components (i.e., x, y, or z), x_{ijk} is the kth measurement for the ith fiducial point, and k is the number of measurements for each fiducial point. For the calibrations of the image and robot, $n = 10$ and $\overline{x}_{ij}$ is the mean measurement for the jth component of the ith fiducial point.

1.10.1.2 Fiducial Registration Error

The exact positions of N fiducials, which are given by $P = [\mathbf{p}_j; j = 1, ..., N]$ in the coordinate system of the transducer, are known for either the image calibration phantom or the robot calibration plates. For the calibration of the image, we measured the positions of N fiducials (the intersection of the nylon strings), which are given by $Q = [\mathbf{q}_j; j = 1, ..., N]$, in the 3-D coordinate system of the image. For the calibration of the robot, we measured the positions of N fiducials (the small divots), which are given by $Q = [\mathbf{q}_j; j = 1, ..., N]$, in the coordinate system of the robot by moving it so that it would touch the small divot holes. The FRE was calculated as the error of the rms distance between the corresponding fiducial positions before and after the registration process. The FRE is given by:

$$\mathrm{FRE} = \sqrt{\frac{\sum_{j=1}^{N}\left\| q_j = F(p_j) \right\|^2}{N}} \qquad (1.5)$$

The rigid-body transformation that registers the exact fiducial positions P with the measured fiducial positions Q is given by F.

1.10.1.3 Target Registration Error

TRE is defined as the distance between the corresponding points (not the fiducial points) and is calculated using (1.5) before and after the registration process. We used four targets in the image calibration phantom to determine the TRE; four other markers on the plates were used to determine the TRE for robot calibration.

1.10.2 Results of Calibration

Our ability to localize the intersections of the nylon strings in the 3-D TRUS image was analyzed along the x, y, and z directions and tabulated in Table 1.1 as the average FLE. As shown in Table 1.1, the FLE for localizing the intersection of the strings is similar in the x or y directions and larger in the z direction. These values are attributed to errors of measurement in the z direction, which has the poorest resolution in 3-D TRUS images, corresponding to the elevation (i.e., out-of-plane) direction of the acquired 2-D images [116].

For the calibration of the robot, the FLE for localizing the divots on the two orthogonal plates is approximately the same in all three orthogonal directions. This error, which is caused by the flexibility of the arm and the backlash in the joints of

Table 1.1 FLE for Identification of the Fiducials Used for Both Image and Robot Calibration (in Millimeters)

	Image Calibration			Robot Calibration		
	x	y	z	x	y	z
Mean	0.05	0.06	0.11	0.17	0.21	0.20
Standard deviation	0.01	0.02	0.01	0.07	0.11	0.10

the arm, is greater than the localization of the intersection of the strings; therefore, this error will dominate the calibration errors of the system.

As shown in Table 1.2, the mean FRE and TRE for the calibration of the robot were calculated to be 0.52 ± 0.18 mm and 0.68 ± 0.29 mm, respectively. Because the FRE is larger for the calibration of the robot than the FRE for the calibration of the image, both the mean FRE and TRE are larger for the calibration of the robot than the FRE and the TRE for the calibration of the image. Because system errors will result from both calibrations of the image and the robot, the errors in the calibration of the robot will dominate the accuracy of integration for the two coordinate systems.

1.10.3   Accuracy of Needle Placement

The accuracy of the robot's needle placement was determined by moving the tip of the needle to nine locations on a 5 cm × 5 cm grid that represented the skin of the patient (i.e., a 3 × 3 grid of targeting points). The tip of the needle was then located by using a three-axis stage (Parker Hannifin Co., Irwin, Pennsylvania) with a measuring accuracy of 2 μm. The displacement error, ε_d, between the measured position and targeted position of the needle tip was calculated using (1.6):

$$\varepsilon_d = \sqrt{\left(x - x_i\right)^2 + \left(y - y_i\right)^2 + \left(z - z_i\right)^2} \qquad (1.6)$$

where (x, y, z) are the coordinates for the targeted point and (x_i, y_i, z_i) are the coordinates for the ith measured point. The mean error of the needle placement, $\bar{\varepsilon}_d$, and the standard deviation (STD) from 10 measurements at each position were found to be 0.15 ± 0.06 mm.

Table 1.2 FRE and TRE for the Image and Robot Calibration of the 3-D US-Guided Brachytherapy System (in Millimeters)

	Image Calibration		Robot Calibration	
	FRE	TRE	FRE	TRE
Mean	0.12	0.23	0.52	0.68
STD	0.07	0.11	0.18	0.29

Source: [45].

1.10.4   Accuracy of Needle Angulation

The accuracy of the robot's needle angulation was determined by attaching a small plate to the needle holder. The robot was then used to angulate the plate vertically and laterally (0°, 5°, 10°, 15°). We measured the orientation of the plate using the three-axis stage and determined the angulation error by comparing the angle of the measured plate and of the expected plate.

As shown in Table 1.3, the mean angular differences between the measured plate and the expected plate were less than 0.12° with a mean angle of 0.07°. However, angulation errors will propagate displacement errors as the distance from the needle insertion is increased.

A needle was inserted 10 cm from the needle guide and the displacement error was estimated. Using the mean and maximum angulation errors, we calculated the mean and maximum displacement errors for a 10-cm insertion to be ±0.13 and ±0.50 mm, respectively.

1.10.5   Accuracy of Needle Targeting

The accuracy of the needle targeting system was evaluated using two tissue-mimicking phantoms constructed from agar [131] and contained in a box. One side of the box was removed to allow the insertion of the needle into the agar gel (Figure 1.13).

As shown in Figure 1.13, each phantom contained two rows of stainless steel beads with diameters of 0.8 mm to provide four different bead-targeting configurations: two different needle insertion depths and two different distances from the ultrasound transducer. The bead configuration formed a 4 cm × 4 cm × 4 cm cube that simulated the approximate size of a large prostate. The targeting experiments involved: (1) producing a 3-D TRUS image, (2) identifying a bead to be targeted, (3) choosing a trajectory, (4) positioning the robot to allow insertion of the needle to the target, and (5) inserting the needle into the target. The accuracy of the needle targeting system was calculated by determining the deviations made by the tip of the needle from the location of the preinsertion bead in the 3-D TRUS image. By averaging the results from all targeting experiments, we calculated the mean error to be 0.74 ± 0.24 mm. In addition, we plotted the results of the accuracy of the needle targeting system as a 3-D scatterplot of the needle tips relative to the target (Figure 1.14).

1.10.6   Accuracy of Needle Tracking

Because the parallel trajectory of the needle insertion to the axis of the US transducer can be verified by observing the insertion of the needle in the set of real-time 2-D

Table 1.3   Guidance Error Analysis of Needle Angulation by the Robot for Vertical and Lateral Angulation of the Needle (in Degrees)

	Vertical Angulation			Lateral Angulation		
Targeted angles	5°	10°	15°	5°	10°	15°
Mean angle difference $\bar{\varepsilon}_a$	0.02°	0.04°	0.06°	0.07°	0.10°	0.12°
Standard deviation	0.02°	0.06°	0.05°	0.05°	0.03°	0.08°
Maximum angle difference	0.08°	0.22°	0.18°	0.15°	0.15°	0.28°

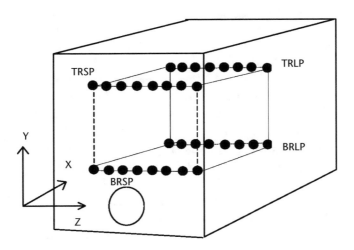

Figure 1.13 Schematic diagram of the prostate phantom. The four rows of circles represents the four different bead configurations. The needle entered the phantom from the front, parallel to the *x*-axis and was inserted into the cylindrical channel representing the rectum. TRLP = top row, long penetration; TRSP = top row, short penetration; BRLP = bottom row, long penetration; BRSP = bottom row, short penetration.

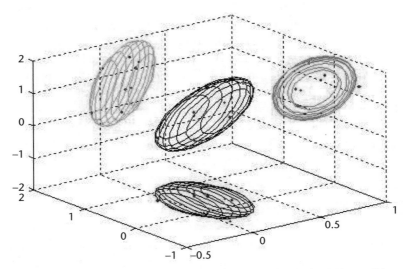

Figure 1.14 Needle targeting accuracy is displayed in a 3-D plot as the 95% confidence ellipsoid. The origin of the coordinate system represents the target, and the needle tip positions after insertion (relative to the targets) are represented by the squares. The projections of the needle tip positions and the ellipsoid (on the three orthogonal planes) are also shown. These results are for the targets near the transducer and for a short penetration as shown in Figure 1.13. (*From:* [45]. © 2004 American Association of Physicists in Medicine. Reprinted with permission.)

TRUS images, we compared the results from the oblique needle tracking to the results obtained from the parallel trajectory tracking. We used the robot to guide the insertion of a needle at different angles ($\pm5°$, $\pm10°$, $\pm15°$) with respect to the trajectory of parallel insertion. For each oblique trajectory, the needle was automatically tracked and its angle of insertion was automatically determined and compared with the angle of parallel insertion. This test, which used an agar gel phantom and a chicken tissue phantom [3, 132], was repeated five times for each angle in order to determine the mean tracking error as well as the standard deviation.

Table 1.4 shows the evaluation results of the needle-tracking algorithm. Using the chicken tissue phantom, the average execution time was 0.13 ± 0.01 second, and the average angulation error was $0.54° \pm 0.16°$. Using the agar gel phantoms, the average execution time was 0.12 ± 0.01 second, and the average angulation error was $0.58° \pm 0.36°$.

The results shown in Table 1.4 demonstrate that the insertion of the needle can be tracked in near real time, and that the tracking error does not significantly depend on the angle of insertion. Figure 1.14 shows the views provided to the physician during needle tracking.

1.10.7 Accuracy of Seed Implantation

The accuracy of seed implantation and the guidance of 3-D TRUS were evaluated using a brachytherapy needle to implant spherical steel beads with diameters of 0.8 mm. The beads were implanted into several agar gel phantoms at different positions and different needle trajectories ($0°$, $\pm5°$, $\pm10°$, $\pm15°$) with respect to the axis of the TRUS probe. These tests were carried out by: (1) choosing a trajectory and position in the 3-D TRUS image, (2) positioning the robot at the chosen trajectory, (3) guiding a needle to its target, and (4) implanting a bead. The error of implantation was determined by: (1) obtaining a 3-D TRUS image after the implantation, (2) identifying the bead in the 3-D TRUS image, and (3) calculating the distance between the actual positions of the bead to the planned position.

Table 1.5 shows the accuracy of the bead implantation. The mean error was calculated to be 2.59 ± 0.76 mm. The largest error was observed in the y direction (in the vertical plane), corresponding to the bevel direction of the needle.

1.11 Conclusions

We have described the principles of 3-D ultrasound imaging and its use for image-guided prostate brachytherapy. We have also demonstrated the technical feasibility of a robot-aided prostate brachytherapy procedure using 3-D TRUS guid-

Table 1.4 Needle Segmentation Results for Chicken Tissue and Agar Phantoms

Angle		$-15°$	$-10°$	$-5°$	$5°$	$10°$	$15°$
Test results for chicken tissue phantom	Time (sec)	0.13	0.11	0.12	0.12	0.12	0.14
	Accuracy (deg)	0.50	0.51	0.43	0.37	0.74	0.74
Test results for agar phantoms	Time (sec)	0.12	0.12	0.12	0.11	0.12	0.13
	Accuracy (deg)	0.30	0.71	0.48	0.68	0.42	0.86

Table 1.5 Brachytherapy Seed Implantation Error for Different
Angulations of the Needle (in Millimeters)

Angle	$-20°$	$-15°$	$-10°$	$-5°$	$0°$	$5°$	$10°$	$15°$	$20°$
x_i	0.62	0.33	0.34	0.78	0.41	0.54	0.65	0.40	0.51
y_i	2.79	2.01	1.02	2.97	1.58	2.97	3.27	3.03	1.99
z_i	0.89	0.61	0.79	0.75	0.53	0.75	0.98	0.63	0.87
Total error	2.99	2.13	1.33	3.16	1.72	3.11	3.48	3.12	2.23
Mean ± STD	2.59 ± 0.76 mm								

Note: The error is shown in each of the 3-D orthogonal directions as well as the total error.
Source: [29].

ance. Using robotic assistance, our experimental results indicate that the brachytherapy needle can be guided accurately and consistently to target points in the 3-D TRUS image. We expect that the introduction of a robotic system and the implementation of 3-D TRUS tools for automatic needle detection of oblique insertion, localization of the implanted seeds, and monitoring of prostate changes will produce an effective intraoperative prostate brachytherapy procedure.

In our robot-assisted and 3-D TRUS-guided system, the robot and 3-D TRUS image coordinate systems were registered using specially designed calibration phantoms to position the hole of the needle guidance in relation to the 3-D TRUS image. This approach allowed us to remove the constraints of the parallel needle trajectory that are present in current prostate brachytherapy procedures. Rather than having it fixed onto the US transducer holder, the needle guide was mounted onto a robotic arm. This mount provided greater flexibility over the needle's trajectory and reduced interference with the pubic arch.

In registering the coordinate systems of the robot and the imaging system, fiducial points in the 3-D TRUS image must be determined. Errors in locating these fiducial points will propagate and result in targeting errors, which lead to systematic errors. By comparing Table 1.2 with Table 1.1, we can see that the FLE for the robot fiducials is greater than the FLE for the 3-D TRUS fiducials; therefore, errors in calibrating the robot coordinate system will dominate and deteriorate the overall accuracy of the system.

By comparing the results obtained in the evaluation of needle-targeting accuracy using a rigid rod with the results obtained for bead implantation accuracy with a brachytherapy needle, we can see that the errors observed in the latter are larger and dominate the overall performance of the system. Due to the bevel at the needle tip, the large implantation error was primarily caused by the deflection of the needle from the planned trajectory. Solving this source of error will greatly improve the performance of the system [115].

Acknowledgments

The authors acknowledge the help of Kerry Knight in the preparation of this manuscript. The authors also gratefully acknowledge the financial support of the Canadian Institutes of Health Research, and the Ontario R&D Challenge Fund. The first author holds a Canada Research Chair and acknowledges the support of the Canada Research Chair program and the Canadian Foundation for Innovation.

References

[1]   Comeau, R. M., A. Fenster, and T. M. Peters, "Intraoperative US in Interactive Image-Guided Neurosurgery," *Radiographics,* Vol. 18, No. 4, 1998, pp. 1019–1027.

[2]   Downey, D. B., A. Fenster, and J. C. Williams, "Clinical Utility of Three-Dimensional US," *Radiographics,* 2000, Vol. 20, No. 2, pp. 559–571.

[3]   Smith, W. L., et al., "Three-Dimensional Ultrasound-Guided Core Needle Breast Biopsy," *Ultrasound Med. Biol.,* Vol. 27, No. 8, 2001, pp. 1025–1034.

[4]   Chin, J. L., et al., "Three-Dimensional Prostate Ultrasound and Its Application to Cryosurgery," *Tech. Urol.,* Vol. 2, No. 4, 1996, pp. 187–193.

[5]   Chin, J. L., et al., "Three-Dimensional Transrectal Ultrasound Guided Cryoablation for Localized Prostate Cancer in Nonsurgical Candidates: A Feasibility Study and Report of Early Results," *J. Urol.,* Vol. 159, No. 3, 1998, pp. 910–914.

[6]   Brinkley, J. F., et al., "In Vitro Evaluation of an Ultrasonic Three-Dimensional Imaging and Volume System," *Ultrasonic Imaging,* Vol. 4, No. 2, 1982, pp. 126–139.

[7]   Fenster, A., and D. B. Downey, "Three-Dimensional Ultrasound Imaging: A Review," *IEEE Eng. Med. Biol.,* Vol. 15, 1996, pp. 41–51.

[8]   Ghosh, A., N. C. Nanda, and G. Maurer, "Three-Dimensional Reconstruction of Echo-Cardiographic Images Using the Rotation Method," *Ultrasound Med. Biol.,* Vol. 8, No. 6, pp. 1982, 655–661.

[9]   Greenleaf, J. F., et al., "Multidimensional Visualization in Echocardiography: An Introduction," *Mayo Clin. Proc.,* Vol. 68, No. 3, 1993, pp. 213–220.

[10] King, D. L., et al., "Three-Dimensional Echocardiography," *Am. J. Cardiac Imaging,* Vol. 7, No. 3, 1993, pp. 209–220.

[11] Nelson, T. R., and D. H. Pretorius, "Three-Dimensional Ultrasound of Fetal Surface Features," *Ultrasound Obstet. Gynecol.,* Vol. 2, 1992, pp. 166–174.

[12] Rankin, R. N., et al., "Three-Dimensional Sonographic Reconstruction: Techniques and Diagnostic Applications," *AJR Am. J. Roentgenol.,* Vol. 161, No. 4, 1993, pp. 695–702.

[13] Smith, W. L., and A. Fenster, "Optimum Scan Spacing for Three-Dimensional Ultrasound by Speckle Statistics," *Ultrasound Med. Biol.,* Vol. 26, No. 4, 2000, pp. 551–562.

[14] Fenster, A., D. B. Downey, and H. N. Cardinal, "Three-Dimensional Ultrasound Imaging," *Phys. Med. Biol.,* Vol. 46, No. 5, 2001, pp. R67–R99.

[15] Downey, D. B., and A. Fenster, "Vascular Imaging with a Three-Dimensional Power Doppler System," *AJR Am. J. Roentgenol.,* Vol. 165, No. 3, 1995, pp. 665–668.

[16] Fenster A., et al., "Three-Dimensional Ultrasound Imaging," *Physics of Medical Imaging, Proc. SPIE,* Vol. 2432, 1995, pp. 176–184.

[17] Picot, P., et al., "Three-Dimensional Colour Doppler Imaging of the Carotid Artery," *Medical Imaging, Proc. SPIE,* Vol. 1444, 1991, pp. 206–213.

[18] Picot, P. A., et al., "Three-Dimensional Colour Doppler Imaging," *Ultrasound Med. Biol.,* Vol. 19, No. 2, 1993, pp. 95–104.

[19] Pretorius, D. H., T. R. Nelson, and J. S. Jaffe, "3-Dimensional Sonographic Analysis Based on Color Flow Doppler and Gray Scale Image Data: A Preliminary Report," *J. Ultrasound Med.,* Vol. 11, No. 5, 1992, pp. 225–232.

[20] Downey, D. B., and A. Fenster, "Three-Dimensional Power Doppler Detection of Prostate Cancer [letter]," *American J. of Roentgenology,* Vol. 165, No. 3, 1995, p. 741.

[21] Landry, A., and A. Fenster, "Theoretical and Experimental Quantification of Carotid Plaque Volume Measurements Made by 3-D Ultrasound Using Test Phantoms," *Medical Physics,* Vol. 29, No. 10, 2002, pp. 2319–2327.

[22] Landry, A., J. D. Spence, and A. Fenster, "Measurement of Carotid Plaque Volume by 3-Dimensional Ultrasound," *Stroke,* Vol. 35, No. 4, 2004, pp. 864–869.

[23] Landry, A., J. D. Spence, and A. Fenster, "Quantification of Carotid Plaque Volume Measurements Using 3-D Ultrasound Imaging," *Ultrasound Med. Biol.,* Vol. 31, No. 6, 2005, pp. 751–762.

[24] Ainsworth, C. D., et al., "3-D Ultrasound Measurement of Change in Carotid Plaque Volume: A Tool for Rapid Evaluation of New Therapies," *Stroke,* Vol. 35, 2005, pp. 1904–1909.

[25] Bamber, J. C., et al., "Data Processing for 3-D Ultrasound Visualization of Tumour Anatomy and Blood Flow," *Proc. SPIE,* Vol. 1808, 1992, pp. 651–663.

[26] Carson, P. L., et al., "Approximate Quantification of Detected Fractional Blood Volume and Perfusion from 3-D Color Flow and Doppler Power Signal Imaging," *Proc. 1993 Ultrasonics Symp.,* 1993, pp. 1023–1026.

[27] King, D. L., D. L. J. King, and M. Y. Shao, "Evaluation of In Vitro Measurement Accuracy of a Three-Dimensional Ultrasound Scanner," *J. Ultrasound Med.,* Vol. 10, No. 2, 1991, pp. 77–82.

[28] Delabays, A., et al., "Transthoracic Real-Time Three-Dimensional Echocardiography Using a Fan-Like Scanning Approach for Data Acquisition: Methods, Strengths, Problems, and Initial Clinical Experience," *Echocardiography,* Vol. 12, No. 1, 1995, pp. 49–59.

[29] Downey, D. B., D. A. Nicolle, and A. Fenster, "Three-Dimensional Orbital Ultrasonography," *Can. J. Ophthalmol.,* Vol. 30, No. 7, 1995, pp. 395–398.

[30] Downey, D. B., D. A. Nicolle, and A. Fenster, "Three-Dimensional Ultrasound of the Eye," *Admin. Radiol. J.,* Vol. 14, 1995, pp. 46–50.

[31] Gilja, O. H., et al., "In Vitro Evaluation of Three-Dimensional Ultrasonography in Volume Estimation of Abdominal Organs," *Ultrasound Med. Biol.,* Vol. 20, No. 2, 1994, pp. 157–165.

[32] Sohn, C. H., et al., "Three-Dimensional Ultrasound Imaging of Benign and Malignant Breast Tumors—Initial Clinical Experiences," *Geburtshilfe Frauenheilkd,* Vol. 52, No. 9, 1992, pp. 520–525.

[33] Benacerraf, B. R., et al., "Three- and 4-Dimensional Ultrasound in Obstetrics and Gynecology: Proceedings of the American Institute of Ultrasound in Medicine Consensus Conference," *J. Ultrasound Med.,* Vol. 24, No. 12, 2005, pp. 1587–1597.

[34] Dolkart, L., M. Harter, and M. Snyder, "Four-Dimensional Ultrasonographic Guidance for Invasive Obstetric Procedures," *J. Ultrasound Med.,* Vol. 24, No. 9, 2005, pp. 1261–1266.

[35] Goncalves, L. F., et al., "Three- and 4-Dimensional Ultrasound in Obstetric Practice: Does It Help?" *J. Ultrasound Med.,* Vol. 24, No. 12, 2005, pp. 1599–1624.

[36] Lee, W., "3-D Fetal Ultrasonography," *Clin. Obstet. Gynecol.,* Vol. 46, No. 4, 2003, pp. 850–867.

[37] Devore, G. R., "Three-Dimensional and Four-Dimensional Fetal Echocardiography: A New Frontier," *Curr. Opin. Pediatr.,* Vol. 17, No. 5, 2005, pp. 592–604.

[38] Blake, C. C., et al., "Variability and Accuracy of Measurements of Prostate Brachytherapy Seed Position In Vitro Using Three-Dimensional Ultrasound: An Intra- and Inter-Observer Study," *Medical Physics,* Vol. 27, No. 12, 2000, pp. 2788–2795.

[39] Tong, S., et al., "A Three-Dimensional Ultrasound Prostate Imaging System," *Ultrasound Med. Biol.,* Vol. 22, No. 6, 1996, pp. 735–746.

[40] Tong, S., et al., "Intra- and Inter-Observer Variability and Reliability of Prostate Volume Measurement Via Two-Dimensional and Three-Dimensional Ultrasound Imaging," *Ultrasound Med. Biol.,* Vol. 24, No. 5, 1998, pp. 673–681.

[41] Elliot, T. L., et al., "Accuracy of Prostate Volume Measurements In Vitro Using Three-Dimensional Ultrasound," *Acad. Radiol.,* Vol. 3, No. 5, 1996, pp. 401–406.

[42] Downey, D. B., J. L. Chin, and A. Fenster, "Three-Dimensional US-Guided Cryosurgery," *Radiology,* Vol. 197, 1995, p. 539.

[43] Chin, J. L., et al., "Three Dimensional Transrectal Ultrasound Imaging of the Prostate: Clinical Validation," *Can. J. Urol.,* Vol. 6, No. 2, 1999, pp. 720–726.

[44] Onik, G. M., D. B. Downey, and A. Fenster, "Three-Dimensional Sonographically Monitored Cryosurgery in a Prostate Phantom," *J. Ultrasound Med.,* Vol. 15, No. 3, 1996, pp. 267–270.

[45] Wei, Z., et al., "Robot-Assisted 3-D-TRUS Guided Prostate Brachytherapy: System Integration and Validation," *Med. Phys.,* Vol. 31, No. 3, 2004, pp. 539–558.

[46] Wei, Z., et al., "Oblique Needle Segmentation and Tracking for 3-D TRUS Guided Prostate Brachytherapy," *Med. Phys.,* Vol. 32, No. 9, 2005, pp. 2928–2941.

[47] Pagoulatos, N., D. R. Haynor, and Y. Kim, "A Fast Calibration Method for 3-D Tracking of Ultrasound Images Using a Spatial Localizer," *Ultrasound Med. Biol.,* Vol. 27, No. 9, 2001, pp. 1219–1229.

[48] Geiser, E. A., et al., "A Mechanical Arm for Spatial Registration of Two-Dimensional Echocardiographic Sections," *Catheterization Cardiovasc. Diagnosis,* Vol. 8, No. 1, 1982, pp. 89–101.

[49] Brinkley, J. F., et al., "Fetal Weight Estimation from Lengths and Volumes Found by Three-Dimensional Ultrasonic Measurements," *J. Ultrasound Med.,* Vol. 3, No. 4, 1984, pp. 163–168.

[50] King, D. L., D. L. J. King, and M. Y. Shao, "Three-Dimensional Spatial Registration and Interactive Display of Position and Orientation of Real-Time Ultrasound Images," *J. Ultrasound Med.,* Vol. 9, No. 9, 1990, pp. 525–532.

[51] King, D. L., et al., "Ultrasound Beam Orientation During Standard Two-Dimensional Imaging: Assessment by Three-Dimensional Echocardiography," *J. Am. Soc. Echocardiography,* Vol. 5, No. 6, 1992, pp. 569–576.

[52] Levine, R. A., et al., "Three-Dimensional Echocardiographic Reconstruction of the Mitral Valve, with Implications for the Diagnosis of Mitral Valve Prolapse," *Circulation,* Vol. 80, No. 3, 1989, pp. 589–598.

[53] Moritz, W. E., et al., "An Ultrasonic Technique for Imaging the Ventricle in Three Dimensions and Calculating Its Volume," *IEEE Trans. on Biomed. Eng.,* Vol. 30, No. 8, 1983, pp. 482–492.

[54] Rivera, J. M., et al., "Three-Dimensional Reconstruction of Ventricular Septal Defects: Validation Studies and In Vivo Feasibility," *J. Am. Coll. Cardiol.,* Vol. 23, No. 1, 1994, pp. 201–208.

[55] Weiss, J. L., et al., "Accuracy of Volume Determination by Two-Dimensional Echocardiography: Defining Requirements Under Controlled Conditions in the Ejecting Canine Left Ventricle," *Circulation,* Vol. 67, No. 4, 1983, pp. 889–895.

[56] Nelson, T. R., and T. T. Elvins, "Visualization of 3-D Ultrasound Data," *IEEE Computer Graphics Appl.,* November 1993, pp. 50–57.

[57] Bonilla-Musoles, F., et al., "Use of Three-Dimensional Ultrasonography for the Study of Normal and Pathologic Morphology of the Human Embryo and Fetus: Preliminary Report," *J. Ultrasound Med.,* Vol. 14, No. 10, 1995, pp. 757–765.

[58] Detmer, P. R., et al., "3-D Ultrasonic Image Feature Localization Based on Magnetic Scanhead Tracking: In Vitro Calibration and Validation," *Ultrasound Med. Biol.,* Vol. 20, No. 9, 1994, pp. 923–936.

[59] Ganapathy, U., and A. Kaufman, "3-D Acquisition and Visualization of Ultrasound Data," *Visualization in Biomedical Computing, Proc. SPIE,* Vol. 1808, 1992, pp. 535–545.

[60] Hodges, T. C., et al., "Ultrasonic Three-Dimensional Reconstruction: In Vitro and In Vivo Volume and Area Measurement," *Ultrasound Med. Biol.,* Vol. 20, No. 8, 1994, pp. 719–729.

[61]   Hughes, S. W., et al., "Volume Estimation from Multiplanar 2-D Ultrasound Images Using a Remote Electromagnetic Position and Orientation Sensor," *Ultrasound Med. Biol.*, Vol. 22, No. 5, 1996, pp. 561–572.

[62]   Leotta, D. F., P. R. Detmer, and R. W. Martin, "Performance of a Miniature Magnetic Position Sensor for Three-Dimensional Ultrasound Imaging," *Ultrasound Med. Biol.*, Vol. 23, No. 4, 1997, pp. 597–609.

[63]   Gilja, O. H., et al., "Intragastric Distribution and Gastric Emptying Assessed by Three-Dimensional Ultrasonography," *Gastroenterology*, Vol. 113, No. 1, 1997, pp. 38–49.

[64]   Nelson, T. R., and D. H. Pretorius, "Visualization of the Fetal Thoracic Skeleton with Three-Dimensional Sonography: A Preliminary Report," *AJR Am. J. Roentgenol.*, Vol. 164, No. 6, 1995, pp. 1485–1488.

[65]   Ohbuchi, R., D. Chen, and H. Fuchs, "Incremental Volume Reconstruction and Rendering for 3-D Ultrasound Imaging," *Visualization in Biomedical Computing, Proc. SPIE*, Vol. 1808, 1992, pp. 312–323.

[66]   Pretorius, D. H., and T. R. Nelson, "Prenatal Visualization of Cranial Sutures and Fontanelles with Three-Dimensional Ultrasonography," *J. Ultrasound Med.*, Vol. 13, No. 11, 1994, pp. 871–876.

[67]   Raab, F. H., et al., "Magnetic Position and Orientation Tracking System," *IEEE Trans. on Aerospace and Electronic Syst.*, Vol. AES-15, 1979, pp. 709–717.

[68]   Riccabona, M., et al., "Distance and Volume Measurement Using Three-Dimensional Ultrasonography," *J. Ultrasound Med.*, Vol. 14, No. 12, 1995, pp. 881–886.

[69]   Mercier, L., et al., "A Review of Calibration Techniques for Freehand 3-D Ultrasound Systems," *Ultrasound Med. Biol.*, Vol. 31, No. 4, 2005, pp. 449–471.

[70]   Lindseth, F., et al., "Probe Calibration for Freehand 3-D Ultrasound," *Ultrasound Med. Biol.*, Vol. 29, No. 11, 2003, pp. 1607–1623.

[71]   Rousseau, F., P. Hellier, and C. Barillot, "Confhusius: A Robust and Fully Automatic Calibration Method for 3-D Freehand Ultrasound," *Med. Image Anal.*, Vol. 9, No. 1, 2005, pp. 25–38.

[72]   Leotta, D. F., "An Efficient Calibration Method for Freehand 3-D Ultrasound Imaging Systems," *Ultrasound Med. Biol.*, Vol. 30, No. 7, 2004, pp. 999–1008.

[73]   Gooding, M. J., S. H. Kennedy, and J. A. Noble, "Temporal Calibration of Freehand Three-Dimensional Ultrasound Using Image Alignment," *Ultrasound Med. Biol.*, Vol. 31, No. 7, 2005, pp. 919–927.

[74]   Dandekar, S., et al., "A Phantom with Reduced Complexity for Spatial 3-D Ultrasound Calibration," *Ultrasound Med. Biol.*, Vol. 31, No. 8, 2005, pp. 1083–1093.

[75]   Poon, T. C., and R. N. Rohling, "Comparison of Calibration Methods for Spatial Tracking of a 3-D Ultrasound Probe," *Ultrasound Med. Biol.*, Vol. 31, No. 8, 2005, pp. 1095–1108.

[76]   Gee, A. H., et al., "A Mechanical Instrument for 3-D Ultrasound Probe Calibration," *Ultrasound Med. Biol.*, Vol. 31, No. 4, 2005, pp. 505–518.

[77]   Smith, W., and A. Fenster, "Statistical Analysis of Decorrelation-Based Transducer Tracking for Three-Dimensional Ultrasound," *Med. Phys.*, Vol. 30, No. 7, 2003, pp. 1580–1591.

[78]   Friemel, B. H., et al., "Speckle Decorrelation Due to Two-Dimensional Flow Gradients," *IEEE Trans. on Ultrasonics Ferroelectrics and Frequency Control*, Vol. 45, No. 2, 1998, pp. 317–327.

[79]   Tuthill, T. A., et al., "Automated Three-Dimensional US Frame Positioning Computed from Elevational Speckle Decorrelation," *Radiology*, Vol. 209, No. 2, 1998, pp. 575–582.

[80]   Huang, Q. H., et al., "3-D Measurement of Body Tissues Based on Ultrasound Images with 3-D Spatial Information," *Ultrasound Med. Biol.*, Vol. 31, No. 12, 2005, pp. 1607–1615.

[81]   Shattuck, D. P., et al., "Explososcan: A Parallel Processing Technique for High Speed Ultrasound Imaging with Linear Phased Arrays," *J. Acoust. Soc. Am.*, Vol. 75, No. 4, 1984, pp. 1273–1282.

[82]   Smith, S. W., H. G. Pavy, Jr., and O. T. von Ramm, "High-Speed Ultrasound Volumetric Imaging System. Part I. Transducer Design and Beam Steering," *IEEE Trans. on Ultrasonics Ferroelectrics and Frequency Control,* Vol. 38, 1991, pp. 100–108.

[83]   Smith, S. W., G. E. Trahey, and O. T. von Ramm, "Two-Dimensional Arrays for Medical Ultrasound," *Ultrasonic Imaging,* Vol. 14, No. 3, 1992, pp. 213–233.

[84]   Snyder, J. E., J. Kisslo, and O. von Ramm, "Real-Time Orthogonal Mode Scanning of the Heart. I. System Design," *J. Am. Coll. Cardiol.,* Vol. 7, No. 6, 1986, pp. 1279–1285.

[85]   Turnbull, D. H., and F. S. Foster, "Beam Steering with Pulsed Two-Dimensional Transducer Arrays," *IEEE Trans. on Ultrasonics Ferroelectrics and Frequency Control,* Vol. 38, 1991, pp. 320–333.

[86]   von Ramm, O. T., and S. W. Smith, "Real Time Volumetric Ultrasound Imaging System," *Proc. SPIE,* Vol. 1231, 1990, pp. 15–22.

[87]   von Ramm, O. T., S. W. Smith, and H. G. Pavy, Jr., "High-Speed Ultrasound Volumetric Imaging System. Part II. Parallel Processing and Image Display," *IEEE Trans. on Ultrasonics Ferroelectrics and Frequency Control,* Vol. 38, 1991, 109–115.

[88]   Oralkan, O., et al., "Volumetric Acoustic Imaging Using 2-D CMUT Arrays," *IEEE Trans. on Ultrasonics Ferroelectrics and Frequency Control,* Vol. 50, No. 11, 2003, pp. 1581–1594.

[89]   Prakasa, K. R., et al., "Feasibility and Variability of Three Dimensional Echocardiography in Arrhythmogenic Right Ventricular Dysplasia/Cardiomyopathy," *Am. J. Cardiol.,* Vol. 97, No. 5, 2006, pp. 703–709.

[90]   Xie, M. X., et al., "Real-Time 3-Dimensional Echocardiography: A Review of the Development of the Technology and Its Clinical Application," *Prog. Cardiovascular Disease,* Vol. 48, No. 3, 2005, pp. 209–225.

[91]   Sklansky, M., "Advances in Fetal Cardiac Imaging," *Pediatr. Cardiol.,* Vol. 25, No. 3, 2004, pp. 307–321.

[92]   Robb, R. A., *Three-Dimensional Biomedical Imaging: Principles and Practice,* New York: VCH Publishers, 1995.

[93]   Nelson, T. R., et al., "Three-Dimensional Echocardiographic Evaluation of Fetal Heart Anatomy and Function: Acquisition, Analysis, and Display," *J. Ultrasound Med.,* Vol. 15, No. 1, 1996, pp. 1–9.

[94]   Gee, A., et al., "Processing and Visualizing Three-Dimensional Ultrasound Data," *Br. J. Radiol.,* Vol. 77, special issue, No. 2, 2004, pp. S186–S193.

[95]   Nelson, T. R., et al., *Three-Dimensional Ultrasound,* Philadelphia, PA: Lippincott-Raven, 1999.

[96]   Fenster, A., and D. Downey, "Three-Dimensional Ultrasound Imaging," *Proc. SPIE,* Vol. 4549, 2001, pp. 1–10.

[97]   Fenster, A., and D. B. Downey, "Three-Dimensional Ultrasound Imaging," in *Handbook of Medical Imaging, Volume I, Physics and Psychophysics,* Bellingham, WA: SPIE Press, 2000, pp. 735–746.

[98]   Pretorius, D. H., and T. R. Nelson, "Fetal Face Visualization Using Three-Dimensional Ultrasonography," *J. Ultrasound Med.,* Vol. 14, No. 5, 1995, pp. 349–356.

[99]   Deng, J., and C. H. Rodeck, "Current Applications of Fetal Cardiac Imaging Technology," *Curr. Opin. Obstet. Gynecol.,* Vol. 18, No. 2, 2006, pp. 177–184.

[100]   Levoy, M., "Volume Rendering, a Hybrid Ray Tracer for Rendering Polygon and Volume Data," *IEEE Computer Graphics Appl.,* Vol. 10, 1990, pp. 33–40.

[101]   Mroz, L., A. Konig, and E. Groller, "Maximum Intensity Projection at Warp Speed," *Computers and Graphics,* Vol. 24, No. 3, 1992, pp. 343–352.

[102]   Fruhauf, T., "Raycasting Vector Fields," *IEEE Visualization,* 1996, pp. 115–120.

[103]   Kniss, J., G. Kindlmann, and C. Hansen, "Multidimensional Transfer Functions for Interactive Volume Rendering," *IEEE Trans. on Visualization and Computer Graphics,* Vol. 8, No. 3, 2002, pp. 270–285.

[104] Sun, Y., and D. L. Parker, "Performance Analysis of Maximum Intensity Projection Algorithm for Display of MRA Images," *IEEE Trans. on Med. Imaging*, Vol. 18, No. 12, 1999, pp. 1154–1169.

[105] Mroz, L., H. Hauser, and E. Groller, "Interactive High-Quality Maximum Intensity Projection," *Computer Graphics Forum*, Vol. 19, No. 3, 2000, pp. 341–350.

[106] Kirby, R. S., "Pre-Treatment Staging of Prostate Cancer: Recent Advances and Future Prospects," *Prostate Cancer and Prostatic Diseases*, Vol. 1, No. 1, 1997, pp. 2–10.

[107] Litwin, M. S., "Health-Related Quality-of-Life After Treatment for Localized Prostate-Cancer," *Cancer*, Vol. 75, No. 7, 1995, pp. 2000–2003.

[108] Sylvester, J., et al., "Interstitial Implantation Techniques in Prostate Cancer," *J. Surg. Oncol.*, Vol. 66, No. 1, 1997, pp. 65–75.

[109] Nag, S., *Principles and Practice of Brachytherapy*, Armonk, NY: Futura Publishing Company, 1997.

[110] Stokes, S. H., et al., "Transperineal Ultrasound-Guided Radioactive Seed Implantation for Organ-Confined Carcinoma of the Prostate," *Int. J. Radiat. Oncol. Biol. Phys.*, Vol. 37, No. 2, 1997, pp. 337–341.

[111] Pouliot, J., et al., "Optimization of Permanent 125I Prostate Implants Using Fast Simulated Annealing. *Int. J. Radiat. Oncol. Biol. Phys.*, Vol. 36, No. 3, 1996, pp. 711–720.

[112] Parker, S. H., et al., "Percutaneous Large-Core Breast Biopsy: A Multi-Institutional Study," *Radiology*, Vol. 193, No. 2, 1994, pp. 359–364.

[113] Tong, S., et al., "Analysis of Linear, Area and Volume Distortion in 3-D Ultrasound Imaging," *Ultrasound Med. Biol.*, Vol. 24, No. 3, 1998, pp. 355–373.

[114] Fenster, A., et al., "3-D Ultrasound Imaging: Applications in Image-Guided Therapy and Biopsy," *Computers and Graphics*, Vol. 26, No. 4, 2002, pp. 557–568.

[115] Wan, G., et al., "Brachytherapy Needle Deflection Evaluation and Correction," *Med. Phys.*, Vol. 32, No. 4, 2005, pp. 902–909.

[116] Fenster, A., D. B. Downey, and H. N. Cardinal, "Three-Dimensional Ultrasound Imaging," *Phys. Med. Biol.*, Vol. 46, No. 5, 2001, pp. R67–R99.

[117] Ding, M., and A. Fenster, "A Real-Time Biopsy Needle Segmentation Technique Using Hough Transform," *Med. Phys.*, Vol. 30, No. 8, 2003, pp. 2222–2233.

[118] Ding, M., H. N. Cardinal, and A. Fenster, "Automatic Needle Segmentation in Three-Dimensional Ultrasound Images Using Two Orthogonal Two-Dimensional Image Projections," *Med. Phys.*, Vol. 30, No. 2, 2003, pp. 222–234.

[119] Ladak, H. M., et al., "Prostate Boundary Segmentation from 2-D Ultrasound Images," *Med. Phys.*, Vol. 27, No. 8, 2000, pp. 1777–1788.

[120] Hu, N., et al., "Prostate Boundary Segmentation from 3-D Ultrasound Images," *Med. Phys.*, Vol. 30, 2003, pp. 1648–1659.

[121] Ladak, H. M., et al., "Testing and Optimization of a Semiautomatic Prostate Boundary Segmentation Algorithm Using Virtual Operators," *Med. Phys.*, Vol. 30, No. 7, 2003, pp. 1637–1647.

[122] Wang, Y., et al., "Semiautomatic Three-Dimensional Segmentation of the Prostate Using Two-Dimensional Ultrasound Images," *Med. Phys.*, Vol. 30, No. 5, 2003, pp. 887–897.

[123] Lobregt, S., and M. A. Viergever, "A Discrete Dynamic Contour Model," *IEEE Trans. on Med. Imaging*, Vol. 14, No. 1, 1995, pp. 12–24.

[124] McInerney, T., and D. Terzopoulos, "A Dynamic Finite Element Surface Model for Segmentation and Tracking in Multidimensional Medical Images with Application to Cardiac 4-D Image Analysis," *Comput. Med. Imaging Graph.*, Vol. 19, No. 1, 1995, pp. 69–83.

[125] Nath, R., et al., "Dosimetry of Interstitial Brachytherapy Sources: Recommendations of the AAPM Radiation Therapy Committee Task Group No. 43, American Association of Physicists in Medicine [published erratum appears in *Med. Phys.*, Vol. 23, No. 9, September 1996, p. 1579]," *Med. Phys.*, Vol. 22, No. 2, 1995, pp. 209–234.

[126] Arun, K. S., T. S. Huang, and S. D. Blostein, "Least-Squares Fitting of Two 3-D Point Sets," *IEEE Trans. on Pattern Anal. Machine Intell.,* No. 9, 1987, pp. 698–700.

[127] Ding, M., and A. Fenster, "Projection-Based Needle Segmentation in 3-D Ultrasound Images," *Computer-Aided Surgery,* Vol. 9, No. 5, 2004, pp. 193–201.

[128] Maurer, C. R. J., R. J. Maciunas, and J. M. Fitzpatrick, "Registration of Head CT Images to Physical Space Using a Weighted Combination of Points and Surfaces," *IEEE Trans. on Med. Imaging,* Vol. 17, 1998, pp. 753–761.

[129] Fitzpatrick, J. M., J. West, and C. R. Maurer, "Predicting Error in Rigid-Body, Point-Based Registration," *IEEE Trans. on Med. Imaging,* Vol. 17, 1998, pp. 694–702.

[130] Maurer, C. R. J., J. J. McCrory, and J. M. Fitzpatrick, "Estimation of Accuracy in Localizing Externally Attached Markers in Multimodal Volume Head Images," *Proc. SPIE,* Vol. 1898, 1993, pp. 43–51.

[131] Rickey, D. W., et al., "A Wall-Less Vessel Phantom for Doppler Ultrasound Studies," *Ultrasound Med. Biol.,* Vol. 21, No. 9, 1995, pp. 1163–1176.

[132] Smith, W., et al., "A Comparison of Core Needle Breast Biopsy Techniques: Free-Hand Versus 3-D Ultrasound Guidance," *Academic Radiology,* Vol. 9, 2002, pp. 541–550.

CHAPTER 2

Despeckle Filtering in Ultrasound Imaging of the Carotid Artery

C. P. Loizou

2.1 Introduction

The use of ultrasound in the diagnosis and assessment of arterial disease is well established because of its noninvasive nature, its low cost, and the continuing improvements in image quality [1]. Speckle, a form of locally correlated multiplicative noise, corrupts medical ultrasound imaging, making visual observation difficult [2, 3]. The presence of speckle noise in ultrasound images has been documented since the early 1970s when researchers such as Burckhardt [2], Wagner [3], and Goodman [4] described the fundamentals and the statistical properties of speckle noise.

Speckle is not truly a "noise" in the typical engineering sense, because its texture often carries useful information about the image being viewed. Speckle is the main factor that limits contrast resolution in diagnostic ultrasound imaging, thereby limiting the detectability of small, low-contrast lesions and making ultrasound images generally difficult for the nonspecialist to interpret [2, 3, 5, 6]. In fact, due to the presence of speckle, even ultrasound experts with sufficient experience may not be able to draw useful conclusions from the images [6]. Speckle noise also limits the effective application of image processing and analysis algorithms (i.e., edge detection, segmentation) and displays in two dimensions and volume rendering in three dimensions. Therefore, speckle is most often considered a dominant source of noise in ultrasound imaging and should be filtered out [2, 5, 6] without affecting important features of the image.

The objective of this work was to carry out a comparative evaluation of despeckle filtering techniques based on texture analysis, image quality evaluation metrics, and visual assessment by experts on 440 ultrasound images of a carotid artery bifurcation. Results of this study have also been published in [7–9]. This chapter was also published in a longer paper in [9].

The wide spread of mobile and portable telemedicine ultrasound scanning instruments has necessitated the need for better image processing techniques that offer a clearer image to the medical practitioner. This makes the use of efficient despeckle filtering a very important task. Early attempts to suppress speckle noise were implemented by averaging of uncorrelated images of the same tissue recorded under different spatial positions [5, 10, 11]. Although these methods are effective for speckle reduction, they require multiple images of the same object to be obtained [12]. Speckle-reducing filters originated from the synthetic aperture radar (SAR) community [10]. Since the early 1980s, these filters have been applied to ultrasound imaging [13]. Filters that are used widely in both SAR and ultrasound imaging include the Frost [14], Lee [10, 15, 16], and Kuan [12, 17] filters.

Table 2.1 summarizes the despeckle filtering techniques that are investigated in this study, grouped under the following categories: local statistics, median, homogeneity, geometric, homomorphic, anisotropic diffusion, and wavelet filtering. Furthermore, in Table 2.1 the main investigators, the methodology used, and the corresponding filter names are given. These filters are briefly introduced in this section, and presented in greater detail in Section 2.2.

Some of the local statistic filters are the Lee [10, 15, 16], the Frost [14], and the Kuan [12, 17] filters. The Lee and Kuan filters have the same structure, although the Kuan filter is a generalization of the Lee filter. Both filters form the output image by computing the central pixel intensity inside a filter-moving window, which is calculated from the average intensity values of the pixels and a coefficient of variation inside the moving window. Kuan considered a multiplicative speckle model and designed a linear filter, based on the minimum mean-square-error (MMSE) criterion, which provides optimal performance when the histogram of the image intensity is Gaussian distributed. The Lee filter [10] is a particular case of the Kuan filter based on a linear approximation made for the multiplicative noise model. The Frost filter [14] provides a balance between the averaging and the all-pass filters. It was

Table 2.1 Overview of Despeckle Filtering Techniques

Speckle Reduction Technique	Investigator	Method	Filter Name
Local statistics	[7–17]	Moving window utilizing local statistics:	
	[7–15]	(a) Mean (m), variance (σ^2)	lsmv
	[34]	(b) Mean, variance, third and fourth moments (higher moments) and entropy	
	[2–16]	(c) Homogeneous mask area filters	lsminsc
		(d) Wiener filtering	wiener
Median	[35]	Median filtering	median
Homogeneity	[8]	Based on the most homogeneous neighborhood around each pixel	homog
Geometric	[11]	Nonlinear iterative algorithm	gf4d
Homomorphic	[2, 18, 19]	The image is logarithmically transformed, the fast Fourier transform (FFT) is calculated, denoised, the inverse FFT is calculated, and finally exponentially transformed back	homo
Anisotropic diffusion	[2, 5, 13, 14, 20–25]	Nonlinear filtering technique for simultaneously performing contrast enhancement and noise reduction. Exponential damp kernel filters utilizing diffusion	ad
		Coherence enhancing diffusion	nldif
Wavelet	[16, 26–30, 37]	Only the useful wavelet coefficients are utilized	waveltc

Source: [9].

designed as an adaptive Wiener filter that assumes an autoregressive exponential model for the image.

In the homogeneity group, the filtering is based on the most homogeneous neighborhood around each image pixel [8, 9]. Geometric filters [11] are based on nonlinear iterative algorithms, which increment or decrement the pixel values in a neighborhood based on their relative values. The method of homomorphic filtering [18, 19] is similar to the logarithmic point operations used in histogram improvement, where dominant bright pixels are de-emphasized. In homomorphic filtering, the FFT of the image is calculated, then denoised, and then the inverse FFT is calculated.

Some other despeckle filtering methods, such as anisotropic diffusion [2, 20–24], speckle reducing anisotropic diffusion [5], and coherence anisotropic diffusion [25] presented in the literature, are nonlinear filtering techniques for simultaneously performing contrast enhancement and noise reduction by utilizing the coefficient of variation [5]. Furthermore, in the wavelet category, filters for suppressing the speckle noise were documented. These filters are making use of a realistic distribution of the wavelet coefficients [2, 16, 26–31] where only the useful wavelet coefficients are utilized. Different wavelet shrinkage approaches were investigated, usually based on Donoho's work [30].

The majority of speckle reduction techniques have certain limitations that can be briefly summarized as follows:

1. They are sensitive to the size and shape of the window. The use of different window sizes greatly affects the quality of the processed images. If the window is too large oversmoothing will occur, subtle details of the image will be lost in the filtering process, and edges will be blurred. On the other hand, a small window will decrease the smoothing capability of the filter and will not reduce speckle noise, thus making the filter ineffective.

2. Some of the despeckle methods based on window approaches require thresholds to be used in the filtering process, which have to be estimated empirically. The inappropriate choice of a threshold may lead to average filtering and noisy boundaries, thus leaving the sharp features unfiltered [7, 11, 15].

3. Most of the existing despeckle filters do not enhance the edges; they merely inhibit smoothing near the edges. When an edge is contained in the filtering window, the coefficient of variation will be high and smoothing will be inhibited. Therefore, speckle in the neighborhood of an edge will remain after filtering. Such filters are not directional in the sense that in the presence of an edge, all smoothing is precluded. Instead of inhibiting smoothing in directions perpendicular to the edge, smoothing in directions parallel to the edge is allowed.

4. Different evaluation criteria for evaluating the performance of despeckle filtering are used by different studies. Although most of the studies use quantitative criteria such as the mean-square-error (MSE) and speckle index (c), there are additional quantitative criteria, such as texture analysis and classification, image quality evaluation metrics, and visual assessment by experts, that could be investigated.

To the best of our knowledge, only two studies have investigated despeckle fil-tering on ultrasound images of the carotid artery. In [5], speckle reducing anisotropic diffusion as the most appropriate method was proposed. This technique was compared with the Frost filter [14], Lee filter [15], and homomorphic filtering [19] and documented that anisotropic diffusion performed better. In [32] a prepro-cessing procedure was first applied to the ultrasound carotid artery image, which modifies the image so that the noise becomes very close to white Gaussian. Then fil-tering methods based on an additive model may be applied. Three different filters, namely, the wavelet, the total variation filter, and anisotropic diffusion, were applied before and after image modification, and they showed that the quality of the image was improved when they were applied after the noise was modified to Gaussian.

In this study, we compare the performance of 10 despeckle filters on 440 ultra-sound images of carotid artery bifurcations. The performance of these filters was evaluated using texture analysis, the kNN classifier, image quality evaluation met-rics, and visual evaluation by two experts. The results of our study show that despeckle filtering improves the class separation between asymptomatic and symp-tomatic ultrasound images of the carotid artery.

In the following section, a brief overview of despeckle filtering techniques is pre-sented. In Section 2.3 we present the methodology and describe the material, record-ing of the ultrasound images, texture and statistical analyses, the kNN classifier, image quality evaluation metrics and the experiment carried out for visual evalua-tion. Sections 2.4 and 2.5 present the results and discussion, respectively.

2.2   Despeckle Filtering

To be able to derive an efficient despeckle filter, a speckle noise model is needed. The speckle noise model may be approximated as multiplicative if the envelope signal that is received at the output of the beam former of the ultrasound imaging system is captured before logarithmic compression. It can be defined as follows:

$$y_{i,j} = x_{i,j} n_{i,j} + a_{i,j} \qquad (2.1)$$

where $y_{i,j}$ represents the noisy pixel in the middle of the moving window, $x_{i,j}$ repre-sents the noise-free pixel, $n_{i,j}$ and $a_{i,j}$ represent the multiplicative and additive noise, respectively, and i, j are the indices of the spatial locations that belong in the 2-D space of real numbers, $i, j \in \Re^2$. Logarithmic compression is applied to the enve-lope-detected echo signal in order to fit it in the display range [25, 33]. It has been shown that the logarithmic compression affects the speckle noise statistics in such a way that the local mean becomes proportional to the local variance rather than the standard deviation [8, 25, 27, 29, 33]. More specifically, logarithmic compression affects the high-intensity tail of the Rayleigh and Rician probability density func-tion (PDF) more than the low-intensity part. As a result, the speckle noise becomes very close to white Gaussian noise corresponding to the uncompressed Rayleigh sig-nal [33]. Because the effect of additive noise is considerably smaller compared with that of multiplicative noise, (2.1) may be written as:

$$y_{i,j} \approx x_{i,j} n_{i,j} \tag{2.2}$$

Thus, the logarithmic compression transforms the model in (2.2) into the classical signal in additive noise form as:

$$\log(y_{i,j}) = \log(x_{i,j}) + \log(n_{i,j}) \tag{2.3a}$$

$$g_{i,j} = f_{i,j} + nl_{i,j} \tag{2.3b}$$

For the rest of this work, the term $\log(y_{i,j})$, which is the observed pixel on the ultrasound image display after logarithmic compression, is denoted as $g_{i,j}$; and the terms $\log(x_{i,j})$ and $\log(n_{i,j})$, which are the noise-free pixel and noise component after logarithmic compression, are denoted as $f_{i,j}$ and $nl_{i,j}$, respectively [see (2.3b)].

2.2.1 Local Statistics Filtering

Most of the techniques for speckle reduction filtering reviewed in the literature use local statistics. Their working principle may be described by a weighted average calculation using subregion statistics to estimate statistical measures over different pixel windows varying from 3×3 up to 15×15. All of these techniques assume that the speckle noise model has a multiplicative form as given in (2.2) [7–17, 25, 27].

2.2.1.1 First-Order Statistics Filtering (lsmv, wiener)

The filters utilizing the first-order statistics such as the variance and the mean of the neighborhood may be described with the model as in (2.3). Hence, the algorithms in this class may be traced back to the following equation [5, 7–18]:

$$f_{i,j} = \bar{g} + k_{i,j}(g_{i,j} - \bar{g}) \tag{2.4}$$

where $f_{i,j}$ is the estimated noise-free pixel value, $g_{i,j}$ is the noisy pixel value in the moving window, $\bar{g}$ is the local mean value of an $N_1 \times N_2$ region surrounding and including pixel $g_{i,j}$, $k_{i,j}$ is a weighting factor, where $k \in [0..1]$ and i, j are the pixel coordinates. The factor $k_{i,j}$ is a function of the local statistics in a moving window. It can be found in the literature [9, 10, 12, 15] and may be derived in different forms as follows:

$$k_{i,j} = (1 - \bar{g}^2 \sigma^2)/(\sigma^2(1 + \sigma_n^2)) \tag{2.5}$$

$$k_{i,j} = \sigma^2/(\bar{g}^2 \sigma_n^2 + \sigma^2) \tag{2.6}$$

$$k_{i,j} = (\sigma^2 - \sigma_n^2)/\sigma^2 \tag{2.7}$$

The values σ^2 and σ_n^2 represent the variance in the moving window and the variance of noise in the whole image, respectively. The noise variance may be calculated for the logarithmically compressed image by computing the average noise variance over a number of windows with dimensions considerably larger than the filtering window. In each window the noise variance is computed as:

$$\sigma_n^2 = \sum_{i=1}^{p} \sigma_p^2 / \overline{g}_p \qquad (2.8)$$

where σ_p^2 and $\overline{g}_p$ are the variance and mean of the noise in the selected windows, respectively, and p is the index covering all windows in the whole image [9, 25, 26, 33]. If the value of $k_{i,j}$ is 1 (in edge areas), this will result in an unchanged pixel, whereas a value of 0 (in uniform areas) replaces the actual pixel by the local average, $\overline{g}$, over a small region of interest [see (2.4)]. In this study the filter *lsmv* uses (2.5). The filter *wiener* uses a pixel-wise adaptive Wiener method [2–6, 14], implemented as given in (2.4), with the weighting factor $k_{i,j}$, as given in (2.7). For both despeckle filters *lsmv* and *wiener* the moving window size was 5×5.

2.2.1.2 Homogeneous Mask Area Filtering (*lsminsc*)

The *lsminsc* filter is a 2-D filter operating in a 5×5 pixel neighborhood by searching for the most homogeneous neighborhood area around each pixel, using a 3×3 sub-set window [9, 34]. The middle pixel of the 5×5 neighborhood is substituted with the average gray level of the 3×3 mask with the smallest speckle index C, where C for log-compressed images is given by:

$$C = \sigma_s^2 / \overline{g}_s \qquad (2.9)$$

where σ_s^2 and $\overline{g}_s$ represent the variance and mean of the 3×3 window. The window with the smallest C is the most homogeneous semiwindow, which presumably does not contain any edge. The filter is applied iteratively until the gray levels of almost all pixels in the image do not change.

2.2.2 Median Filtering (*median*)

The *median* filter [35] is a simple nonlinear operator that replaces the middle pixel in the window with the median value of its neighbors. The moving window for the *median* filter was 7×7.

2.2.3 Maximum Homogeneity over a Pixel Neighborhood Filtering (*homog*)

The *homog* filter is based on an estimation of the most homogeneous neighborhood around each image pixel [9, 36]. The filter takes into consideration only pixels that belong in the processed neighborhood (7×7 pixels) using (2.10), under the assumption that the observed area is homogeneous. The output image is then given by:

$$f_{i,j} = \left(c_{i,j} g_{i,j} \right) \Big/ \sum_{i,j} c_{i,j}, \text{ with } \quad c_{i,j} = 1 \text{ if } \left(1 - 2\sigma_n\right)\overline{g} \le g_{i,j} \le \left(1 + 2\sigma_n\right)\overline{g}$$

$$\text{or } c_{i,j} = 0 \text{ otherwise} \qquad (2.10)$$

The *homog* filter does not require any parameters or thresholds to be tuned, thus making the filter suitable for automatic interpretation.

2.2.4   Geometric Filtering (*gf4d*)

The concept behind geometric filtering is that speckle appears in the image as narrow walls and valleys. The geometric filter, through iterative repetition, gradually tears down the narrow walls (bright edges) and fills up the narrow valleys (dark edges), thus smearing the weak edges that need to be preserved.

The *gf4d* filter [11] investigated in this study uses a nonlinear noise reduction technique. It compares the intensity of the central pixel in a 3×3 neighborhood with those of its eight neighbors and, based on the neighborhood pixel intensities, it increments or decrements the intensity of the central pixel such that it becomes more representative of its surroundings. The operation of the geometric filter *gf4d* is described in Figure 2.1 and has the following form:

1. Select direction and assign pixel values:
   Select the direction to be NS and the corresponding three consecutive pixels to be *a*, b, *c* [see Figure 2.1(a, b), respectively].
2. Carry out central pixel adjustments:
   Do the following intensity adjustments [see Figure 2.1(b)]:
   if $a \geq b + 2$, then $b = b + 1$;
   if $a > b$ and $b \leq c$, then $b = b + 1$;
   if $c > b$ and $b \leq a$, then $b = b + 1$;
   if $c \geq b + 2$, then $b = b + 1$;
   if $a \leq b - 2$, then $b = b - 1$;
   if $a < b$ and $b \geq c$, then $b = b - 1$;
   if $c < b$ and $b \geq a$, then $b = b - 1$;
   if $c \leq b - 2$, then $b = b - 1$.
3. Repeat.
4. Repeat steps 1 and 2 for the west-east (WE) direction, west-north to south-east (WN-SE) direction, and north-east to west-south (NE to WS) direction [see Figure 2.1(a)].

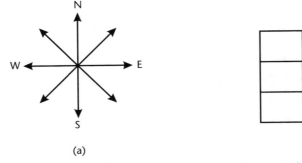

(a)   (b)

Figure 2.1   (a) Directions of implementation of the *gf4d* geometric filter, and (b) pixels selected for the NS direction where the intensity of central pixel *b* is adjusted based on the intensities values for pixels *a, b,* and *c.*

2.2.5   Homomorphic Filtering (*homo*)

The *homo* filter performs homomorphic filtering for image enhancement by calculating the FFT of the logarithmic compressed image, applying a denoising homomorphic filter function $H(.)$, and then performing the inverse FFT of the image [18, 19]. The homomorphic filter function $H(.)$ can be constructed either using a bandpass Butterworth or a high-boost Butterworth filter. In this study, a high-boost Butterworth filter was used with the homomorphic function [18]:

$$H_{u,v} = \gamma_L + \frac{\gamma_H}{1 + \left(D_0/D_{u,v}\right)^2} \qquad (2.11a)$$

with

$$D_{u,v} = \sqrt{\left(u - N/2\right)^2 + \left(v - N/2\right)^2} \qquad (2.11b)$$

where $D_0 = 1.8$ is the cutoff frequency of the filter; $\gamma_L = 0.4$ and $\gamma_H = 0.6$, are the gains for the low and high frequencies, respectively; u and v are the spatial coordinates of the frequency transformed image; and N the dimensions of the image in the u, v space. This form of filtering sharpens features and flattens speckle variations in an image.

2.2.6   Diffusion Filtering

Diffusion filters remove noise from an image by modifying the image with a partial differential equation (PDE). The smoothing is carried out depending on the image edges and their directions. Anisotropic diffusion is an efficient nonlinear technique for simultaneously performing contrast enhancement and noise reduction. It smooths homogeneous image regions but retains image edges [5, 23, 24] without requiring any information from the image power spectrum. It may thus be applied directly to logarithmic compressed images. Consider applying the isotropic diffusion equation given by $d_{gi,j,t}/dt = div(d\nabla g)$ using the original noisy image $g_{i,j,t=0}$ as the initial condition, where $g_{i,j,t=0}$ is an image in the continuous domain, i, j specifies spatial position, t, is an artificial time parameter, d is the diffusion constant, and ∇g is the image gradient. Modifying the image according to this linear isotropic diffusion equation is equivalent to filtering the image with a Gaussian filter. In this section we present conventional anisotropic diffusion (*ad*) and coherent nonlinear anisotropic diffusion (*nldif*).

2.2.6.1   Anisotropic Diffusion Filtering (*ad*)

Perona and Malik [24] replaced the classical isotropic diffusion equation, as described earlier in this chapter, by introducing a function $d_{i,j,t} = f(|\nabla g|)$, which smooths the original image while trying to preserve brightness discontinuities with:

$$\frac{dg_{i,j,t}}{dt} = div\left[d_{i,j,t}\nabla g_{i,j,t}\right] = \left[\frac{d}{di}d_{i,j,t}\frac{d}{di}g_{i,j,t}\right] + \left[\frac{d}{dj}d_{i,j,t}\frac{d}{dj}g_{i,j,t}\right] \qquad (2.12a)$$

where $|\nabla g|$ is the gradient magnitude, and $d(|\nabla g|)$ is an edge stopping function, which is chosen to satisfy $d \to 0$ when $|\nabla g| \to \infty$ so that the diffusion is stopped across edges. This function, called the diffusion coefficient $d(|\nabla g|)$, which is a monotonically decreasing function of the gradient magnitude $|\nabla g|$, yields intraregion smoothing not inter-region smoothing [20, 21, 23, 24] by impeding diffusion at image edges. It increases smoothing parallel to the edge and stops smoothing perpendicular to the edge, as the highest gradient values are perpendicular to the edge and dilated across edges.

The choice of $d(|\nabla g|)$, can greatly affect the extent to which discontinuities are preserved. For example, if $d(|\nabla g|)$, is constant at all locations, then smoothing progresses in an isotropic manner. If $d(|\nabla g|)$, is allowed to vary according to the local image gradient, then we have anisotropic diffusion. A basic anisotropic PDE is given in (2.12a). Two different diffusion coefficients were proposed in [24] and also derived in [23]. The diffusion coefficients suggested were:

$$d\left(|\nabla g|\right) = \frac{1}{1 + \left(|\nabla g_{i,j}|/K\right)^2}$$ (2.12b)

where K in (2.12b) is a positive gradient threshold parameter, known as diffusion or flow constant [23]. In our study the diffusion coefficient in (2.12b) was used because it was found to perform better in our carotid artery images.

A discrete formulation of the anisotropic diffusion in (2.12a) is [2, 23, 24]:

$$\frac{dg_{i,j}}{dt} = \frac{\lambda}{|\eta_s|} \left\{ \begin{array}{l} d_{i+1,j,t}\left[g_{i+1,j} - g_{i,j}\right] + d_{i-1,j,t}\left[g_{i-1,j} - g_{i,j}\right] \\ + d_{i,j+1,t}\left[g_{i,j+1} - g_{i,j}\right] + d_{i,j-1,t}\left[g_{i,j-1} - g_{i,j}\right] \end{array} \right\}$$ (2.13a)

where the new pixel gray value, $f_{i,j}$, at location i, j, is:

$$f_{i,j} = g_{i,j} + \frac{1}{4}\frac{dg_{i,j}}{dt}$$ (2.13b)

where $d_{i+1,j,t}$, $d_{i-1,j,t}$, $d_{i,j+1,t}$, and $d_{i,j-1,t}$ are the diffusion coefficients for the west, east, north, and south pixel directions, in a four-pixel neighborhood, around the pixel i, j, where diffusion is computed, respectively. The coefficient of variation leads to the largest diffusion where the nearest-neighbor difference is largest (largest edge), while the smallest diffusion is calculated where the nearest-neighbor difference is smallest (the weakest edge). The constant $\lambda \in \Re^+$ is a scalar that determines the rate of diffusion, η_s represents the spatial neighborhood of pixel i, j, and $|\eta_s|$ is the number of neighbors (usually four except at the image boundaries). Perona and Malik [24] linearly approximated the directional derivative in a particular direction as $g_{i,j} = g_{i+1,j} - g_{i,j}$ (for the east direction of the central pixel i, j). Modifying the image according to (2.13), which is a linear isotropic diffusion equation, is equivalent to filtering the image with a Gaussian filter. The parameters for the anisotropic diffusion filter used in this study were $\lambda = 0.25$, $\eta_s = 8$, and the parameter $K = 30$, which was used for the calculation of the edge stopping function $d(|\nabla g|)$, in (2.12b).

2.2.6.2 Coherent Nonlinear Anisotropic Diffusion Filtering (nldif)

The applicability of the *ad* filter of (2.12) is restricted to smoothing with edge enhancement, where $|\nabla g|$, has a higher magnitude at its edges. In general, the function $d(|\nabla g|)$, in (2.12) can be put into a tensor form that measures local coherence of structures such that the diffusion process becomes more directional in both the gradient and the contour directions, which represent the directions of maximum and minimum variations, respectively. Therefore, the *nldif* filter will take the form:

$$\frac{dg_{i,j,t}}{dt} = div[D\nabla g] \tag{2.14a}$$

where $D \in \Re^{2\times2}$ is a symmetric, positive, semidefinite diffusion tensor representing the required diffusion in both the gradient and contour directions and, hence, enhancing coherent structures as well as edges. The design of D, as well as the derivation of the coherent nonlinear anisotropic diffusion model, can be found in [25], and it is given as follows:

$$D = (\omega_1\omega_2)\begin{pmatrix}\lambda_1 & 0 \\ 0 & \lambda_2\end{pmatrix}\begin{pmatrix}\omega_1^T \\ \omega_2^T\end{pmatrix} \tag{2.14b}$$

with

$$\lambda_1 = \begin{cases}\alpha\left(1 - \dfrac{(\mu_1 - \mu_2)^2}{s^2}\right) & \text{if } (\lambda_1 - \lambda_2)^2 \le s^2 \\ 0, & \text{else } \lambda_2 = \alpha\end{cases} \tag{2.14c}$$

where the eigenvectors ω_1, ω_2 and the eigenvalues λ_1, λ_2 correspond to the directions of maximum and minimum variations and the strength of these variations, respectively. The flow at each point is affected by the local coherence, which is measured by $(\mu_1 - \mu_2)$ in (2.14c).

The parameters used in this study for the *nldif* filter were $s^2 = 2$ and $\alpha = 0.9$, which were used for the calculation of the diffusion tensor D, and the parameter step size $m = 0.2$, which defined the number of diffusion steps performed. The local coherence is close to zero in very noisy regions and diffusion must become isotropic $(\mu_1 = \mu_2 = \alpha = 0.9)$, whereas in regions with lower speckle noise the local coherence must correspond to $(\mu_1 - \mu_2)^2 > s^2$ [25].

2.2.7 Wavelet Filtering (waveltc)

Speckle reduction filtering in the wavelet domain, as used in this study, is based on the idea of the Daubenchies Symlet wavelet and on soft-thresholding denoising. It was first proposed by Donoho [30] and also investigated by [26, 27, 37–40]. The Symlets family of wavelets, although not perfectly symmetrical, was designed to have the least asymmetry and highest number of vanishing moments for a given compact support [30]. The *waveltc* filter implemented in this study is described as follows:

1. Estimate the variance of the speckle noise, σ_n^2, from the logarithmic transformed noisy image using (2.8).
2. Compute the discrete wavelet transform (DWT) using the Symlet wavelet for two scales.
3. For each sub-band:
   Compute a threshold [28, 30]:

$$T = \begin{cases} (T_{max} - \alpha(j-1))\sigma_n & \text{if } T_{max} - \alpha(j-1) > T_{min} \\ T_{min}\sigma_n & \text{else} \end{cases} \qquad (2.15)$$

   where α is a decreasing factor between two consecutive levels, T_{max} is a maximum factor for σ_n, and T_{min} is a minimum factor. The threshold T is primarily calculated using σ_n and a decreasing factor, $T_{max} - \alpha(j-1)$.
   Apply the thresholding procedure in (2.15) on the wavelet coefficients in step 2.
4. Invert the multiscale decomposition to reconstruct the despeckled image, f.

2.3 Methodology

2.3.1 Material

A total of 440 ultrasound images of carotid artery bifurcations—220 asymptomatic and 220 symptomatic—were investigated in this study. Asymptomatic images were recorded from patients at risk of atherosclerosis in the absence of clinical symptoms, whereas symptomatic images were recorded from patients at risk of atherosclerosis who had already developed clinical symptoms, such as a stroke episode.

2.3.2 Recording of Ultrasound Images

The ultrasound images of the carotid artery bifurcations for this study were acquired using an ATL HDI-3000 ultrasound scanner. The ATL HDI-3000 ultrasound scanner is equipped with 64 elements, fine pitch, high resolution, a 38-mm broadband array, a multielement ultrasound scanning head with an operating frequency range of 4 to 7 MHz, a 10×8 mm acoustic aperture, and a transmission focal range of 0.8 to 11 cm [41]. In this work all images were recorded as they are displayed in the ultrasound monitor, after logarithmic compression. The images were recorded digitally on a magneto-optical drive, with a resolution of 768×756 pixels with 256 gray levels. The image resolution was 16.66 pixels/mm.

2.3.3 Despeckle Filtering

Ten despeckle filters were investigated as presented in Section 2.2, and were applied on the 440 logarithmically compressed ultrasound images.

2.3.4 Texture Analysis

Texture provides useful information for the characterization of atherosclerotic plaque [42]. In this study a total of 56 different texture features were extracted both from the original and the despeckled images as follows [42, 43]:

2.3.4.1 Statistical Features (SF)

1. Mean;
2. Median;
3. Variance (σ^2);
4. Skewness (σ^3);
5. Kurtosis (σ^4);
6. Speckle index (σ/m).

2.3.4.2 Spatial Gray-Level Dependence Matrices (SGLDMs)

SGLDM as proposed by Haralick et al. [43]:

1. Angular second moment;
2. Contrast;
3. Correlation;
4. Sum of squares: variance;
5. Inverse difference moment;
6. Sum average;
7. Sum variance;
8. Sum entropy;
9. Entropy;
10. Difference variance;
11. Difference entropy;
12. Information measures of correlation.

Each feature was computed using a distance of one pixel. Also, for each feature the mean values and the range of values were computed and used as two different feature sets.

2.3.4.3 Gray Level Difference Statistics (GLDS) [44]

1. Contrast;
2. Angular second moment;
3. Entropy;
4. Mean.

2.3.4.4 Neighborhood Gray Tone Difference Matrix (NGTDM) [45]

1. Coarseness;

   2. Contrast;

   3. Business;

   4. Complexity;

   5. Strength.

2.3.4.5 Statistical Feature Matrix (SFM) [46]

   1. Coarseness;

   2. Contrast;

   3. Periodicity;

   4. Roughness.

2.3.4.6 Laws Texture Energy Measures (TEM) [46]

For the laws TEM extraction, vectors of length $l = 7$, $L = (1, 6, 15, 20, 15, 6, 1)$, $E = (-1, -4, -5, 0, 5, 4, 1)$ and $S = (-1, -2, 1, 4, 1, -2, -1)$ were used, where L performs local averaging, E acts as an edge detector, and S acts as a spot detector. The following TEM features were extracted:

   1. LL: texture energy (TE) from LL kernel;

   2. EE: TE from EE kernel;

   3. SS: TE from SS kernel;

   4. LE: average TE from LE and EL kernels;

   5. ES: average TE from ES and SE kernels;

   6. LS: average TE from LS and SL kernels.

2.3.4.7 Fractal Dimension Texture Analysis (FDTA) [46]

The Hurst coefficient, $H^{(k)}$, for resolutions with $k = 1, 2, 3, 4$.

2.3.4.8 Fourier Power Spectrum (FPS) [46]

   1. Radial sum;

   2. Angular sum.

2.3.5 Distance Measures

To identify the most discriminant features separating asymptomatic and symptomatic ultrasound images before and after despeckle filtering, the following distance measure was computed for each feature [42]:

$$dis_{zc} = |m_{za} - m_{zs}| / \sqrt{\sigma_{za}^2 + \sigma_{zs}^2} \qquad (2.16)$$

where z is the feature index, c if o indicates the original image set and if f indicates the despeckled image set, m_{za} and m_{zs} are the mean values, and σ_{za} and σ_{zs} are the standard deviations of the asymptomatic and symptomatic classes, respectively.

The most discriminant features are the ones with the highest distance values [42]. If the distance after despeckle filtering is increased, that is:

$$dis_{zf} > dis_{zo} \qquad (2.17)$$

then it can be derived that the classes may be better separated.

For each feature, a percentage distance was computed as follows:

$$feat\_dis_z = \left(dis_{zf} - dis_{zo}\right)100 \qquad (2.18)$$

For each feature set, a score distance was computed as:

$$Score\_Dis = (1/N)\sum_{z=1}^{N}\left(dis_{zf} - dis_{zo}\right)100 \qquad (2.19)$$

where N is the number of features in the feature set. Note that for all features a larger feature distance shows improvement.

2.3.6 Univariate Statistical Analysis

The Wilcoxon rank sum test was used to detect if for each texture feature a significant (S) or not significant (NS) difference exists between the original and the despeckled images at $p < 0.05$.

2.3.7 kNN Classifier

The statistical k-nearest-neighbor (kNN) classifier using the Euclidean distance with $k = 7$ was also used to classify a plaque image as asymptomatic or symptomatic [42]. The leave-one-out method was used for evaluating the performance of the classifier, where each case is evaluated in relation to the rest of the cases. This procedure is characterized by no bias concerning the possible training and evaluation bootstrap sets. The kNN classifier was chosen because it is simple to implement and computationally very efficient. This is highly desired due to the many feature sets and filters tested [46].

2.3.8 Image Quality Evaluation Metrics

Differences between the original images, $g_{i,j}$, and the despeckled images, $f_{i,j}$, were evaluated using the image quality evaluation metrics presented in Section 4.3 of this book and in [47].

2.3.9 Visual Evaluation by Experts

Visual evaluation can be broadly categorized as the ability of an expert to extract useful anatomic information from an ultrasound image. The visual evaluation varies of course from expert to expert and is subject to the observer's variability [48]. The visual evaluation, in this study, was carried out according to the ITU-R recommendations with the Double Stimulus Continuous Quality Scale (DSCQS) procedure [49].

A total of 100 ultrasound images of carotid artery bifurcations (50 asymptomatic and 50 symptomatic) were evaluated visually by two vascular experts (a cardiovascular surgeon and a neurovascular specialist) before and after despeckle filtering. For each case, the original and the despeckled images (despeckled with filters *lsmv, lsminsc, median, wiener, homog, gf4d, homo, ad, nldif,* and *waveltc*) were presented without labeling at random to the two experts. The experts were asked to assign a score on a scale of 1 to 5 corresponding to low and high subjective visual perception criteria, with a score of 5 being given to an image with the best visual perception. Therefore, the maximum score for a filter was 500 if the expert assigned a score of 5 for all 100 images. For each filter, the score was divided by 5 so it could be expressed in percentage format. The experts were allowed to give equal scores to more than one image in each case. For each class and for each filter the average score was computed.

The two vascular experts evaluated the area around the distal common carotid, 2 to 3 cm before the bifurcation and the bifurcation. It is known that measurements taken from the far wall of the carotid artery are more accurate than those taken from the near wall [50]. Furthermore, the experts were examining the image in the lumen area, in order to identify the existence of plaque or not.

2.4   Results

In this section we present the results of the 10 despeckle filters described in Section 2.2, applied on 220 asymptomatic and 220 symptomatic ultrasound images of carotid artery bifurcations. A total of 56 texture features were computed, and the most discriminant ones are presented. Furthermore, the performance of these filters is investigated for discriminating between asymptomatic and symptomatic images using the statistical kNN classifier. Moreover, nine different image quality evaluation metrics were computed, as well as visual evaluation scores carried out by two experts.

2.4.1   Evaluation of Despeckle Filtering on a Symptomatic Ultrasound Image

Figure 2.2 shows an ultrasound image of the carotid together with the despeckled images. The best visual results as assessed by the two experts were obtained by the filters *lsmv* and *lsminsc*, whereas the filters *gf4d, ad,* and *nldif* also showed good visual results, but smoothed the image considerably, which means that edges and subtle details could be lost. Filters that showed a blurring effect were the *median, wiener, homog,* and *waveltc* filters. The *wiener, homog,* and *waveltc* filters showed poorer visual results.

2.4.2   Texture Analysis: Distance Measures

Despeckle filtering and texture analysis were carried out on 440 ultrasound images of the carotid. Table 2.2 tabulates the results of *feat_dis$_z$* (2.18), and *Score_dis*

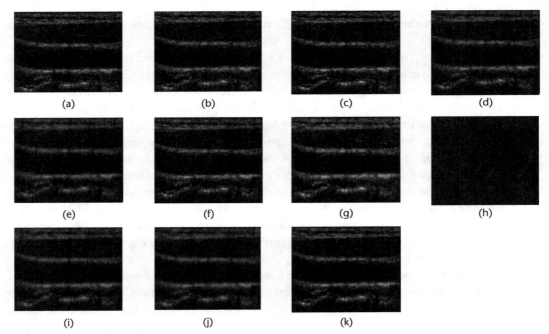

Figure 2.2 (a) Original ultrasound image of a carotid artery (2- to 3-cm proximal to bifurcation) and (b–k) the despeckled filtered images. (a) Original image. (b) *lsmv.* (c) *lsminsc.* (d) *median.* (e) *wiener.* (f) *homog.* (g) *gf4d.* (h) *homo.* (i) *ad.* (j) *nldif.* (k) *waveltc.* (*From:* [9]. © 2005 IEEE. Reprinted with permission.)

(2.19), for the SF, SGLDM range of values, and NGTDM feature sets for the 10 despeckle filters. Results are presented only for these feature sets because they were the ones that provided the best performance. The filters are categorized as local statistics, median, maximum homogeneity (HF), geometric (GF), homomorphic (HM), diffusion, and wavelet filters, as introduced in Sections 2.1 and 2.2. Also the number of iterations (Nr. of It.) for each filter is given, which was selected based on C and on the visual evaluation of the two experts. When C began to change minimally, the filtering process was stopped. The bold values represent the values that showed an improvement after despeckle filtering compared to the original. The last row in each subtable shows the *Score_dis* for all features, where the highest value indicates the best filter in the subtable. Additionally, a total score distance, *Score_dis_T* was computed for all feature sets and is shown in the last row of Table 2.2. Some of the despeckle filters shown in Table 2.2 change a number of texture features by increasing the distance between the two classes (the positive values in Table 2.2); this makes the identification and separation between asymptomatic and symptomatic plaques more feasible. A positive feature distance shows improvement after despeckle filtering, whereas a negative distance shows deterioration.

In the first part of Table 2.2 the results for the SF features are presented, where the best *Score_dis* is given for the filter *homo* followed by the *lsminsc, lsmv, homog, nldif, waveltc, median,* and *wiener;* the worst *Score_dis* is given by *gf4d.* All filters reduced speckle index C. Almost all filters significantly reduced the variance σ^2 and the kurtosis σ^3 of the histogram, as can be seen from the bold values in the first part of Table 2.2.

Table 2.2 Feature Distance (2.18) and Score Distance (2.19) Results for the SF, SGLDM Range of Values, and NGTDM Texture Feature Sets Between Asymptomatic and Symptomatic Carotid Plaque Ultrasound Images

Feature	Local Statistics			Median	HF	GF	HM	Diffusion		Wavelet
	lsmv	lsminsc	wiener	median	homog	gf4d	homo	ad	nldif	waveltc
Nr. of It.	4	1	2	2	1	3	2	20	5	5
SF–Statistical Features										
Mean	14	22	19	4	11	3	164	18	5	15
Median	−5	−17	−26	−5	−5	−15	110	−29	−6	−15
σ^2	18	38	18	7	13	−2	140	9	7	18
σ^3	12	16	5	9	7	−0.1	149	17	7	8
σ^4	−12	−14	−7	−6	−4	−3	117	−21	6	−9
C	0.4	0.3	0.3	0.4	0.3	0.4	0.08	0.3	0.4	0.3
Score_dis	27	45	9	9	22	−17	680	−6	19	17
SGLDM Range of Values–Spatial Gray Level Dependence Matrix										
ASM	−21	−0.5	−29	2	−4	−8	−47	−25	−17	−20
Contrast	47	107	14	64	32	−3	165	104	13	22
Correlation	12	59	15	24	−5	2	10	54	−4	−4
SOSV	9	40	18	10	16	−2	101	9	8	20
IDM	−50	−11	−48	2	−29	−8	94	−54	−34	−43
SAV	17	24	23	7	15	3	169	22	6	18
ΣVar	19	38	18	9	15	−2	90	9	8	20
ΣEntr	−34	−14	−49	3	−19	−4	−11	−47	−30	−36
Score_dis	−1	243	−38	121	21	−22	571	72	−50	−23
NGTDM–Neighborhood Gray Tone Difference Matrix										
Coarseness	30	87	4	9	−16	−7	72	−36	−37	−33
Contrast	7	−0.3	−9	8	0.4	−4	105	5	−27	−15
Busyness	17	26	−30	8	1	−4	48	−14	−39	8
Completion	64	151	21	53	80	2	150	63	18	27
Score_dis	118	264	−14	78	66	−13	375	18	−85	−13
Score_dis – T	144	551	−43	208	108	−52	1,626	84	−116	−19

Key: ASM: angular second moment; SOSV: sum of squares variance; IDM: inverse difference moment; SAV: sum a verage; Σ*Var:* sum variance; HF: homogeneity; GF: geometric; HM: homomorphic. Bold values show improvement after despeckle filtering. *Source:* [9].

In the second part of Table 2.2, the results for the SGLDM range-of-values feature set are tabulated. The filters with the highest *Score_dis* in the SGLDM range-of-values features set were the *homo, lsminsc, median, ad,* and *homog* filters. All other filters (*nldif, wiener, waveltc, gf4d, lsmv*) presented a negative *Score_dis*. The texture features that improved in most of the filters were the contrast, correlation, sum of squares variance, sum average, and sum variance features.

In the third part of Table 2.2, which shows the results for the NGTDM feature set, almost all filters showed an improvement in *Score_dis*. The best filters in the NGTDM feature set were the *homo, lsminsc, homog,* and *lsmv* filters. The texture features that were improved the most were the completion, coarseness, and contrast features. The completion feature of the image was increased by all filters.

Finally, in the last row of Table 2.2, the total score distance, *Score_dis_T*, for all feature sets is shown. The best values were obtained by the *homo, lsminsc, median, lsmv, homog*, and *ad* filters.

2.4.3 Texture Analysis: Univariate Statistical Analysis

Table 2.3 shows the results of the Wilcoxon rank sum test, which was performed on the SGLDM range-of-values features set of Table 2.2 for the 10 despeckle filters. The test was performed to check if significant differences exist between the features computed on the 440 original and the 440 despeckled images. Filters that resulted in the most significant number of features after despeckle filtering as listed in the "Score" column of Table 2.3 were as follows: *lsmv, gf4d, lsminsc*, and *nldif*. The rest of the filters gave a lower number of significantly different features. Features that showed a significant difference after filtering were the inverse difference moment (IDM) angular second moment (ASM), sum of entropy, contrast, correlation, sum of squares variance (SOSV), and sum variance, ΣVar. These features were mostly affected after despeckle filtering and they were significantly different.

2.4.4 Texture Analysis: kNN Classifier

Table 2.4 shows the percentage of correct classification scores for the kNN classifier with $k = 7$ for classifying a subject as asymptomatic or symptomatic. The classifier was evaluated using the leave-one-out method [46] on 220 asymptomatic and 220 symptomatic images of the original and despeckled images. The percentage of correct classifications score is given for the feature sets listed in the table and defined at the bottom of the table. Filters that showed an improvement in classification success scores compared to those of the original image set were the *homo* (3%), *gf4d* (1%), and *lsminsc* (1%) filters (see the last row labeled "Average" in Table 2.4).

The feature sets that benefitted the most from the despeckle filtering were the SF, GLDS, NGTDM, and TEM sets when counting the number of cases in which the correct classification score was improved. Less improvement was observed for the FDTA, SFM, SGLDMm, FPS, and SGLDMr feature sets. For the SGLDMr feature set better results are given for the *lsminsc* filter with an improvement of 2%. This is the only filter that showed an improvement for this class of features. For the TEM feature set the filter *lsmv* showed the best improvement at 9%, whereas for the FPS feature set the filter *lsminsc* gave the best improvement at 5%. The *lsminsc* filter showed improvements for the GLDS and NGTDM feature sets, whereas the *lsmv* filter showed improvements for the SF and TEM feature sets.

2.4.5 Image Quality Evaluation Metrics

Table 2.5 tabulates the image quality evaluation metrics presented in Section 4.3 for the 220 asymptomatic and 220 symptomatic ultrasound images between the original and the despeckled images, respectively. The best values were obtained for the *nldif, lsmv,* and *waveltc* filters with lower MSE, RMSE, Err3 and Err4 and with

Table 2.3 Wilcoxon Rank Sum Test for the SGLDM Range-of-Values Texture Features Applied on the 440 Ultrasound Images of Carotid Plaque Before and After Despeckle Filtering

Feature	Local Statistics			Median	HF	GF	HM	Diffusion		Wavelet	Score
	lsmv	*lsminsc*	*wiener*	*median*	*homog*	*gf4d*	*homo*	*ad*	*nldif*	*waveltc*	
ASM	S	S	NS	NS	S	S	NS	S	S	S	7
Contrast	S	NS	NS	NS	NS	S	NS	NS	S	NS	3
Correlation	S	S	NS	NS	NS	S	NS	NS	NS	NS	3
SOSV	S	NS	NS	NS	NS	S	NS	NS	NS	NS	2
IDM	S	S	NS	S	S	S	S	NS	S	S	8
SAV	NS	NS	NS	NS	NS	NS	NS	NS	NS	NS	0
ΣVar	S	S	NS	NS	NS	NS	NS	NS	NS	NS	2
ΣEntropy	S	S	NS	NS	NS	S	NS	NS	S	S	5
Score	7	5	0	1	2	6	1	1	4	3	

Key: ASM: angular second moment; SOSV: sum of squares variance; IDM: inverse difference moment; SAV: sum average; ΣVar: sum variance; HM: homomorphic; HF: homogeneity; GF: geometric; HM: homomorphic. Score: illustrates the number of S.
At $p < 0.05$, "S" means the feature is significantly different after filtering and "NS" means the feature is not significantly different after filtering.
Source: [9].

Table 2.4 Percentage of Correct Classifications Score for the kNN Classifier with $k = 7$ for the Original and the Filtered Image Sets

Feature Set	No. of Features	Original	Local Statistics			Median	HF	GF	HM	Diffusion		Wavelet
			lsmv	*lsminsc*	*wiener*	*median*	*homog*	*gf4d*	*homo*	*ad*	*nldif*	*waveltc*
SF	5	59	**62**	**61**	**61**	57	**63**	59	**65**	**60**	52	**61**
SGLDMm	13	65	63	64	62	63	**69**	67	**68**	61	**66**	63
SGLDMr	13	70	66	**72**	64	66	65	70	69	64	65	65
GLDS	4	64	63	**66**	61	**69**	64	**66**	**72**	59	58	62
NGTDM	5	64	63	**68**	60	**69**	63	**65**	57	60	61	62
SFM	4	62	62	60	62	58	55	**65**	**68**	59	56	55
TEM	6	59	**68**	52	**60**	59	**66**	**60**	**65**	53	**60**	**60**
FDTA	4	64	63	**66**	53	**68**	53	62	**73**	55	54	62
FPS	2	59	54	**64**	59	58	59	59	59	52	48	55
Average		63	63	**64**	60	63	62	**64**	**66**	58	58	61

SF: statistical features; SGLDMm: spatial gray-level dependence matrix mean values; SGLDMr: spatial gray-level dependence matrix range of values; GLDS: gray-level difference statistics; NGTDM: neighborhood gray-tone difference matrix; SFM: statistical feature matrix; TEM: laws texture energy measures; FDTA: fractal dimension texture analysis; FPS: Fourier power spectrum; HF: homogeneity; GF: geometric; HM: homomorphic.
Bold values indicate improvement after despeckling.
Source: [9].

higher SNR and PSNR (see bottom of Table 2.5 for definitions). The GAE was 0.00 for all cases, and this can be attributed to the fact that the information between the original and the despeckled images remains unchanged. Best values for the universal quality index, *Q*, and the structural similarity index, SSIN, were obtained for the *lsmv* and *nldif* filters.

Table 2.5 Image Quality Evaluation Metrics Computed for the 220 Asymptomatic and 220 Symptomatic Images

Feature Set	Local Statistics			Median	HF	GF	HM	Diffusion		Wavelet
	lsmv	lsminsc	wiener	median	homog	gf4d	homo	ad	nldif	waveltc
Asymptomatic Images										
MSE	13	86	19	131	42	182	758	132	8	11
RMSE	3	9	4	10	6	13	27	11	2	3
Err3	7	17	5	25	14	25	38	21	5	4
Err4	11	26	7	41	24	40	49	32	10	5
GAE	0	0	0	0	0	0	0	0	0	0
SNR	25	17	23	16	21	14	5	14	28	25
PSNR	39	29	36	29	34	27	20	28	41	39
Q	0.83	0.78	0.74	0.84	0.92	0.77	0.28	0.68	0.93	0.65
SSIN	0.97	0.88	0.92	0.94	0.97	0.88	0.43	0.87	0.97	0.9
Symptomatic Images										
MSE	33	374	44	169	110	557	1452	374	8	23
RMSE	5	19	6	13	10	23	37	19	3	5
Err3	10	33	9	25	20	43	51	31	5	6
Err4	16	47	11	38	30	63	64	43	7	8
GAE	0	0	0	0	0	0	0	0	0	0
SNR	24	13	22	16	17	12	5	12	29	25
PSNR	34	23	33	26	28	21	17	23	39	36
Q	0.82	0.77	0.7	0.79	0.87	0.75	0.24	0.63	0.87	0.49
SSIN	0.97	0.85	0.89	0.81	0.94	0.85	0.28	0.81	0.97	0.87

MSE: mean square error; RMSE: randomized MSE; Err3, Err4: Minowski metrics; GAE: geometric average error; SNR: signal-to-noise radio; PSNR: peak SNR; Q: universal quality index; SSIN: structural similarity index.
Source: [9].

2.4.6 Visual Evaluation by Experts

Table 2.6 shows the results of the visual evaluation of the original and despeckled images made by two experts, a cardiovascular surgeon and a neurovascular specialist. The last row of Table 2.6 presents the overall average percentage score assigned by both experts for each filter.

For the cardiovascular surgeon, the average score showed that the best despeckle filter was the *lsmv* filter with a score of 62%, followed by *gf4d, median, homog,* and *original* with scores of 52%, 50%, 45%, and 41%, respectively. For the neurovascular specialist, the average score showed that the best filter was the *gf4d* with a score of 72%, followed by *lsmv, original, lsminsc,* and *median* with scores of 71%, 68%, 68%, and 66%, respectively. The overall average percentage score (last row of Table 2.6) shows that the highest score was given to the *lsmv* filters (67%), followed by *gf4d* (62%), *median* (58%), and *original* (54%). It should be emphasized that the *lsmv* despeckle filter is the only filter that was graded with a higher score than the original by both experts for the asymptomatic and symptomatic image sets.

Table 2.6 Percentage Scoring for Visual Evaluations by Experts of the Original and Despeckled Images: 50 Asymptomatic (A) and 50 Symptomatic (S)

Experts	A/S	Original	Local Statistics		Median	HF	GF	HM	Diffusion	Wavelet
			lsmv	Lsminsc	median	homog	gf4d	homo	nldif	waveltc
Cardiovascular surgeon	A	33	75	33	43	47	61	19	43	32
	S	48	49	18	57	43	42	20	33	22
Average %		41	62	26	50	45	52	19	38	27
Neurovascular specialist	A	70	76	73	74	63	79	23	52	29
	S	66	67	63	58	45	65	55	41	28
Average %		68	71	68	66	54	72	39	47	28
Overall Average %		54	67	47	58	50	62	29	43	28

HF: homogeneity; GF: geometric; HM: homomorphic.
Source: [9].

The reader might observe a difference in the scorings between the two vascular specialists. This is because the cardiovascular surgeon is primarily interested in the plaque composition and texture evaluation, whereas the neurovascular specialist is interested in evaluating the degree of stenosis and the lumen diameter in order to identify the plaque contour. The *lsmv* and *gf4d* filters were identified as the best despeckle filters by both specialists because they improved visual perception with overall average scores of 67% and 62%, respectively. The filters *waveltc* and *homo* were scored by both specialists with the lowest overall average scores of 28% and 29%, respectively.

By examining the visual results of Figure 2.2, the statistical results of Tables 2.2 to 2.5, and the visual evaluation of Table 2.6, we can conclude that the best filters are the *lsmv* and *gf4d* filters, which can be used for both plaque composition enhancement and plaque texture analysis, whereas the *lsmv*, *gf4d*, and *lsminsc* filters are more appropriate for identifying the degree of stenosis and, therefore, may be used when the primary interest is in outlining the plaque borders.

2.5 Discussion

Despeckle filtering is an important operation in the enhancement of ultrasound images of the carotid artery, both in the case of texture analysis and in the case of image quality evaluation and visual evaluation by the experts. In this study 10 despeckle filters were comparatively evaluated on 440 ultrasound images of carotid artery bifurcations, and the validation results are summarized in Table 2.7.

As given in Table 2.7, the *lsmv*, *lsminsc*, and *homo* filters improved the class separation between the asymptomatic and the symptomatic classes (see also Table 2.2). The *lsmv*, *lsminsc*, and *gf4d* filters gave a high number of significantly different features (see Table 2.3). The *lsminsc*, *gf4d*, and *homo* filters provided only a marginal improvement in the percentage of correct classifications (see Table 2.4). Moreover, the *lsmv*, *nldif*, and *waveltc* filters gave better image quality evaluation results (see Table 2.5). The *lsmv* and *gf4d* filters improved the visual assessment carried out by the experts (see Table 2.6). It is clearly shown that the *lsmv* filters gave the best performance, followed by the *lsminsc* and *gf4d* filters (see Table 2.7). The *lsmv* or

Table 2.7 Summary of Despeckle Filtering Findings in Ultrasound Imaging of the Carotid Artery

Despeckle Filter	Statistical and Texture Features (Table 2.2)	Statistical Analysis (Table 2.3)	kNN Classifier (Table 2.4)	Image Quality Evaluation (Table 2.5)	Optical Perception Evaluation (Table 2.6)
Local Statistics					
lsmv	✓	✓		✓	✓
lsminsc	✓	✓	✓		
Geometric Filtering					
gf4d		✓	✓		✓
Homomorphic Filtering					
homo	✓		✓		
Diffusion Filtering					
nldif				✓	
Wavelet Filtering					
waveltc				✓	

Source: [9].

gf4d filter could be used for despeckling asymptomatic images when the expert is interested mainly in the plaque composition and texture analysis. The *lsmv, gf4d,* or *lsmnsc* filter could be used for despeckling of symptomatic images when the expert is interested in identifying the degree of stenosis and the plaque borders. The *homo, nldif,* and *waveltc* filters gave poorer performance.

The *lsmv* filters gave a very good performance with respect to: (1) preserving the mean and the median as well as decreasing the variance and the speckle index of the image; (2) increasing the distance of the texture features between the asymptomatic and the symptomatic classes; (3) significantly changing the SGLDM range-of-values texture features after filtering based on the Wilcoxon rank sum test; (4) marginally improving the classification success rate of the kNN classifier for the classification of asymptomatic and symptomatic images in the cases of the SF, SMF, and TEM feature sets; and (5) improving the quality of the image. The *lsmv* filter, which is a simple filter, is based on local image statistics. It was first introduced in [10, 15, 16] by Jong-Sen Lee and coworkers and it was tested on a few SAR images with satisfactory results. It was also used for SAR imaging in [14] and image restoration in [17], again with satisfactory results.

The *lsminsc* filter gave the best performance with respect to: (1) preserving the mean, as well as decreasing the variance and the speckle index and increasing the contrast of the image; (2) increasing the distance of the texture features between the asymptomatic and the symptomatic classes; (3) significantly changing the SGLDM texture features after filtering based on the Wilcoxon rank sum test; and (4) improving the classification success rate of the kNN classifier for the classification of asymptomatic and symptomatic images in the cases of the SF, SGLDMr, GLDS, NGTDM, FDTA and FPS feature sets.

The *lsminsc* filter was originally introduced by Nagao in [34] and was tested on an artificial and a SAR image with satisfactory performance. In this study the filter was modified by using the speckle index instead of the variance value for each subwindow [as described in (2.9) in Section 2.2.1.2].

The *gf4d* filter gave a very good performance with respect to: (1) decreasing the speckle index; (2) marginally increasing the distance of the texture features between the asymptomatic and the symptomatic classes; (3) significantly changing the SGLDM range-of-values texture features after filtering based on the Wilcoxon rank sum test; and (4) improving the classification success rate of the kNN classifier for the classification of asymptomatic and symptomatic images in the cases of the SGLDMm, GLDS, NGTDM, SFM, and TEM feature sets. The geometric filter *gf4d* was introduced by Crimmins [11], and was tested visually on a few SAR images with satisfactory results.

Filters used for speckle reduction in ultrasound imaging by other investigators include *median* [35] *wiener* [14], *homog* [8], *homo* [18, 19], *ad* [5], and *waveltc* [30, 51]. However, these filters were evaluated on a small number of images, and their performance was tested based only on the mean, median, standard deviation, and speckle index of the image before and after filtering.

The *median* and *wiener* filters were originally used by many researchers for suppressing the additive and later for suppressing the multiplicative noise in different types of images [2–10, 14, 35]. The results of this study showed that the *wiener* and *median* filters were not able to remove the speckle noise and produced blurred edges in the filtered image (see Figure 2.2). In this study the *median* filter performed poorer as shown in Tables 2.2, 2.3, and 2.4.

The *homog* [8] and *homo* [2, 18, 19] filters, were recently used by some researchers for speckle reduction but our results in Tables 2.2, 2.3, and 2.5 and the visual evaluation of the experts in Table 2.6 showed poor performance, especially for the *homo* filter.

Anisotropic diffusion is an efficient nonlinear technique for simultaneously performing contrast enhancement and noise reduction. It smooths homogeneous image regions but retains image edges [24]. Anisotropic diffusion filters usually require many iteration steps compared with the local statistic filters. In a recent study [5], speckle-reducing anisotropic diffusion filtering was proposed as the most appropriate filter for ultrasound images of the carotid artery. However, in this study, *ad*, as shown in Tables 2.2, 2.3, 2.4, 2.5, and 2.6 did not perform as well as the *lsmv*, *gf4d* and *lsminsc* filters.

Furthermore, wavelet filtering proposed by Donoho in [30], was investigated for suppressing the speckle noise in SAR images [16, 37], real-world images [26], and ultrasound images [27] with favorable results. This study showed that the *waveltc* filter gave poorer performance for removing the speckle noise from the ultrasound images of the carotid artery (Tables 2.2 and 2.3).

Anisotropic diffusion and wavelet filtering were also investigated in [32] on carotid artery ultrasound images, where a simple preprocessing procedure based on spectrum equalization and a nonlinear shrinkage procedure was applied, in order to suppress the spiky component to the logarithmic transformed speckle and to modify the noise component to approximately white Gaussian. Then anisotropic diffusion and wavelet filtering, which are based on an additive noise model, were applied. The preprocessing procedure does not modify the structures of a specific filtering method, but rather alters the noise in such a way that it becomes very similar to white Gaussian, so that existing powerful methods for additive noise reduction may be applied. A wavelet multiscale normalized modulus-based wavelet diffusion method was also recently proposed in [51] that utilizes the properties of the wavelet and the edge enhancement feature of the nonlinear diffusion. The performance of this algorithm was tested on cardiac ultrasound images and showed superior performance when compared to the speckle-reducing anisotropic method [5]. In [52] the effect of despeckle filtering on lossy compression was investigated, where compression and then despeckle filtering were applied on ultrasound images of a renal cell carcinoma. The images were simultaneously denoised and compressed in a single step, and it was shown that the Laplacian distribution compression scheme performed better.

In conclusion, despeckle filtering is an important operation in the enhancement of ultrasonic imaging of the carotid artery. In this study it was shown that simple filters based on local statistics (*lsmv* and *lsminsc*) and geometric filtering (*gf4d*) could be used successfully for the processing of these images. In this context, despeckle filtering can be used as a preprocessing step for the automated segmentation of the IMT [53] and the carotid plaque, followed by the carotid plaque texture analysis, and classification. This field is currently being investigated by our group [54]. Initial findings show promising results; however, further work is required to evaluate the performance of the suggested despeckle filters at a larger scale as well as their impact in clinical practice. In addition, the usefulness of the proposed despeckle filters in portable ultrasound systems and in wireless telemedicine systems still has to be investigated.

Acknowledgments

This work was partly funded through the Integrated System for the Support of the Diagnosis for the Risk of Stroke (IASIS) project of the Fifth Annual Program for the Financing of Research of the Research Promotion Foundation of Cyprus 2002–2005, as well as, through the Integrated System for the Evaluation of Ultrasound Imaging of the Carotid Artery (TALOS) project of the Program for Research and Technological Development 2003–2005 of the Research Promotion Foundation of Cyprus.

References

[1] Lamont, D., et al., "Risk of Cardiovascular Disease Measured by Carotid Intima-Media Thickness at Age 49–51: Life Course Study," *BMJ*, Vol. 320, January 2000, pp. 273–278.

[2] Burckhardt, C. B., "Speckle in Ultrasound B-Mode Scans," *IEEE Trans. on Sonics and Ultrasonics*, Vol. SU-25, No. 1, 1978, pp. 1–6.

[3] Wagner, R. F., et al., "Statistics of Speckle in Ultrasound B-Scans," *IEEE Trans. on Sonics and Ultrasonics*, Vol. 30, 1983, pp. 156–163.

[4] Goodman, J. W., "Some Fundamental Properties of Speckle," *J. of Optical Society of America*, Vol. 66, No. 11, 1976, pp. 1145–1149.

[5] Yongjian, Y., and S. T. Acton, "Speckle Reducing Anisotropic Diffusion," *IEEE Trans. on Image Processing*, Vol. 11, No. 11, November 2002, pp. 1260–1270.

[6] Prager, R. W., et al., *Speckle Detection in Ultrasound Images Using First Order Statistics*, GUED/F-INFENG/TR 415, University of Cambridge Dept. of Engineering Report, July 2002, pp. 1–17.

[7] Loizou, C. P., et al., "Speckle Reduction in Ultrasound Images of Atherosclerotic Carotid Plaque," *DSP 2002: Proc. IEEE 14th Int. Conf. on Digital Signal Processing*, Vol. 2, July 2002, pp. 525–528.

[8] Christodoulou, C. I., et al., "Despeckle Filtering in Ultrasound Imaging of the Carotid Artery," *Proc. Second Joint EMBS/BMES Conference*, Houston, TX, October 23–26, 2002, pp. 1027–1028.

[9] Loizou, C. P., et al. "Comparative Evaluation of Despeckle Filtering in Ultrasound Imaging of the Carotid Artery," *IEEE Trans. on Ultrasonics Ferroelectrics and Frequency Control*, Vol. 52, No. 10, 2005, pp. 1653–1669.

[10] Lee, J. S., "Speckle Analysis and Smoothing of Synthetic Aperture Radar Images," *Computer Graphics and Image Processing*, Vol. 17, 1981, pp. 24–32.

[11] Busse, L., T. R. Crimmins, and J. R. Fienup, "A Model Based Approach to Improve the Performance of the Geometric Filtering Speckle Reduction Algorithm," *IEEE Ultrasonic Symposium*, 1995, pp. 1353–1356.

[12] Kuan, D. T., et al., "Adaptive Restoration of Images with Speckle," *IEEE Trans. on Acoustic Speech and Signal Processing*, Vol. ASSP-35, 1987, pp. 373–383.

[13] Insana, M., et al., "Progress in Quantitative Ultrasonic Imaging," *Proc. SPIE Medical Imaging III, Image Formation*, Vol. 1090, 1989, pp. 2–9.

[14] Frost, V. S., et al., "A Model for Radar Images and Its Application for Adaptive Digital Filtering of Multiplicative Noise," *IEEE Trans. on Pattern Analysis and Machine Intelligence*, Vol. 4, No. 2, 1982, pp. 157–165.

[15] Lee, J. S., "Digital Image Enhancement and Noise Filtering by Using Local Statistics," *IEEE Trans. on Pattern Analysis and Machine Intelligence*, PAMI-2, No. 2, 1980, pp. 165–168.

[16] Lee, J. S., "Refined Filtering of Image Noise Using Local Statistics," *Computer Graphics and Image Processing*, Vol. 15, 1981, pp. 380–389.

[17] Kuan, D. T., and A. A. Sawchuk, "Adaptive Noise Smoothing Filter for Images with Signal Dependent Noise," *IEEE Trans. on Pattern Analysis and Machine Intelligence*, Vol. PAMI-7, No. 2, 1985, pp. 165–177.

[18] Solbo, S., and T. Eltoft, "Homomorphic Wavelet Based-Statistical Despeckling of SAR Images," *IEEE Trans. on Geosc. Remote Sensing*, Vol. 42, No. 4, 2004, pp. 711–721.

[19] Saniie, J., T. Wang, and N. Bilgutay, "Analysis of Homomorphic Processing for Ultrasonic Grain Signal Characterization," *IEEE Trans. on Ultrasonics, Ferroelectrics and Frequency Control*, Vol. 3, 1989, pp. 365–375.

[20] Jin, S., Y. Wang, and J. Hiller, "An Adaptive Non-Linear Diffusion Algorithm for Filtering Medical Images," *IEEE Trans. on Information Technology in Biomedicine*, Vol. 4, No. 4, December 2000, pp. 298–305.

[21] Weickert, J., B. Romery, and M. Viergever, "Efficient and Reliable Schemes for Nonlinear Diffusion Filtering," *IEEE Trans. on Image Processing*, Vol. 7, 1998, pp. 398–410.

[22] Rougon, N., and F. Preteux, "Controlled Anisotropic Diffusion," *Proc. Conf. on Nonlinear Image Processing VI*, San Jose, CA, February 5–10, 1995, pp. 1–12.

[23] Black, M., et al., "Robust Anisotropic Diffusion," *IEEE Trans. on Image Processing*, Vol. 7, No. 3, March 1998, pp. 421–432.

[24] Perond, P., and J. Malik, "Scale-Space and Edge Detection Using Anisotropic Diffusion," *IEEE Trans. on Pattern Analysis and Machine Intelligence*, Vol. 12, No. 7, July 1990, pp. 629–639.

[25] Abd-Elmoniem, K., A.-B. Youssef, and Y. Kadah, "Real-Time Speckle Reduction and Coherence Enhancement in Ultrasound Imaging Via Nonlinear Anisotropic Diffusion," *IEEE Trans. on Biomed. Eng.*, Vol. 49, No. 9, September 2002, pp. 997–1014.

[26] Zhong, S., and V. Cherkassky, "Image Denoising Using Wavelet Thresholding and Model Selection," *Proc. IEEE Int. Conf. on Image Processing*, Vancouver, Canada, November 2000, pp. 1–4.

[27]   Achim, A., A. Bezerianos, and P. Tsakalides, "Novel Bayesian Multiscale Method for Speckle Removal in Medical Ultrasound Images," *IEEE Trans. on Medical Imaging*, Vol. 20, No. 8, 2001, pp. 772–783.

[28]   Zong, X., A. Laine, and E. Geiser, "Speckle Reduction and Contrast Enhancement of Echocardiograms Via Multiscale Nonlinear Processing," *IEEE Trans. on Medical Imaging*, Vol. 17, No. 4, 1998, pp. 532–540.

[29]   Hao, X., S. Gao, and X. Gao, "A Novel Multiscale Nonlinear Thresholding Method for Ultrasonic Speckle Suppressing," *IEEE Trans. on Medical Imaging*, Vol. 18, No. 9, 1999, pp. 787–794.

[30]   Donoho, D. L., "Denoising by Soft Thresholding," *IEEE Trans. on Inform. Theory*, Vol. 41, 1995, pp. 613–627.

[31]   Wink, A. M., and J. B. T. M. Roerdink, "Denoising Functional MR Images: A Comparison of Wavelet Denoising and Gaussian Smoothing," *IEEE Trans. on Medical Imaging*, Vol. 23, No. 3, 2004, pp. 374–387.

[32]   Michailovich, O. V., and A. Tannenbaum, "Despeckling of Medical Ultrasound Images," *IEEE Trans. on Ultrasonics, Ferroelectrics and Frequency Control*, Vol. 53, No. 1, January 2006, pp. 64–78.

[33]   Dutt, V., "Statistical Analysis of Ultrasound Echo Envelope," Ph.D. Dissertation, Rochester, MN: Mayo Graduate School, 1995.

[34]   Nagao, M., and T. Matsuyama, "Edge Preserving Smoothing," *Computer Graphic and Image Processing*, Vol. 9, 1979, pp. 394–407.

[35]   Huang, T., G. Yang, and G. Tang, "A Fast Two-Dimensional Median Filtering Algorithm," *IEEE Trans. on Acoustics, Speech and Signal Processing*, Vol. 27, No. 1, 1979, pp. 13–18.

[36]   Ali, S. M., and R. E. Burge, "New Automatic Techniques for Smoothing and Segmenting SAR Images," *Signal Processing*, Vol. 14, 1988, pp. 335–346.

[37]   Medeiros, F. N. S., et al., "Edge Preserving Wavelet Speckle Filtering," *Proc. 5th IEEE Southwest Symposium on Image Analysis and Interpretation*, Santa Fe, NM, April 7–9, 2002, pp. 281–285.

[38]   Moulin, P., "Multiscale Image Decomposition and Wavelets," in *Handbook of Image and Video Processing*, A. Bovik, (ed.), Boston, MA: Academic Press, 2000, pp. 289–300.

[39]   Scheunders, P., "Wavelet Thresholding of Multivalued Images," *IEEE Trans. on Image Processing*, Vol. 13, No. 4, 2004, pp. 475–483.

[40]   Gupta, S., R. C. Chauhan, and S. C. Sexana, "Wavelet-Based Statistical Approach for Speckle Reduction in Medical Ultrasound Images," *Medical and Biological Engineering and Computing*, Vol. 42, 2004, pp. 189–192.

[41]   Philips Medical System Company, *Comparison of Image Clarity, SonoCT Real-Time Compound Imaging Versus Conventional 2-D Ultrasound Imaging*, ATL Ultrasound, Report, 2001.

[42]   Christodoulou, C. I., et al., "Texture-Based Classification of Atherosclerotic Carotid Plaques," *IEEE Trans. on Medical Imaging*, Vol. 22, No. 7, 2003, pp. 902–912.

[43]   Haralick, R. M., K. Shanmugam, and I. Dinstein, "Texture Features for Image Classification," *IEEE Trans. on Systems, Man, and Cybernetics*, Vol. SMC-3, November 1973, pp. 610–621.

[44]   Weszka, J. S., C. R. Dyer, and A. Rosenfield, "A Comparative Study of Texture Measures for Terrain Classification," *IEEE Trans. on Systems, Man, and Cybernetics*, Vol. SMC-6, April 1976, pp. 269–285.

[45]   Amadasun, M., and R. King, "Textural Features Corresponding to Textural Properties," *IEEE Trans. on Systems, Man, and Cybernetics*, Vol. 19, No. 5, September 1989, pp. 1264–1274.

[46]   Wu, C. M., Y. C. Chen, and K.-S. Hsieh, "Texture Features for Classification of Ultrasonic Images," *IEEE Trans. on Med. Imaging*, Vol. 11, June 1992, pp. 141–152.

[47] Chen, T. J., et al., "A Novel Image Quality Index Using Moran I Statistics," *Physics in Medicine and Biology*, Vol. 48, 2003, pp. 131–137.

[48] Loizou, C. P., et al., "Quantitative Quality Evaluation of Ultrasound Imaging in the Carotid Artery," *Med. Biol. Eng. Comput.*, Vol. 44, No. 5, 2006, pp. 414–426.

[49] Winkler, S., "Vision Models and Quality Metrics for Image Processing Applications," Ph.D. Dissertation, University of Lausanne–Switzerland, December 21, 2000.

[50] Elatrozy, T., et al., "The Effect of B-Mode Ultrasonic Image Standardization of the Echodensity of Symptomatic and Asymptomatic Carotid Bifurcation Plaque," *Int. Angiology*, Vol. 17, No. 3, September 1998, pp. 179–186.

[51] Yue, Y., et al., "Nonlinear Multiscale Wavelet Diffusion for Speckle Suppression and Edge Enhancement in Ultrasound Images," *IEEE Trans. on Medical Imaging*, Vol. 25, No. 3, March 2006, pp. 297–311.

[52] Loizou, C. P., et al., "Snakes Based Segmentation of the Common Carotid Artery Intima Media," *Med. Biol. Eng. Comput.*, Vol. 45, No. 1, 2007, pp. 35–49.

[53] Pattichis, C. S., et al., "Cardiovascular: Ultrasound Imaging in Vascular Cases," in *Wiley Encyclopaedia of Biomedical Engineering*, M. Akay, (ed.), New York: Wiley, 2006.

[54] Gupta, N., M. N. S. Swamy, and E. Plotkin, "Despeckling of Medical Ultrasound Images Using Data and Rate Adaptive Lossy Compression," *IEEE Trans. on Medical Imaging*, Vol. 24, No. 6, June 2005, pp. 743–754.

PART II
2-D/3-D Ultrasound in Vascular Imaging

CHAPTER 3

3-D US Imaging of the Carotid Arteries

Aaron Fenster, Grace Parraga, Anthony Landry, Bernard Chiu, Michaela Egger, and J. David Spence

3.1 Introduction

Determining the severity of carotid atherosclerotic stenosis has been an important step in establishing patient management pathways and identifying patients who can benefit from carotid endarterectomy versus those who should be treated using lifestyle and pharmaceutical interventions. Recently a number of research groups have developed phenotypes other than carotid stenosis using noninvasive imaging. Monitoring carotid plaque progression/regression and identifying vulnerable or high-risk plaques that can lead to thrombogenic events using noninvasive imaging tools now involve multiple disciplines and multiple modalities, including image processing.

Morphological characterization of carotid plaques has been used for risk stratification and genetic research [1, 2], evaluation of a patient's response to therapy [3], evaluation of new risk factors [4, 5], and quantification of the effects of new therapies [6]. Although 1-D measurements, such as the intima-media thickness, have been important for measuring the progression of carotid end-organ disease, 2-D measurements of a plaque area have been shown to have significant advantages because the plaque area progresses more than twice as quickly as does the thickness [4]. The use of conventional ultrasound in measuring plaque size (cross-sectional area) in a single longitudinal view has several limitations.

Conventional 2-D US has been used to quantify and correlate plaque morphology and composition with the risk of stroke; however, the results have been mixed. Some investigators have reported high accuracies for identifying features such as intraplaque hemorrhage [7, 8], but others have found only moderate sensitivity and specificity [9]. Some of these discrepancies are related to the lack of standardization and the variability of conventional ultrasound exams. However, it is generally agreed that several ultrasound techniques must be improved in order to more accurately and precisely quantify and characterize plaques.

Conventional 2-D US has also been useful in the assessment of plaque morphology. However, due to insufficient image contrast and to the variability of conventional 2-D US exams, accurate assessments of morphological plaque changes (of volume and surface irregularity) are difficult. Because it is difficult to relocate the 2-D image of the plaque, monitoring the changes in the development of the plaque over long periods of time using conventional 2-D US is problematic. In addition, the reconstruction of a 3-D impression of the vascular anatomy and the plaque using multiple 2-D images is time consuming and prone to variability and inaccuracy [10, 11].

Three-dimensional ultrasound is a recent development that promises to improve the visualization and quantification of complex anatomy and pathology to monitor the progression of atherosclerosis [11]. Noninvasive and reproducible 3-D imaging techniques that allow for the direct visualization and quantification of plaque development are becoming more important in serial monitoring of disease progression and regression [12–15]. Furthermore, the characterization of carotid plaques in three dimensions will potentially improve investigations into the changes of surface morphology, plaque geometry, and plaque distribution. These investigations can provide important information about the effects of antiatherosclerotic therapies.

3.1.1 Limitations of Conventional Carotid Ultrasound Imaging

In conventional carotid ultrasound examinations, the transducer is manipulated manually over the arteries by a sonographer or a radiologist. This approach works well when using ultrasound to *diagnosis* or *detect* the presence of carotid disease; however, quantitative measurements are subject to variability. Obtaining the required information from a series of 2-D images is largely dependent on the skill and experience of the operator in performing these tasks. This approach is suboptimal due to the following limitations:

1. Conventional ultrasound images are two dimensional; hence, the diagnostician must mentally transform multiple images in order to reconstruct a 3-D impression of the complex 3-D structure of the vessel and plaque. This may lead to variable and incorrect quantification, size, and extent of plaque morphology.

2. Conventional 2-D US images are difficult to reproduce; therefore, conventional 2-D US is nonoptimal for quantitative prospective studies, particularly where small changes in the plaque due to therapy are being followed over the course of time.

3. The patient's anatomy sometimes restricts the angle of the image, resulting in the inaccessibility of the optimal image plane necessary for the diagnosis of carotid disease and the assessment of plaque morphology.

4. Diagnostic and therapeutic decisions, at times, require accurate measurements of the volumes of the lesion. Several techniques for improving the management of atherosclerosis are now available; however, current 2-D US volume measuring techniques, which assume an idealized shape of the lesion to calculate the volume, use only simple measures of the width in a few views. This method potentially leads to inaccurate results and operator variability.

3.2   3-D US Imaging of the Carotid Arteries

Recent research has shown that 3-D US of the carotid arteries not only allows for the visualization, measurement (volume), and characterization of the carotid plaque but also provides the capability to monitor plaque progression and regression as well as to identify vulnerable plaques [13, 16–18]. The ability to monitor the progression and regression (i.e., the changes in volume and morphology) of carotid plaques quantitatively provides researchers with important information about the plaque's response to therapy (e.g., medication or diet mediated lowering of lipids) and the plaque's natural history [15, 18–21].

In the following sections, we review the methods developed and utilized for acquiring 3-D carotid ultrasound images and the tools required for visualizing and measuring plaque volume. The tools for monitoring the changes in the plaque's volume and the plaque's surface features will also be discussed. For further information about the technical and computational aspects of 3-D US, readers should refer to Chapter 1 of this book and to the recent review articles and books on the subject [11, 22–28].

3.2.1   3-D Carotid Ultrasound Scanning Techniques

Because images of the carotid arteries require at least a 4-cm scanning length, real-time 3-D (four-dimensional) systems cannot be used effectively. Thus, all 3-D US systems that are currently used to acquire images of the carotid arteries are conventional 1-D US transducers that produce 2-D US images. Different methods are used to locate the positions and orientations of 2-D images within the volume. Two 3-D US approaches have been successfully used to image the carotid arteries: mechanical linear scanners and magnetically tracked free-hand scanners. These approaches are summarized in the following sections.

3.2.1.1   Mechanical Linear 3-D Carotid Ultrasound Imaging

Linear scanners use a motorized mechanism to translate the transducer linearly along the neck of the patient as shown in Figure 3.1. While the transducer is moved, transverse 2-D images of the carotid arteries are acquired at regular spatial intervals. Each image in the set of acquired 2-D images is spaced equally so that all subsequent images are uniform and parallel to the preceding image. The length of the scan, which is the image length of the carotid arteries in 3-D, depends on the length of the mechanical scanning mechanism. The resolution of the image in the 3-D scanning direction (i.e., along the artery) can be optimized by varying the translating speed and sampling interval in order to match the sampling rate to the frame rate of the ultrasound machine and to match the sampling interval to (half) the elevational resolution of the transducer [29]. Typically, 2-D US images are acquired every 0.2 mm. If the 2-D US images are acquired at 30 frames per second, a 4-cm length will require 200 2-D images, which can be scanned in 6.7 seconds without cardiac gating.

The simple predefined geometry of the acquired 2-D US images can be reconstructed to form a 3-D image [29]. Using this approach, a 3-D image can be viewed

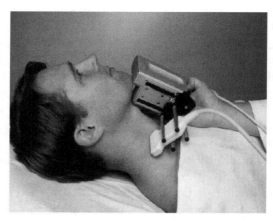

Figure 3.1 Photograph of a mechanical linear scanner mechanism used to acquire 3-D US images of the carotid arteries. The transducer is translated along the arteries, while conventional 2-D US images are acquired by a computer and reconstructed into a 3-D image.

immediately after scanning to determine if additional 3-D scans are necessary. The specific advantages of immediate review of 3-D images after a scan are that it significantly shortens the examination time and reduces digital storage requirements because a decision can be made not to store unnecessary 3-D images.

Because the 3-D carotid ultrasound image is produced from a series of conventional 2-D images, the resolution in the 3-D image will not be isotropic. In the direction parallel to the acquired 2-D US image planes, the resolution of the reconstructed 3-D image will be equal to the original 2-D images; however, in the direction of the 3-D scan along the arteries, the resolution of the reconstructed 3-D image will depend on the elevational resolution of the transducer and the interslice spacing [29]. Because the elevational resolution is less than the in-plane resolution of the 2-D US images, the resolution of the 3-D US image will be the lowest in the 3-D scanning direction (i.e., elevation). Therefore, a transducer with good elevational resolution should be used to obtain optimal results.

The scanning parameters can be adjusted depending on the following two types of ultrasound images: 3-D B-mode and 3-D power Doppler. The 3-D B-mode is used for imaging vessel walls and plaque; one to three focal zones result in a 2-D image acquisition rate of about 30 to 15 frames per second and a 3-D scan of 6.7 to 13.4 seconds for the acquisition of 200 2-D US images. For 3-D power Doppler imaging, which is used to display blood flow, an increased persistence with no cardiac gating results in an acquisition rate of 7.5 images per second and a 3-D scan time of 20 seconds.

The 3-D mechanical scanning offers three advantages: short imaging times, high-quality 3-D images, and fast reconstruction times. However, its bulkiness and weight sometimes make it inconvenient to use. Linear scanning has been successfully implemented in many vascular imaging applications using B-mode and color Doppler images of the carotid arteries [30–34], vascular test phantoms [35–37], and power Doppler images [30–36]. An example of a mechanical scanner is shown in Figure 3.1, and several examples of linearly scanned 3-D US images of carotid arteries with complex plaques are shown in Figure 3.2.

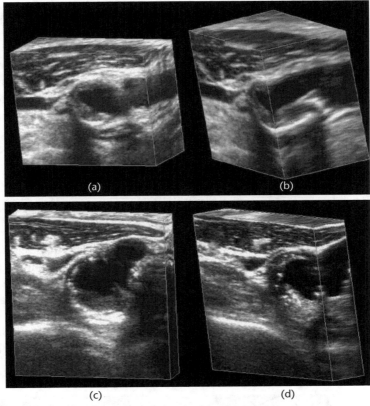

Figure 3.2   Two 3-D US views of two different patients with complex and ulcerated carotid plaques. For each patient, (a, c) a transverse and (b, d) a longitudinal view are shown side by side.

3.2.1.2   Magnetically Tracked Free-Hand 3-D Carotid Ultrasound Imaging

To overcome size and weight issues related to mechanical scanning, free-hand scanning techniques have been developed that do not require motorized mechanisms [37–42]. To measure position and orientation, a sensor is attached to the transducer. As the transducer is manipulated across the neck, a set of transverse 2-D images of the carotid arteries is digitally acquired by a computer, and the position and orientation of each image is obtained and recorded. This information is then used to reconstruct the 3-D image. Because the relative positions of the acquired 2-D images are not predefined, the operator must ensure that the set of 2-D images does not have any significant gaps and that the desired region is scanned. In addition, the ability to scan the neck freely allows longer regions of the neck to be imaged.

Although several free-hand scanning approaches have been developed (e.g., articulated arms, acoustic sensing, magnetic field sensing, and image-based sensing), the most successful approach for imaging the carotid arteries uses a sensor with six degrees of freedom to measure the strength of a magnetic field. In this approach, a transmitter is used to produce a spatially varying magnetic field; a small sensor detects the magnetic field in order to determine the relative position and orientation of the sensor. Typically, the transmitter is placed beside the patient and the sensor is mounted onto the transducer.

The magnetic field sensors are small and unobtrusive devices that allow the transducer to be tracked with fewer constraints than any other mechanical scanning approach. However, electromagnetic interference (e.g., ac power cabling and motors) can compromise the accuracy of tracking. Geometric distortions in the final 3-D US image can occur if ferrous or highly conductive metals are located nearby. By ensuring that the environment near the patient is free of metals and electrical interference, high-quality 3-D images of the carotid arteries can be produced. Figure 3.3 shows a 3-D carotid image that has been obtained with a magnetic field sensor.

3.2.2 Reconstruction of 3-D US Images

In the 3-D reconstruction procedure, the set of acquired 2-D images is used to reconstruct a 3-D (voxel-based) volume. Each of the 2-D US images is placed in its correct location within the volume. The gray-scale values of any volume element (or voxel) not sampled by the 2-D images are then calculated by interpolating between the appropriate images. As a result, all 2-D image information is preserved, allowing for the viewing of the original 2-D planes as well as the other types of views. The advent of desktop computers has made this process very efficient by allowing 3-D reconstructions to occur while 2-D images are being acquired. Thus, it is possible to view

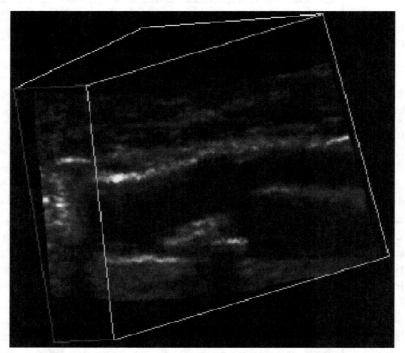

Figure 3.3 A 3-D B-mode ultrasound image of carotid arteries obtained with the free-hand scanning approach using a magnetic position and orientation measurement. The 3-D image has been "sliced" to reveal a plaque at the entrance of the internal carotid artery. The plaque is heterogenous, with calcified regions casting a shadow.

the complete 3-D US image immediately after the acquisition of the images has been completed.

3.2.3   Viewing of 3-D US Images

After a 3-D image has been reconstructed, it is ready to be viewed. The *cube-view approach* is based on multiplanar rendering using texture mapping. In this technique, a 3-D image is displayed as a polyhedron, and the appropriate image for each plane is "painted" on the face of the cube (texture mapped). Users not only can rotate the polyhedron in order to obtain the desired orientation of the 3-D image, but also can move any of the surfaces (i.e., by slicing of the 3-D image parallel or obliquely) to the original, while the appropriate data are texture mapped in real time onto the new face. As a result, users always have 3-D image-based cues, which relate the plane being manipulated to the rest of the anatomy. These visual cues allow users to efficiently identify the desired structures [11, 22, 26, 27]. Examples of this approach are shown in Figures 3.2 and 3.3.

In the *orthogonal plane view,* three perpendicular planes are extracted from the volume data and displayed simultaneously on a computer screen as 2-D images. Graphical cues are added to the screen and overlaid onto the 2-D images to indicate their relative orientations. Users can then select and move single or multiple planes within the 3-D image volume to provide a cross-sectional view at any location or orientation, including oblique views.

The *volume rendering technique* presents a display of the entire 3-D image after it has been projected onto a 2-D plane. The projection of the image is typically accomplished via ray-casting techniques [43]. The image voxels that intersect each ray are weighted and summed. The desired result is displayed in the rendered image. Although this approach has been used primarily to display fetal anatomy, it can also be used to display vascular anatomy [30] and views of carotid plaques, as shown in Figure 3.4.

3.3   Measurement of Plaque Volume

3.3.1   Plaque Segmentation by Manual Planimetry

Carotid disease can be measured using either of two manual planimetry techniques: the plaque volume technique or the vessel wall volume technique. When using the plaque volume technique, each 3-D image is "sliced" transverse to the vessel axis, starting from one end of the plaque using an interslice difference (ISD) of 1.0 mm [Figure 3.5(a)]. Using software developed in our laboratory, we can contour the plaque in each cross-sectional image using a cross-haired cursor (Figure 3.5). As the contours are manually outlined, the visualization software calculates the area of the contours automatically. Sequential areas of the contours are averaged and multiplied by the ISD in order to calculate the incremental volume. A summation of incremental volumes provides a measure of the total plaque volume. After measuring a complete plaque volume, the 3-D US image can be viewed in multiple orientations in order to verify that the set of contours has outlined the entire plaque

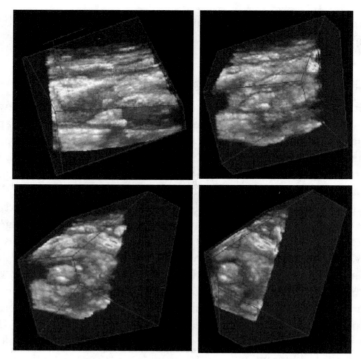

Figure 3.4 Four views of the 3-D carotid ultrasound image shown in Figure 3.3. Each panel has been volume rendered to display the appearance of the plaque in the common carotid artery. This approach enhances the appearance of the plaque as image speckle is suppressed.

volume. A typical plaque having 10 to 30 slices requires approximately 15 minutes of measuring time.

The vessel wall volume technique, which is commonly used for analyzing MR images, is an alternative method for quantifying disease in the carotid arteries. In this approach, the lumen (blood–intima boundary) and the vessel wall (intima–adventia boundary) are segmented in each slice. The area inside the lumen boundary is subtracted from the area inside the vessel wall boundary. Sequential areas are averaged and multiplied by the ISD to give the incremental vessel wall volume. The summation of incremental volumes provides a measure of the total vessel wall volume (Figure 3.6).

3.3.2   Studies of Plaque Volume

During the past few years, a number of investigators have explored the use of 3-D US imaging in measurements of carotid plaque volume [15, 44–47]. These in vivo and in vitro investigations have also reported on the variability of plaque volume measurements for plaques ranging in size and geometry. Delcker et al. [45–49] have measured carotid plaque volumes ranging from 2 to 200 mm^3. The intraobserver and interobserver variability for the full series of plaque sizes were found to be 2.8% and 3.8%, respectively, demonstrating that 3-D US imaging can be used to obtain reproducible measurements of carotid plaque volumes [36, 45, 49]. Delcker et al. have further demonstrated that in vivo measurements of carotid plaques have bene-

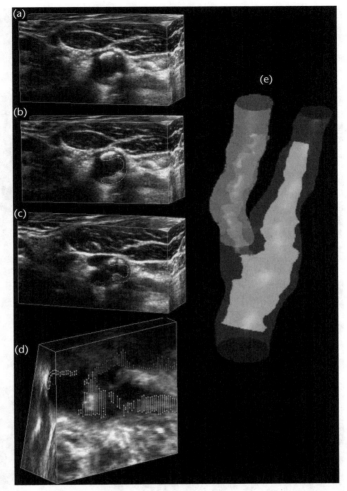

Figure 3.5 Steps used in measurement of total plaque volume from 3-D US images. (a) First, the 3-D image is "sliced" to obtain a transverse view. (b, c) Using a mouse-driven cross-haired cursor, the plaque is outline in successive image "slices" until all the plaques have been traversed. (d) The vessel can be sliced to reveal a longitudinal view with the outlines of the plaques. (e) After outlining all of the plaques, the total volume can be calculated and a mesh fitted to provides a view of the plaque surface together with the boundary of the vessel.

fitted from ECG-triggered data acquisition. The intraobserver variability, the interobserver variability, and the follow-up variability were shown to decrease significantly when compared to the non-ECG-triggered measurements [49]. Palombo et al. [47] measured carotid plaque volume ranging from 7 to 450 mm^3 and obtained reliability coefficients that were close to 1 for this in vivo intraobserver and interobserver study.

The monitoring of carotid plaque progression and regression requires accurate and reproducible techniques to measure and analyze carotid plaque volume and morphology. Thus, reproducible segmentations of the vessel lumen, plaque surface, and carotid wall borders are required. Although a number of studies have demonstrated the potential utility of 3-D US, few have attempted to quantify the accuracy

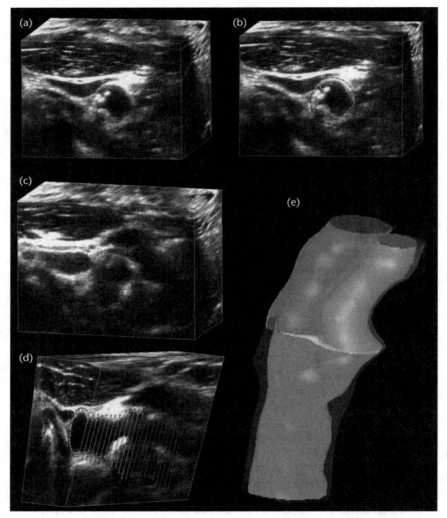

Figure 3.6 Steps used in measurement of vessel wall plus plaque volume from 3-D US images.
(a) First, the 3-D image is "sliced" to obtain a transverse view. (b, c) Using a mouse-driven
cross-haired cursor, the vessel boundary and the lumen boundary plaque are outlined separately in
successive image "slices" until all the slices have been traversed (typically 1.5 cm above and below
the carotid bifurcation). (d) The vessel can be sliced to reveal a longitudinal view with the outlines
and correct any errors. (e) After outlining has been completed, the vessel wall plus plaque volume
can be calculated and a mesh fitted to provide a view of the vessel and the lumen boundaries.

of these measurements by determining the actual volume of each plaque. In addition, the variability in the measurement of plaque volume has generally been reported over a large range of plaque volumes. In the analysis of plaque progression, it is important to understand the relationship between plaque volume and plaque volume measurement variability so that plaque changes at different stages of plaque progression can be interpreted appropriately. In the following sections we review both the accuracy and the variability of plaque measurement as a function of plaque volume.

3.3.3   The Effects of Interslice Distance

The value of the ISD between the slices of the acquired 3-D US images will affect the measurement of the plaque volume. If the value of the ISD is chosen to be small (in order to obtain results that are accurate), then the quantification of the volume procedure will be time consuming. If the value of ISD is chosen to be large (in order to reduce the number of analyzed slices and to reduce the time of measurement), then inaccurate and variable results may occur. Thus, we investigated the effects of the ISD on the relative accuracy and variability of the plaque volume measurement by measuring the volume of five plaques, which ranged in size from 42.2 to 604.1 mm^3. These plaques were measured five times using nine separate ISDs, which ranged from 1.0 to 5.0 mm in increments of 0.5 mm. The results are shown in Figure 3.7. These results show that the relative plaque volume (normalized to measurements with an ISD of 1 mm) had remained unchanged for ISDs between 1.0 and 3.0 mm and had decreased to 0.83 mm for larger ISDs. Thus, the plaque volume will be systematically underestimated for ISDs greater than 3.0 mm. The error bars in Figure 3.7 show that the variability of the plaque volume measurement increases as the ISD increases. The coefficient of variation (i.e., the SD divided by the mean), which increased as the ISD increased, was found to be approximately 10% for ISDs of 3 mm or less. Based on these results, we had chosen an ISD of 1 mm for quantifying plaque volume.

3.3.4   The Accuracy of Plaque Volume Measurements

To ascertain the accuracy of the plaque volume measurements, the volume of the measured plaque must be compared to the actual volume of the plaque. Because 3-D US-based measurements of carotid plaques in vivo cannot be compared to the actual plaque volume obtained by any other method, an in vitro study using test phantoms with known "plaque" volumes was conducted by Landry and Fenster [50]. In this in

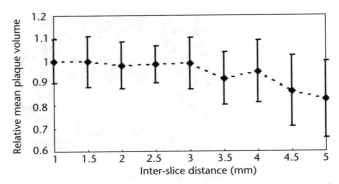

Figure 3.7   Relative mean plaque volume as a function of ISD for five plaque volumes measured by a single observer five times using each ISD investigated. Relative volume is constant for ISDs of 1.0 to 3.0 mm, but then decreases to 0.83 mm for an ISD of 5.0 mm. Plaque volume measurement variability is shown by the error bars that represent ±1 SD. (*From:* [51]. © 2004 American Heart Association. Reprinted with permission.)

vitro study, the vascular phantoms were made from blocks of colloidal agar that contained Sigmacell to scatter the ultrasound beam and to produce the typical ultrasound image speckle; cylindrical channels were used to simulate the vessels. The agar gel was sliced into "plaques" of varying height and length in order to produce several simulated hypoechoic plaques. These "plaques" were then inserted into the cylindrical channels, imaged with 3-D US, and measured.

The results from this study are shown in Figure 3.8. The measurement error of the plaque volume is plotted as a function of plaque volume, which ranges from 68.2 to 285.5 mm^3. The inserted image is a 3-D US sample of the simulated plaque in the vascular phantom. The mean error for the measurement of the plaque volume was found to be 3.1% ± 0.9%. The coefficient of variation (i.e., the SD divided by the mean) in the measurement of the plaque volume was found to be 4.0% ± 1.0% and 5.1% ± 1.4% for intraobserver and interobserver measurements, respectively. Using the manual planimetry method, Landry and Fenster have developed a theoretical description for the variance in the measurement of plaque volume [50]. The root-mean-square (RMS) difference between the experimentally and the theoretically determined values of the coefficient of variation for plaque volume measurement was found to be 9%.

3.3.5 Variability of Plaque Volume Measurements

The variability of plaque volume measurements has been reported by a number of investigators as an average that has been calculated over a wide range of plaque vol-

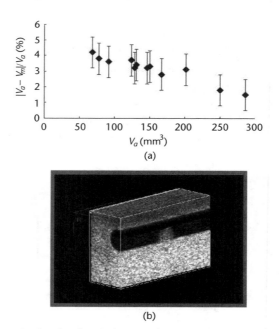

(a)

(b)

Figure 3.8 (a) The error in the simulated plaque volume measurement plotted as the absolute percent difference between the actual (V_a) and measured (V_m) plaque volumes as a function of the average actual volume for each plaque. The error bars represent ±1 SD. (b) A longitudinal view of the 3-D image of the simulated vessel and plaque sliced to reveal the simulated vessel (black) and plaque. (*From:* [50]. © 2002 American Association of Physicists in Medicine. Reprinted with permission.)

umes; however, it is also important to assess the effect of plaque volume on the variability in the measurement of plaque volume. To examine and determine this effect, we conducted a multiple observer study, which determined the intraobserver and interobserver variability in the measurement of plaque volume. Each of the five observers measured the volume of 40 carotid plaques five times during five different sessions using an ISD of 1.0 mm. Using a 3-D mechanical scanning approach, we imaged the carotid plaques from a number of human subjects (23 males and 17 females with an average age of 73.8 ± 6.2 years) who were patients at the Premature Atherosclerosis Clinic and the Stroke Prevention Clinic (London Health Science Centre, University Campus, London, Ontario, Canada). The volume of the plaques ranged from 37.4 to 604.1 mm^3.

Table 3.1 and Figure 3.9 show a summary of the results obtained using Analysis of Variance (ANOVA) [52, 53]. Table 3.1 lists the results for the entire set of data as well as for the five subsets (a, b, c, d, e) consisting of seven to eight plaques, which were grouped together by volume.

For the smallest volumes of plaques, which range from 37.4 to 89.5 mm^3, Table 3.1 shows the interstandard and intrastandard error of measurement (SEM_i) to be $13.1\% \pm 1.8\%$ and $12.5\% \pm 1.9\%$, respectively (subset a). For the largest volumes of plaques, which range from 519.4 to 604.1 mm^3 (subset e), the SEM_{inter} and SEM_{intra} were found to be $4.3\% \pm 1.0\%$ and $3.9\% \pm 1.3\%$, respectively. Figure 3.9 not only shows this effect but also shows that the coefficient of variation in the intraobserver measurement of the plaque volume decreases as the plaque volume increases. For the entire range of plaque volumes measured in this study (37.4 to 604.1 mm^3), we determined the intraobserver coefficient of variation to be 19.2% and 1.9%, respectively. For the same range of plaque volumes, we determined the range of the interobserver coefficient of variation to be from 24.1% to 2.2%.

3.3.6   Automated Segmentation of Plaques from 3-D US Images

The manual planimetry method for measuring plaque volume requires that an operator: (1) "cut" into a 3-D US image, (2) orient the revealed plane appropriately, and (3) outline the plaque surface manually. To obtain the volume of the complete plaque, the operator must "cut" further into the volume and repeat the process until

Table 3.1   Summary of the ANOVA Results of Study in Which the Volumes of 40 Plaques Were Repeatedly Measured by Five Observers

Plaque Subset	Mean V (mm³)	SEM_inter/V (%)	SEM_intra/V (%)
a	65.4	13.1	12.5
b	130.5	6.8	7.0
c	247.4	4.9	4.9
d	426.0	5.5	4.2
e	560.6	4.3	3.9
Global data	276.3	6.9	6.5

The plaques have been grouped in five subsets, and the mean of each subset tabulated. The interobserver and intraobserver coefficient of variation (standard error of measurement divided by the mean) has been tabulated.
Source: [51].

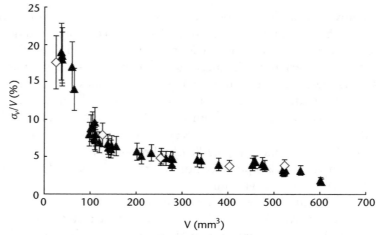

Figure 3.9 The results of the ANOVA analysis of the repeated measurements of 40 plaques made by five observers. The results are plotted as the intraobserver coefficient of variation (standard deviation divided by the mean plaque volume), σ_v/V, as a function of mean plaque volume, V (diamonds). The five plaque volumes measured in the repeat 3-D US scan study are plotted as open diamonds. The error bars represent ±1 SD. (*From:* [51]. © 2004 American Heart Association. Reprinted with permission.)

the complete plaque is traversed. This approach for manually outlining the plaque is not only arduous but also time consuming. Substantial experience is needed by an operator in order to obtain reasonable results.

We have developed an efficient semiautomated segmentation algorithm that is based on deformable models that require minimal user interaction and processing time to segment the boundary of plaques [54–57]. Our approach is based on a deformable model that defines a boundary as a series of vertices connected by a set of vectors, which produces a 3-D mesh. This structure allows for the modeling of forces between the nodes of the mesh and the external forces acting on the mesh, which are based on the image properties (e.g., the image gradients found at the lumen boundary). Using this model, we can calculate the forces at each vertex, which is moved iteratively until the values of the forces acting on the vertices are zero.

The algorithm contains three of the following major steps: (1) the placement of the initial deformable balloon model (i.e., a 3-D mesh composed of triangles) inside the lumen; (2) the inflation of the model toward the plaque and vessel surface; and (3) the automatic localization of the plaque and vessel surface. When equilibrium is reached (under the influence of inflation forces, internal forces, and image forces), the model approximately represents the shape of the surface. Because the surface of the plaque may not be smooth and uniform, we deform the mesh further by using the local image-based forces to localize the surface of the plaque. Equations (3.1) and (3.2) describe the dynamics of the model and the forces acting on the model. Equation (3.1) defines the motion for the ith vertex of the mesh:

$$x_i'(t) + g\big(x_i(t)\big) = f\big(x_i(t)\big) \tag{3.1}$$

The position and velocity of the vertex are given by $x_i(t)$ and $x'_i(t)$, respectively. The resultant surface tension at the vertex and the "driving" force are given by $g(x_i(t))$ and $f(x_i(t))$, respectively. Equilibrium is reached when $x'_i(t)$ becomes zero.

The equilibrium position of each vertex, and hence the final shape of the model, is obtained by iteratively updating the variable of time in (3.1) by Δt. Once the surface of the plaque has been segmented, an image-based force replaces the inflation force. The image force causes the mesh to fit the fine details of the plaque. A 3-D potential function, P, is constructed from the image data, which attracts the model to the 3-D gradients at the surface of the plaque:

$$P\left(x_i(t)\right) = 1/\left[\|\nabla G_\sigma * I\| + \varepsilon\right] \tag{3.2}$$

where G_σ is a 3-D Gaussian smoothing kernel of width σ, and ε is a constant that prevents the denominator of (3.2) from dividing by zero. The minima of the potential function coincide with the plaque surface. The potential function produces a force field, which is used to deform the model as $f(x_i(t) = -k\nabla P(x_i(t)$. The value of k, which must be optimized so that the mesh will conform to the surface of the plaque, is an adjustable term that controls the strength of the force. In addition, the surface tension of the mesh must be also chosen appropriately. If the model is too stiff, then it will not deform into small ulcerations, but if it is too compliant, then the model may extend into echolucent regions.

Figure 3.10 shows the results of segmentation using a 3-D image of a patient's carotid arteries with plaque. The segmentation required only about 10 seconds to generate the lumen surface of the arteries. The mean separation between manual segmentation and semiautomated (algorithm) segmentation was 0.35 mm with a standard deviation of 0.25 mm [56].

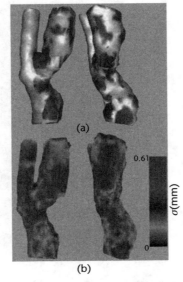

Figure 3.10 Analysis results of the 3-D carotid segmentation. (a) Image showing local agreement between manual and algorithm segmentation. Agreement is shown in white and disagreement in gray. (b) Local standard deviation (variability) in the 3-D algorithm segmented boundary, with the scale on the right. (*From:* [56]. © 2000 American Association of Physicists in Medicine. Reprinted with permission.)

3.3.7   Plaque Surface Morphology

The use of conventional 2-D US to identify characteristics of the plaque–lumen boundary (i.e., surface irregularities, roughness, and ulcerations) remains difficult. However, mounting evidence has not only been showing the importance of plaque surface morphology [58–61] but also generating interest in the development of improved ultrasound techniques for visualization and characterizing of plaque surface features. Using 3-D US images, we have been developing improved quantitative techniques for the analysis of plaque surface morphology. Our approach uses the surface of an outlined plaque, which is obtained by either using manual planimetry or using an automated approach. The outlines of a plaque are then converted to a triangular surface mesh, which is smoothed to remove the variation in the surface due to ultrasound speckle. We analyzed the mesh by calculating the local Gaussian curvature at each node of the mesh. To reduce the variation in the measure of the curvature, we determined the local mean Gaussian curvature value [62–64] by averaging the curvature values of all surface points within a 1-mm radius of each point on the mesh.

Figure 3.11 shows a result of the surface roughness metric measured by the mean Gaussian curvature. The roughness metric has been coded in gray and mapped onto the segmented surfaces of the vessels, showing an increase in the roughness for some regions of the surface associated with the plaque. Although these types of

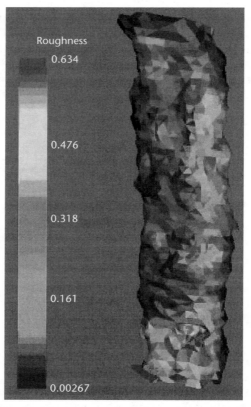

Figure 3.11   Result of analysis of the surface morphology of a section of a common carotid artery. The roughness index is coded in gray and is mapped onto the lumen wall–plaque boundary.

images have yet to be validated, they show the potential of using 3-D US images for monitoring changes in the plaque surface morphology.

3.3.8   Local Distribution and Change of Carotid Wall and Plaque

A number of investigators have studied the spatial distribution of carotid plaque by quantifying the degree of stenosis [20, 65]. Although stenosis profiles are useful in describing the distribution of plaque along a vessel, they do not give the location of the plaque burden within a given transverse slice of the vessel. The following sections describe a metric that can be used to compute the local distribution of carotid disease on a point-by-point basis for each transverse carotid image slice. In addition, we have extended the quantification of plaque burden along the vessel to the quantification of the change of the vessel wall plus plaque thickness of the carotid vessel between two time points.

3.3.8.1   Reconstructing Surfaces from Polygonal Slices

Figures 3.12(a) and (c) show the carotid wall and lumen, which were segmented manually five times by an expert observer, in 2-D transverse image slices that are 1 mm apart. The symmetric correspondence algorithm [66] was used to pair the vertices on adjacent contours. With all of the correspondence vertices defined, the adjacent 2-D curves were connected to form polygonal surfaces. Figure 3.12(b) shows the tessellated surface, which was constructed from the stack of 2-D contours of the carotid wall. Figure 3.12(d) shows the surface constructed for the set of contours representing the carotid lumen.

3.3.8.2   3-D Maps of Average and Standard Deviation Carotid Surfaces

For each of the 3-D carotid US images, an expert observer segmented the carotid wall and lumen five times. After the five surfaces representing the wall and the lumen had been reconstructed using the method described earlier, we obtained the average and standard deviations of the wall and the lumen surfaces using the following steps:

1. For the segmentation of the vessel, the transverse plane was identified, and the average of the normals of the five different transverse planes was calculated. The five segmented surfaces were then resliced at 1-mm intervals using the plane with the average normal, producing five 2-D contours at 1-mm intervals along the vessel.
2. Because the symmetric correspondence algorithm only established a correspondence between a pair of curves, it was applied four times. The smoothest curve (of the five curves obtained in step 1) was chosen, and the symmetric correspondence mappings were established between the smoothest curve and the remaining four curves.
3. Each of the vertices of the smoothest curve (defined in step 2) was linked with correspondence points on four different curves, resulting in a group of

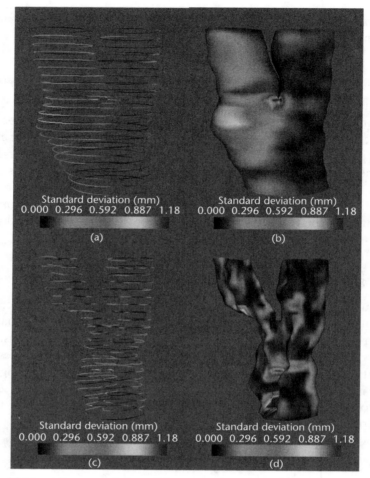

Figure 3.12 (a) Manually segmented carotid artery wall contours. (b) Mean carotid artery wall surface obtained with five manual segmentations. The standard deviation at each location is coded in gray and mapped onto the mean surface. (c) Manually segmented carotid artery lumen contours. (d) Mean carotid artery lumen surface obtained with five manual segmentations. The standard deviation at each location is color coded and mapped onto the mean surface.

five points. The centroid and the standard deviation of the points about this centroid were then computed.

3.3.8.3 Vessel Wall Plus Plaque Thickness Map and Its Standard Deviation

A transverse plane was used to slice the average surfaces of the carotid wall and lumen at 1-mm intervals. The vertices of the resulting 2-D curves were paired according to the symmetric correspondence algorithm [66]. The 3-D thickness map was then computed as the difference between the carotid wall and lumen boundary, which was composed of a stack of 2-D curves, each lying on a transverse plane where the average surfaces of the wall and the lumen were sliced. As shown in Figure 3.13, the resulting thickness map was coded in gray and superimposed on the mean vessel wall 3-D boundary. Figure 3.13 shows the vessel wall plus plaque thickness maps for a patient scanned at baseline and after a 3-month interval, during which he was treated with 80 mg of atorvastatin.

3.3.8.4 3-D Maps of Carotid Wall Plus Plaque Thickness Change

To study carotid plaque change, we generated the vessel wall plus plaque thickness map from 3-D US images of carotid arteries obtained at two different times (see Figures 3.13 and 3.14). To calculate the change in the plaque thickness between the imaging times, we first registered the vessel contours using a slightly modified version of the iterative closest point (ICP) algorithm by Besl et al. [67]. Rather than aligning the centroids of the two surfaces, as proposed in [67], we aligned the bifurcation apex of the carotid vessels. After registering the two thickness maps, we then

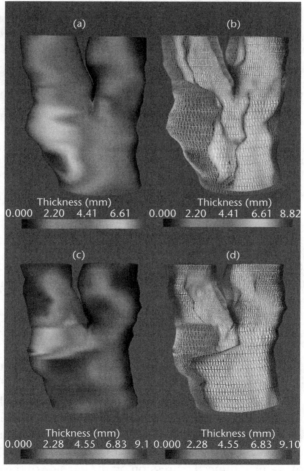

Figure 3.13 (a) Mean carotid artery wall with thickness of vessel wall plus plaque is coded in gray and mapped on its surface (obtained at baseline). (b) The thickness map in part (a) is shown together with the lumen of the carotid artery. (c) Mean carotid artery wall with thickness of vessel wall plus plaque is coded in gray and mapped on its surface (obtained after 3 months of 80 mg of atorvastatin daily treatment). (d) The thickness map in part (c) is shown together with the lumen of the carotid artery.

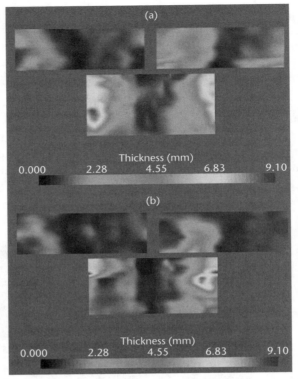

Figure 3.14   The carotid arteries have been cut open and flattened and the thickness of vessel wall plus plaque is coded in gray and mapped on its surface. (a) Results obtained at baseline. (b) Results obtained after 3 months of 80 mg of atorvastatin daily treatment.

calculated the difference between them, coded it in gray, and mapped it onto a mean 3-D boundary that was obtained from the 3-D US image at baseline, as shown in Figure 3.15.

3.3.9   Example Results

Figure 3.13 shows the results for a patient who was treated with 80 mg of atorvastatin daily for 3 months. The results shown in Figure 3.13(a, b) were obtained at baseline, and the results in Figure 3.13(c, d) were obtained after 3 months of treatment. The difference in plaque in the vessel wall plus plaque thickness is also apparent in the flattened map of the carotid arteries shown in Figure 3.14. The difference in the vessel wall plus plaque thickness is shown in Figure 3.15, which shows a pronounced regression on the left side of the CCA with a reduction of plaque thickness of about 7 mm.

The calculation and the display of the distribution of carotid plaque and the distribution of plaque change will help in the study of the natural progression of carotid plaque; more importantly, it will help in the understanding of plaque regression in response to therapy.

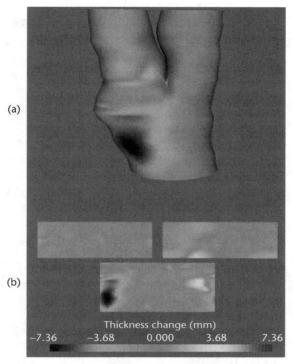

(a)

(b)

Thickness change (mm)

−7.36   −3.68   0.000   3.68   7.36

Figure 3.15   (a) The thickness of the vessel wall plus plaque difference between the results obtained at baseline and after atorvastatin treatment (maps shown in Figure 3.14). A negative value indicates that a plaque thickness reduction was recorded in the second scanning session. (b) The flattened map of the 3-D results shown in part (a).

3.4   Trends and Future Developments

Three-dimensional ultrasound has already demonstrated its clear advantages in obstetrics (e.g., imaging of the fetus), cardiology (e.g., imaging of the cardiac system), and image guidance (e.g., imaging of the prostate during an interventional procedure) [11, 26, 27]. Advances in 3-D US instrumentation and visualization software have advanced sufficiently enough to allow for fast (real-time) 3-D acquisition, real-time reconstruction, and 3-D visualization with real-time image manipulation. Technical advances continue to improve both 3-D US image acquisition and image analysis, promising to make it a routinely available tool on all ultrasound machines.

The availability of improved 3-D US systems is allowing researchers to focus on demonstrating its clinical utility. Although 3-D US has been shown to be useful in imaging the plaques of the carotid arteries, this modality requires further development in order for it to become a routine tool for quantifying the progression and regression of carotid disease. The following four sections discuss the current trends and needed developments in 3-D carotid ultrasound imaging.

3.4.1   The Need for Improved Free-Hand Scanning

Although free-hand scanning techniques are useful, they are still subject to potential artifacts and inaccuracies. Further technical progress is needed in order for these techniques to become standard tools for producing high-quality images of the carotid arteries. Approaches for gating and techniques for ensuring that the carotid arteries are imaged uniformly need improvement. Real-time 3-D US systems (using 2-D arrays) are useful in cardiology. However, further developments are needed in order for them to produce 3-D images of the carotid arteries with an image quality similar to the quality that is currently being produced by conventional 1-D arrays.

3.4.2   The Need for Improved Visualization Tools

Current 3-D US imaging systems use 3-D visualization tools. However, these tools require not only complicated user interfaces but also special training. In order for 3-D carotid imaging to become widely accepted, we need to develop intuitive and automated visualization tools to manipulate the 3-D image so that the user may view the desired section of the carotid arteries using both the texture mapping and volume rendering approaches. The production of volume-rendered images of the carotid arteries and the acquisition of volume-rendered images of the plaque surface require multiple parameters to be manipulated. Most notably, automated techniques are needed to provide immediate optimal rendering without significant user intervention.

3.4.3   Reduction in the Variability of Plaque Volume Measurement

Landry and Fenster [50] have demonstrated that the greatest source of variability in measurement of plaque volume (using manual planimetry) is caused by the variability of locating the initial and final slices. Outlining the plaque contributes to the variability; however, this contribution is small. Figure 3.16 shows a plot of five repeated outlining sessions for one carotid plaque "slice" as well as the standard deviation in determining the boundary. The mean standard deviation in the plaque outline is only 0.28 mm. Thus, to reduce the measurement variability of the plaque volume, we need to improve our strategies for measuring the disease burden.

3.4.4   ECG-Triggered Acquisition of 3-D Carotid US Images

Delcker et al. [49] have shown that ECG-triggered 3-D US acquisition with free-hand scanning resulted in improved reliability of plaque volume measurements. However, triggered acquisition prolongs the scanning time, introduces discomfort for the patient, and adds complexity to the instrumentation. Nonetheless, this approach is promising for improving the quality of the 3-D US images and reducing the variability of plaque volume measurements.

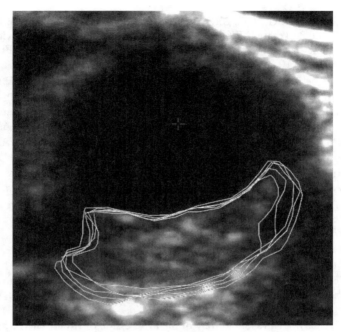

Figure 3.16 Five contours of a plaque obtained by manual segmentation. The figure indicates that the variability in outlining the plaque–lumen interface is small, but the variability in outlining the plaque–vessel wall interface is higher. The greatest variability is at the lateral sides of the vessel.

Acknowledgments

The authors acknowledge the financial support of the Canadian Institutes for Health Research and the Ontario Research and Development Challenge Fund. The first author holds a Canada Research Chair and acknowledges the support of the Canada Research Chair Program.

References

[1]   Hegele, R. A., et al., "Infection-Susceptibility Alleles of Mannose-Binding Lectin Are Associated with Increased Carotid Plaque Area," *J. Investig. Med.*, Vol. 48, No. 3, 2000, pp. 198–202.

[2]   Spence, J. D., M. R. Ban, and R. A. Hegele, "Lipoprotein Lipase (Lpl) Gene Variation and Progression of Carotid Artery Plaque," *Stroke*, Vol. 34, No. 5, 2003, pp. 1176–1180.

[3]   Spence, J. D., et al., "Carotid Plaque Area: A Tool for Targeting and Evaluating Vascular Preventive Therapy," *Stroke*, Vol. 33, No. 12, 2002, pp. 2916–2922.

[4]   Barnett, P. A., et al., "Psychological Stress and the Progression of Carotid Artery Disease," *J. Hypertens.*, Vol. 15, No. 1, 1997, pp. 49–55.

[5]   Spence, J. D., et al., "Plasma Homocyst(E)Ine Concentration, But Not Mthfr Genotype, Is Associated with Variation in Carotid Plaque Area," *Stroke*, Vol. 30, No. 5, 1999, pp. 969–973.

[6]   Hackam, D. G., J. C. Peterson, and J. D. Spence, "What Level of Plasma Homocyst(E)Ine Should Be Treated? Effects of Vitamin Therapy on Progression of Carotid Atherosclerosis in Patients with Homocyst(E)Ine Levels above and Below 14 Umul/L," *AJH*, Vol. 13, 2000, pp. 105–110.

[7]   Zwiebel, W. J., "Duplex Sonography of the Cerebral Arteries: Efficacy, Limitations, and Indications," *AJR Am. J. Roentgenol.*, Vol. 158, No. 1, 1992, pp. 29–36.

[8]   Sterpetti, A. V., et al., "Ultrasonographic Features of Carotid Plaque and the Risk of Subsequent Neurologic Deficits," *Surgery*, Vol. 104, No. 4, 1988, pp. 652–660.

[9]   Widder, B., et al., "Morphological Characterization of Carotid Artery Stenoses by Ultrasound Duplex Scanning," *Ultrasound Med. Biol.*, Vol. 16, No. 4, 1990, pp. 349–354.

[10] Riccabona, M., et al., "In Vivo Three-Dimensional Sonographic Measurement of Organ Volume: Validation in the Urinary Bladder," *J. Ultrasound Med.*, Vol. 15, No. 9, 1996, pp. 627–632.

[11] Fenster, A., D. B. Downey, and H. N. Cardinal, "Three-Dimensional Ultrasound Imaging," *Phys. Med. Biol.*, Vol. 46, No. 5, 2001, pp. R67–R99.

[12] Liapis, C., et al., "Internal Carotid Artery Stenosis: Rate of Progression," *Eur. J. Vasc. Endovasc. Surg.*, Vol. 19, No. 2, 2000, pp. 111–117.

[13] Schminke, U., et al., "Three-Dimensional Power-Mode Ultrasound for Quantification of the Progression of Carotid Artery Atherosclerosis," *J. Neurol.*, Vol. 247, No. 2, 2000, pp. 106–111.

[14] Serena, J., "Ultrasonography of the Progression of Atherosclerotic Plaques," *Rev. Neurol.*, Vol. 29, No. 9, 1999, pp. 851–856.

[15] Ainsworth, C. D., et al., "3-D US Measurement of Change in Carotid Plaque Volume; a Tool for Rapid Evaluation of New Therapies," *Stroke*, Vol. 35, 2005, pp. 1904–1909.

[16] Hatsukami, T. S., et al., "Echolucent Regions in Carotid Plaque: Preliminary Analysis Comparing Three-Dimensional Histologic Reconstructions to Sonographic Findings," *Ultrasound Med. Biol.*, Vol. 20, 1994, pp. 743–749.

[17] Steinke, W., and M. Hennerici, "Three-Dimensional Ultrasound Imaging of Carotid Artery Plaques," *J. Cardiovasc. Technol.*, Vol. 8, 1989, pp. 15–22.

[18] Troyer, A., et al., "Major Carotid Plaque Surface Irregularities Correlate with Neurologic Symptoms," *J. Vasc. Surg.*, Vol. 35, 2002, pp. 741–747.

[19] Zhao, X. Q., et al., "Effects of Prolonged Intensive Lipid-Lowering Therapy on the Characteristics of Carotid Atherosclerotic Plaques In Vivo by MRI: A Case-Control Study," *Arterioscler. Thromb. Vasc. Biol.*, Vol. 21, No. 10, 2001, pp. 1623–1629.

[20] Yao, J., et al., "Three-Dimensional Ultrasound Study of Carotid Arteries Before and After Endarterectomy: Analysis of Stenotic Lesions and Surgical Impact on the Vessel," *Stroke*, Vol. 29, No. 10, 1998, pp. 2026–2031.

[21] Sameshima, T., et al., "Clinical Usefulness of and Problems with Three-Dimensional CT Angiography for the Evaluation of Arteriosclerotic Stenosis of the Carotid Artery: Comparison with Conventional Angiography, MRA, and Ultrasound Sonography," *Surg. Neurol.*, Vol. 51, No. 3, 1999, pp. 301–308; discussion, pp. 308–309.

[22] Nelson, T. R., et al., *Three-Dimensional Ultrasound*, Philadelphia, PA: Lippincott-Raven, 1999.

[23] Nelson, T. R., and D. H. Pretorius, "Three-Dimensional Ultrasound Imaging," *Ultrasound Med. Biol.*, Vol. 24, No. 9, 1998, pp. 1243–1270.

[24] Baba, K., and D. Jurkovic, *Three-Dimensional Ultrasound in Obstetrics and Gynecology*, New York: Parthenon Publishing Group, 1997.

[25] Downey, D. B., and A. Fenster, "Three-Dimensional Ultrasound: A Maturing Technology," *Ultrasound Quarterly*, Vol. 14, No. 1, 1998, pp. 25–40.

[26] Fenster, A., and D. Downey, "Three-Dimensional Ultrasound Imaging," in *Handbook of Medical Imaging, Volume 1, Physics and Psychophysics*, J. Beutel, H. Kundel, and R. Van Metter, (eds.), Bellingham, WA: SPIE Press, 2000, pp. 433–509.

[27] Fenster, A., and D. Downey, "3-D US Imaging," *Annual Review of Biomedical Engineering*, Vol. 2, 2000, pp. 457–475.

[28] Fenster, A., and D. Downey, "Basic Principles and Applications of 3-D Ultrasound Imaging," in *An Advanced Signal Processing Handbook*, S. Stergiopoulos, (ed.), Boca Raton, FL: CRC Press, 2001, pp. 14–34.

[29] Smith, W., and A. Fenster, "Statistical Analysis of Decorrelation-Based Transducer Tracking for Three-Dimensional Ultrasound," *Med. Phys.*, Vol. 30, No. 7, 2003, pp. 1580–1591.

[30] Downey, D. B., and A. Fenster, "Vascular Imaging with a Three-Dimensional Power Doppler System," *AJR Am. J. Roentgenol.*, Vol. 165, No. 3, 1995, pp. 665–658.

[31] Fenster, A., et al., "Three-Dimensional Ultrasound Imaging," *Proc. SPIE Physics of Medical Imaging*, Vol. 2432, 1995, pp. 176–184.

[32] Picot, P. A., et al., "Three-Dimensional Colour Doppler Imaging of the Carotid Artery," *Proc. SPIE Image Capture, Formatting and Display*, Vol. 1444, 1991, pp. 206–213.

[33] Picot, P. A., et al., "Three-Dimensional Colour Doppler Imaging," *Ultrasound Med. Biol.*, Vol. 19, No. 2, 1993, pp. 95–104.

[34] Pretorius, D. H., T. R. Nelson, and J. S. Jaffe, "3-Dimensional Sonographic Analysis Based on Color Flow Doppler and Gray Scale Image Data: A Preliminary Report," *J. Ultrasound Med.*, Vol. 11, No. 5, 1992, pp. 225–232.

[35] Guo, Z., and A. Fenster, "Three-Dimensional Power Doppler Imaging: A Phantom Study to Quantify Vessel Stenosis," *Ultrasound Med. Biol.*, Vol. 22, No. 8, 1996, pp. 1059–1069.

[36] Dabrowski, W., et al., "A Real Vessel Phantom for Flow Imaging: 3-D Doppler Ultrasound of Steady Flow," *Ultrasound Med. Biol.*, Vol. 27, No. 1, 2001, pp. 135–141.

[37] Hughes, S. W., et al., "Volume Estimation from Multiplanar 2-D US Images Using a Remote Electromagnetic Position and Orientation Sensor," *Ultrasound Med. Biol.*, Vol. 22, No. 5, 1996, pp. 561–572.

[38] Leotta, D. F., P. R. Detmer, and R. W. Martin, "Performance of a Miniature Magnetic Position Sensor for Three-Dimensional Ultrasound Imaging," *Ultrasound Med. Biol.*, Vol. 23, No. 4, 1997, pp. 597–609.

[39] Gilja, O. H., et al., "In Vitro Evaluation of Three-Dimensional Ultrasonography Based on Magnetic Scanhead Tracking," *Ultrasound Med. Biol.*, Vol. 24, No. 8, 1998, pp. 1161–1167.

[40] Barratt, D. C., et al., "Optimisation and Evaluation of an Electromagnetic Tracking Device for High-Accuracy Three-Dimensional Ultrasound Imaging of the Carotid Arteries," *Ultrasound Med. Biol.*, Vol. 27, No. 7, 2001, pp. 957–968.

[41] Detmer, P. R., et al., "3-D Ultrasonic Image Feature Localization Based on Magnetic Scanhead Tracking: In Vitro Calibration and Validation," *Ultrasound Med. Biol.*, Vol. 20, No. 9, 1994, pp. 923–936.

[42] Hodges, T. C., et al., "Ultrasonic Three-Dimensional Reconstruction: In Vitro and In Vivo Volume and Area Measurement," *Ultrasound Med. Biol.*, Vol. 20, No. 8, 1994, pp. 719–729.

[43] Levoy, M., "Volume Rendering, a Hybrid Ray Tracer for Rendering Polygon and Volume Data," *IEEE Computer Graphics and Applications*, Vol. 10, 1990, pp. 33–40.

[44] Allott, C. P., et al., "Volumetric Assessment of Carotid Artery Bifurcation Using Freehand-Acquired, Compound 3-D US," *Br. J. Radiol.*, Vol. 72, No. 855, 1999, pp. 289–292.

[45] Delcker, A., and H. C. Diener, "Quantification of Atherosclerotic Plaques in Carotid Arteries by Three-Dimensional Ultrasound," *Br. J. Radiol.*, Vol. 67, No. 799, 1994, pp. 672–678.

[46] Delcker, A., and H. C. Diener, "3-D US Measurement of Atherosclerotic Plaque Volume in Carotid Arteries," *Bildgebung*, Vol. 61, No. 2, 1994, pp. 116–121.

[47] Palombo, C., et al., "Ultrafast Three-Dimensional Ultrasound: Application to Carotid Artery Imaging," *Stroke*, Vol. 29, No. 8, 1998, pp. 1631–1637.

[48] Delcker, A., H. C. Diener, and H. Wilhelm, "Influence of Vascular Risk Factors for Atherosclerotic Carotid Artery Plaque Progression," *Stroke*, Vol. 26, 1995, pp. 2016–2022.

[49] Delcker, A., and C. Tegeler, "Influence of ECG-Triggered Data Acquisition on Reliability for Carotid Plaque Volume Measurements with a Magnetic Sensor Three-Dimensional Ultrasound System," *Ultrasound Med. Biol.*, Vol. 24, No. 4, 1998, pp. 601–605.

[50] Landry, A., and A. Fenster, "Theoretical and Experimental Quantification of Carotid Plaque Volume Measurements Made by 3-D US Using Test Phantoms," *Med. Phys.*, Vol. 29, 2002, pp. 2319–2327.

[51] Landry, A., J. D. Spence, and A. Fenster, "Measurement of Carotid Plaque Volume by 3-Dimensional Ultrasound," *Stroke*, Vol. 35, No. 4, 2004, pp. 864–869.

[52] Eliasziw, M., et al., "Statistical Methodology for the Concurrent Assessment of Interrater and Intrarater Reliability: Using Goniometric Measurements as an Example," *Phys. Ther.*, Vol. 74, No. 8, 1994, pp. 777–788.

[53] Mitchell, J. R., et al., "The Variability of Manual and Computer Assisted Quantification of Multiple Sclerosis Lesion Volumes," *Med. Phys.*, Vol. 23, No. 1, 1996, pp. 85–97.

[54] Mao, F., J. D. Gill, and A. Fenster, "Technique of Evaluation of Semi-Automatic Segmentation Methods," *Proc. SPIE Image Processing*, Vol. 3661, 1999, pp. 1027–1036.

[55] Mao, F., et al., "Segmentation of Carotid Artery in Ultrasound Images: Method Development and Evaluation Technique," *Med. Phys.*, Vol. 27, No. 8, 2000, pp. 1961–1970.

[56] Gill, J. D., et al., "Accuracy and Variability Assessment of a Semiautomatic Technique for Segmentation of the Carotid Arteries from Three-Dimensional Ultrasound Images," *Med. Phys.*, Vol. 27, No. 6, 2000, pp. 1333–1342.

[57] Zahalka, A., and A. Fenster, "An Automated Segmentation Method for Three-Dimensional Carotid Ultrasound Images," *Phys. Med. Biol.*, Vol. 46, No. 4, 2001, pp. 1321–1342.

[58] Bluth, E. I., "Evaluation and Characterization of Carotid Plaque," *Semin. Ultrasound CT MR*, Vol. 18, No. 1, 1997, pp. 57–65.

[59] Fuster, V. E. *The Vulnerable Atherosclerotic Plaque: Understanding, Identification, and Modification*, Armonk, NY: Futura Publishing Company, 1999.

[60] Falk, E., P. K. Shah, and V. Fuster, "Coronary Plaque Disruption," *Circulation*, Vol. 92, No. 3, 1995, pp. 657–671.

[61] Schminke, U., et al., "Three-Dimensional Ultrasound Observation of Carotid Artery Plaque Ulceration," *Stroke*, Vol. 31, 2000, pp. 1651–1655.

[62] Besl, P., and R. Jain, "Invariant Surface Characteristics for 3-D Object Recognition in Range Images," *Comp. Vis. Graph. and Image Proc.*, Vol. 33, 1986, pp. 33–80.

[63] Han, C., T. S. Hatsukami, and C. Yuan, "Accurate Lumen Surface Roughness Measurement Method in Carotid Atherosclerosis," *Proc. of SPIE, Medical Imaging 2001: Image Processing*, Vol. 4322, 2001, pp. 1817–1827.

[64] Stokeyl, E., and S. Wu, "Surface Parameterization and Curvature Measurement of Arbitrary 3-D Objects: Five Practical Methods," *IEEE Trans. on Pattern Analysis and Machine Intelligence*, Vol. 14, 1992, pp. 883–840.

[65] Barratt, D. C., et al., "Reconstruction and Quantification of the Carotid Artery Bifurcation from 3-D US Images," *IEEE Trans. on Medical Imaging*, Vol. 23, No. 5, 2004, pp. 567–583.

[66] Papademetris, X., et al., "Estimation of 3-D Left Ventricular Deformation from Medical Images Using Biomechanical Models," *IEEE Trans. on Medical Imaging*, Vol. 21, No. 7, 2002, pp. 786–800.

[67] Besl, P., et al., "A Method for Registration of 3-D Shapes," *IEEE Trans. on Pattern Analysis and Machine Intelligence*, Vol. 14, No. 2, 1992, pp. 239–256.

CHAPTER 4

Quality Evaluation of Ultrasound Imaging of the Carotid Artery

C. P. Loizou

4.1 Introduction

Ultrasound imaging is a powerful noninvasive diagnostic imaging modality in medicine [1]. However, like all medical imaging modalities that exhibit various image artifacts, ultrasound is subject to a locally correlated multiplicative noise, called speckle, which degrades image quality and compromises diagnostic confidence [1]. For medical images, quality can be objectively defined in terms of performance in clinically relevant tasks such as lesion detection and classification, where typical tasks are the detection of an abnormality, the estimation of some parameters of interest, or the combination of the above [2]. Most studies today have assessed the equipment performance by testing diagnostic performance of multiple experts, which also suffers from intraobserver and interobserver variability. Although this is the most important method of assessing the results of image degradation, few studies have attempted to perform physical measurements of degradation [3]. This chapter was also published in a longer paper in [4].

Image quality is important when evaluating or segmenting atherosclerotic carotid plaques [5, 6] or the intima-media thickness (IMT) in the carotid artery [7], where speckle obscures subtle details [8] in the image. Speckle, which is a multiplicative noise, is the major performance-limiting factor in visual lesion detection in ultrasound imaging that makes the signal or lesion difficult to detect [3, 6–8]. In recent studies [4, 8] we have shown that normalization and speckle reduction improve the visual perception of the expert in the assessment of ultrasound imaging of the carotid artery. Traditionally, suspected plaque formation is confirmed using color blood flow imaging, where the types of the plaque were visually identified, and the delineations of the plaque and IMT were made manually by medical experts [6, 7].

To be able to design accurate and reliable quality metrics, it is necessary to understand what quality means to the expert. An expert's satisfaction when watching an image depends on many factors. One of the most important, of course, is image content and material. Research in the area of image quality has shown that the quality is dependent on many parameters, such as viewing distance, display size, resolution, brightness, contrast, sharpness, colorfulness, naturalness, and other factors [9].

It is also important to note that there is often a difference between fidelity (the accurate reproduction of the original on the display) and perceived quality. Sharp images with high contrast are usually more appealing to the average expert. Likewise, subjects prefer slightly more colorful and saturated images despite realizing that they look somewhat unnatural [10]. For studying visual quality some of the definitions above should be related to the human-visual system. Unfortunately, subjective quality may not be described by an exact figure, due to its inherent subjectivity, it can only be described statistically. Even in psychological threshold experiments, where the task of the expert is to give a yes or no answer, there exists a significant variation between expert's contrast sensitivity functions and other critical low-level visual parameters. When speckle noise is apparent in the image, the expert's differing experiences with noise are bound to lead to different weightings of the artifact [9]. Researchers showed that experts and nonexperts, with respect to image quality, examine different critical image characteristics to form their final opinion [11].

The objective of this study was to investigate the usefulness of image quality evaluation based on image quality metrics and on visual perception in ultrasound imaging of the carotid artery after normalization and speckle reduction filtering. For this task we have evaluated the quality of ultrasound imaging of the carotid artery on two different ultrasound scanners, the HDI ATL-3000 and the HDI ATL-5000, before and after speckle reduction, after image normalization, and after image normalization and speckle reduction filtering. Statistical and texture analyses were carried out on the original and processed images, and these findings were compared with the visual perception review, carried out by two experts.

4.2 Methodology

4.2.1 Ultrasound Imaging Scanners

The images used in this study, were captured using two different ultrasound scanners, the ATL HDI-3000 and the ATL HDI-5000 (Advanced Technology Laboratories, Seattle, Washington).

The ATL HDI-3000 ultrasound scanner is equipped with a 64-element, fine-pitch, high-resolution, 38-mm broadband array, a multielement ultrasound scanning head with an operating frequency range of 4 to 7 MHz, an acoustic aperture of 10×8 mm, and a transmission focal range of 0.8 to 11 cm [12].

The ATL HDI-5000 ultrasound scanner is equipped with a 256-element, fine-pitch, high-resolution, 50-mm linear array and a multielement ultrasound scanning head with an extended operating frequency range of 5 to 12 MHz that offers real spatial compound imaging. The scanner increases the image clarity using SonoCT imaging by enhancing the resolution and borders, and interface margins are better displayed. Several tests made by the manufacturer showed that the ATL HDI-5000 scanner was superior overall to conventional 2-D imaging systems, primarily because of the reduction of speckle, contrast resolution, and tissue differentiation [12].

The settings for the two ultrasound scanners were the same during the acquisition of all images in this study. The images were captured with the ultrasound probe positioned at right angles to the adventitia and the image was magnified or the depth was adjusted so that the plaque would fill a substantial area of the image. Digital images were resolution normalized at 16.66 pixels/mm (see the image normalization section). This was carried out due to the small variations in the number of pixels per millimeter of image depth (i.e., for deeply situated carotid arteries, the image depth was increased and therefore the digital image spatial resolution would have decreased) and in order to maintain uniformity in the digital image's spatial resolution [13]. B-mode scan settings were adjusted at 170 dB, so that the maximum dynamic range was used with a linear postprocessing curve. To ensure that a linear postprocessing curve was used, these settings were preselected (by selecting the appropriate start-up presets from the software) and were included in the start-up settings of the ultrasound scanner.

The position of the probe was adjusted so that the ultrasonic beam was vertical to the artery wall. The time gain compensation (TGC) curve was adjusted (gently sloping) to produce uniform intensity of echoes on the screen, but it was vertical in the lumen of the artery where attenuation in blood was minimal, so that echogenicity of the far wall was the same as that of the near wall. The overall gain was set so that the appearance of the plaque was assessed to be optimal and little noise appeared within the lumen. It was then decreased so that at least some areas in the lumen appeared to be free of noise (black).

4.2.2   Material

A total of 80 symptomatic B-mode longitudinal ultrasound images from identical vessel segments of the carotid artery bifurcation were acquired from each ultrasound scanner. The images were recorded digitally on a magneto-optical drive with a resolution of 768×576 pixels with 256 gray levels.

The images were recorded at the Institute of Neurology and Genetics in Nicosia, Cyprus, from 32 female and 48 male symptomatic patients between 26 and 95 years old, with a mean age of 54 years old. These subjects were at risk of atherosclerosis and had already developed clinical symptoms, such as a stroke or a transient ischemic attack.

In addition, 10 symptomatic ultrasound images of the carotid artery representing different types of atherosclerotic carotid plaque formation with irregular geometry typically found in this blood vessel were acquired from each scanner.

Plaques may be classified into the following types: (1) type I: uniformly echolucent (black), where bright areas occupy less than 15% of the plaque area; (2) type II: predominantly echolucent, where bright echoes occupy 15% to 50% of the plaque area; (3) type III: predominantly echogenic, where bright echoes occupy 50% to 85% of the plaque area; (4) type IV: uniformly echogenic, where bright echoes occupy more than 85% of the plaque area; and (5) type V: calcified cap with acoustic shadow so that the rest of the plaque cannot be visualized [5, 6]. In this study the plaques delineated were of type II, III, and IV because these types of images are easier to manually delineate because the fibrous cap, which is the border between blood and plaque, is more easily identified. Plaque of type I does not have

very visible borers, and plaques of type V produce acoustic shadowing and are also
not very visible.

4.2.3   Speckle Reduction

The linear scaling speckle reduction filter (*lsmv: linear scaling mean variance*) utilizing the mean and the variance of a pixel neighborhood, first introduced in [14] and
implemented by our group, was used in this study. The *lsmv* filter was also used in
other studies for the speckle reduction filtering of ultrasound carotid artery images
and forms an output image as follows [8, 4, 15]:

$$f_{i,j} = \bar{g} + k_{i,j}\left(g_{i,j} - \bar{g}\right) \tag{4.1}$$

where $f_{i,j}$ is the new estimated noise-free pixel value in the moving window, $g_{i,j}$ is the
noisy pixel value in the middle of the moving window, $\bar{g}$ is the local mean value of an
$M \times N$ region surrounding and including pixel $g_{i,j}$, $k_{i,j}$ is a weighting factor with $k_{i,j}$
$\in [0..1]$, and i, j are the absolute pixel coordinates. The factor $k_{i,j}$ is a function of the
local statistics in a moving window and may be derived as [14]:

$$k_{i,j} = \sigma_g^2 \big/ \left(\bar{g}^2 \sigma_g^2 + \sigma_n^2\right) \tag{4.2}$$

The values σ_g^2, and σ_n^2, represent the variance in the moving window and the variance of the noise in the whole image, respectively. The noise variance, σ_n^2, may be
calculated for the logarithmically compressed image by computing the average noise
variance over a number of windows with dimensions that are considerably larger
than those of the filtering window. In each window the noise variance is computed
as [8]:

$$\sigma_n^2 = \sum_{i=1}^{p} \sigma_p^2 \big/ \bar{g}_p \tag{4.3}$$

where σ_p^2 and $\bar{g}_p$ are the variance and mean of the noise in the selected windows,
respectively, and p is the index covering all windows in the whole image [16]. If the
value of $k_{i,j}$ is 1 (in edge areas), this will result in an unchanged pixel, whereas a value
of 0 (in uniform areas) replaces the actual pixel by the local average $\bar{g}$ over a small
region of interest [see (4.1)]. It has been shown that speckle in ultrasound images can
be approximated by the Rayleigh distribution [17–19], which is implicitly contained
by σ_n^2 in (4.2) and (4.3). Speckle reduction filtering was applied on the images using
the *lsmv* filter, which was applied for four times iteratively on the images using a $7 \times$
7 moving pixel window without overlapping because this produced the best results
[8].

4.2.4   Image Normalization

The need for image standardization or postprocessing has been suggested in the
past, and normalization using only blood echogenicity as a reference point has been
applied in ultrasound images of the carotid artery [5]. A brightness adjustment of

the ultrasound images was used in this study, and this has been shown to improve image compatibility by reducing the variability introduced by different gain settings and also to facilitate ultrasound tissue comparability [5, 13].

The images were normalized manually by linearly adjusting the image so that the median gray-level value of the blood was 0 to 5, and the median gray level of the adventitia (artery wall) was 180 to 190. The scale of the gray level of the images ranged from 0 to 255 [20]. This normalization using blood and adventitia as reference points was necessary in order to extract comparable measurements when processing images obtained by different operators or different equipment [20].

The image normalization procedure performed in this study was implemented in MATLAB software (version 6.1.0.450, release 12.1, May 2001, by the MathWorks, Inc.), and tested on a Pentium III desktop computer, running at 1.9 GHz with 512 MB of RAM memory. The same software and computer station were also used for all other methods employed in this study.

4.2.5   Statistical and Texture Analysis

Texture provides useful information that can be used by humans for the interpretation and analysis of many types of images. It may provide useful information about object characterization in ultrasound images [21]. The following statistical and texture features (SF) were extracted from the original and the processed images to evaluate their usefulness based on speckle reduction filtering, image normalization, and visual perception evaluation.

4.2.5.1   Statistical Features [21]

1. Mean;
2. Median;
3. Variance (σ^2);
4. Skewness (σ^3);
5. Kurtosis (σ^4);
6. Speckle index ($c = \sigma^2/\bar{g}$).

4.2.5.2   Spatial Gray-Level Dependence Matrix—Range of Values (SGLDM)

Selected features as proposed by Haralick et al. [22] and measured in four directions, namely, in the east, west, north, and south directions of a pixel neighborhood. The ranges of these four values were computed for each feature where the following features were computed:

1. Entropy;
2. Contrast;
3. Angular second moment (ASM).

The Wilcoxon matched-pairs signed rank sum test was also used to determine if a significant (S) or not significant (NS) difference existed between the results of the visual perception evaluation made by the two experts, and the statistical and texture features, at $p < 0.05$. The test was applied on all 80 images of the carotid artery for the original (NF), speckle reduction (DS), normalized (N), and normalized speckle reduction (NDS) images.

4.3   Image Quality and Evaluation Metrics

Differences between the original, $g_{i,j}$, and the processed, $f_{i,j}$, images were evaluated using the following image quality and evaluation metrics, which were used as statistical measures. The basic idea is to compute a single number that reflects the quality of the processed image. Processed images with higher metrics have been evaluated to be better [23]. The following measures, which are easy to compute and have clear physical meaning, were computed.

4.3.1   Normalized Mean-Square-Error

The mean-square error (MSE) measures the quality change between the original and processed image in an $M \times N$ window [24]:

$$\text{MSE} = \frac{1}{MN} \sum_{i=1}^{M} \sum_{j=1}^{N} \left(\frac{g_{i,j} - f_{i,j}}{lpg_{i,j}} \right)^2 \tag{4.4}$$

where $lpg_{i,j}$ is the lowpass filtered images of the original image, $g_{i,j}$. For the case where $lpg_{i,j}$ is equal to zero, its value is replaced with the smallest gray-level value in the image. The MSE has been widely used to quantify image quality; when is used alone, it does not correlate strongly enough with perceptual quality. It should, therefore, be used together with other quality metrics and visual perception reviews [11, 24].

4.3.2   Normalized Root Mean-Square-Error

The root mean-square-error (RMSE) is the square root of the squared error averaged over an $M \times N$ array [25]:

$$\text{RMSE} = \sqrt{\frac{1}{MN} \sum_{i=1}^{M} \sum_{j=1}^{N} \left(\frac{g_{i,j} - f_{i,j}}{lpg_{i,j}} \right)^2} \tag{4.5}$$

The popularity of the RMSE value arises mostly from the fact that it is, in general, the best approximation of the standard error.

4.3.3   Normalized Error Summation in the Form of the Minkowski Metric

The error summation in the form of the Minkowski metric, Err, is the norm of the dissimilarity between the original and the processed images [3, 24, 26]:

$$Err = \left(\frac{1}{MN} \sum_{i=1}^{M} \sum_{j=1}^{N} \left| \frac{g_{i,j} - f_{i,j}}{lpg_{i,j}} \right|^{\beta} \right)^{1/\beta} \tag{4.6}$$

computed for $\beta = 3$ (Err3) and $\beta = 4$ (Err4). For $\beta = 2$, the RMSE is computed as in (4.5), whereas for $\beta = 1$, the RMSE is computed as the absolute difference, and for $\beta = \infty$ as the maximum difference measure.

4.3.4 Normalized Geometric Average Error

The geometric average error (GAE) is a measure that determines if the transformed image is very bad [27], and it is used to replace or complete the RMSE. It is positive only if every pixel value is different between the original and the transformed image. The GAE is approaching zero if there is a very good transformation (small differences) between the original and the transformed image, and is moving away from zero vice versa. This measure is also used for tele-ultrasound, when transmitting ultrasound images, and is defined as:

$$GAE = \left(\prod_{i=1}^{N} \prod_{j=1}^{M} \sqrt{\frac{g_{i,j} - f_{i,j}}{lpg_{i,j}}} \right)^{1/NM} \tag{4.7}$$

4.3.5 Normalized Signal-to-Noise Ratio

The signal-to-noise ratio (SNR) [28] is given by:

$$SNR = 10 \log_{10} \frac{\displaystyle\sum_{i=1}^{M} \sum_{j=1}^{N} \left(\frac{g_{i,j}^{2} + f_{i,j}^{2}}{lpg_{i,j}} \right)}{\displaystyle\sum_{i=1}^{M} \sum_{j=1}^{N} \left(\frac{g_{i,j} - f_{i,j}}{lpg_{i,j}} \right)^{2}} \tag{4.8}$$

The SNR, RMSE, and *Err* have proven to be very sensitive tests for image degradation, but they are completely nonspecific. Any small change in image noise, filtering, and transmitting preferences would cause an increase in the preceding measures.

4.3.6 Normalized Peak Signal-to-Noise Radio

The peak SNR (PSNR) [28] is computed by:

$$PSNR = -10 \log_{10} \frac{MSE}{s^{2}} \tag{4.9}$$

where s is the maximum intensity in the original image. the PSNR is higher for a better transformed image and lower for a poorly transformed image. It measures image fidelity that determines how closely the transformed image resembles the original image.

4.3.7 Mathematically Defined Universal Quality Index

The universal quality index, Q, models any distortion as a combination of three different factors [3] (loss of correlation, luminance distortion, and contrast distortion) and is derived as follows:

$$Q = \frac{\sigma_{gf}}{\sigma_f \sigma_g} \frac{2\bar{f}\bar{g}}{(\bar{f})^2 + (\bar{g})^2} \frac{2\sigma_f \sigma_g}{\sigma_f^2 + \sigma_g^2}, \quad -1 < Q < 1 \tag{4.10}$$

where $\bar{g}$ and $\bar{f}$ represent the mean of the original and transformed image values, with their standard deviations, σ_g and σ_f, of the original and transformed values of the analysis window, and σ_{gf}, represents the covariance between the original and transformed images. The value of Q is computed for a sliding window of size 8×8 without overlapping. Its highest value is 1 if $g_{i,j} = f_{i,j}$, whereas its lowest value is -1 if $f_{i,j} = 2$ $\bar{g} - g_{i,j}$.

4.3.8 Structural Similarity Index

The structural similarity index (SSIN) between two images [26], which is a generalization of (4.10), is given by:

$$\text{SSIN} = \frac{(2\bar{g}\bar{f} + c_1)(2\sigma_{gf} + c_2)}{(\bar{g}^2 + \bar{f}^2 + c_1)(\sigma_g^2 + \sigma_f^2 + c_2)} \quad -1 < Q < 1 \tag{4.11}$$

where $c_1 = 0.01dr$ and $c_2 = 0.03dr$, with $dr = 255$, which represents the dynamic rage of the ultrasound images. The range of values for the SSIN lies between -1 for a bad and 1 for a good similarity between the original and transformed images, respectively. It is computed in a manner similar to that used for the Q measure for a sliding window of size 8×8 without overlapping.

4.4 Visual Perception Evaluation

Visual evaluation can be broadly categorized as the ability of an expert to extract useful anatomic information from an ultrasound image. The visual evaluation varies of course from expert to expert and is subject to the expert's variability [2]. The visual evaluation in this study was carried out according to the ITU-R recommendations with the Double Stimulus Continuous Quality Scale (DSCQS) procedure [27]. All of the visual evaluation experiments were carried out at the same workstation under indirect fluorescent lighting typical of an office environment. Two vascular experts evaluated the images. The vascular experts, an angiologist and a neurovascular specialist, were allowed to position themselves comfortably with respect to the viewing monitor, where a typical distance of about 50 cm was kept. Experts in real-life applications employ a variety of conscious and unconscious strategies for image evaluation, and it was our intent to create an application environment as close as possible to the real one. The two vascular experts evaluated 80 ultrasound images recorded from each ultrasound scanner before and after speckle reduction, after image normalization, and after normalization and speckle reduction filtering.

The two vascular experts evaluated the area around the distal common carotid, between 2 to 3 cm before the bifurcation, and the bifurcation. It is known that measurements taken from the far wall of the carotid artery are more accurate than those taken from the near wall [5, 17]. Furthermore, the experts were examining the image in the lumen area in order to identify the existence of a plaque or not. The primary interest of the experts was the area around the borders between blood and tissue of the carotid artery, and how much better they could differentiate blood from carotid wall, intima media, or plaque surface.

For each image, an individual expert was asked to assign a score of 1 to 5, corresponding to low and high subjective visual perception criteria. A 5 was given to an image with the best visual perception. Therefore, the maximum score for a procedure was 400 (i.e., the expert assigned a score of 5 to all 80 images). For each procedure, the score was divided by four to allow it to be expressed in a percentage format. The experts were allowed to give equal scores to more than one image in each case. For each preprocessing procedure the average score was computed.

4.5   Results

Figure 4.1 illustrates the NF, DS, N, and NDS images for the two ultrasound image scanners. The figure shows that the images for the ATL HDI-3000 scanner have greater speckle noise compared to the ATL HDI-5000 images. Moreover, the lumen borders and the IMT are more easily identified with the ATL HDI-5000 on the N and NDS images.

Table 4.1 shows the results in percentage format for the visual perception evaluation made by the two vascular experts on the two scanners. It is clearly shown that the highest scores are given for the NDS images, followed by the N, DS, and NF images for both scanners from both experts.

Table 4.1 Visual Perception Evaluation of Image Quality on 80 Images Processed from Each Scanner for the Original (NF), Speckle Reduction (DS), Normalized (N), and Normalized Speckle Reduction (NDS) Images

Visual Perception Score								
Ultrasound Scanner	*ATL HDI-3000*				*ATL HDI-5000*			
Preprocessing procedure	*NF*	*DS*	*N*	*NDS*	*NF*	*DS*	*N*	*NDS*
Angiologist	30	43	69	72	26	42	59	70
Neurovascular specialist	41	56	54	71	49	53	59	72
Average	36	50	62	72	38	48	59	71

Source: [4].

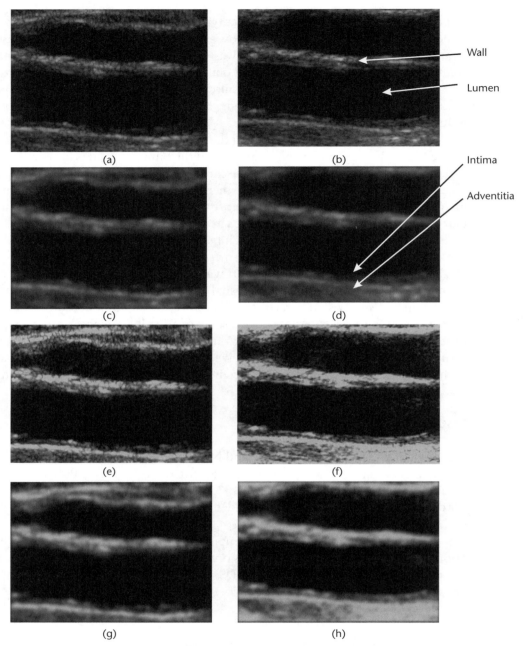

Figure 4.1 Ultrasound carotid artery images of (a, b) the original (NF), (c, d) speckle reduction (DS), (e, f) normalized (N), and (g, h) normalized speckle reduction (NDS) images of the ATL HDI-3000 scanner (left column) and ATL HDI-5000 scanner (right column). (*From:* [8]. © 2005 IEEE. Reprinted with permission.)

Let's look at the results of the Wilcoxon rank sum test for the visual perception evaluation, performed between the NF-DS, NF-N, NF-NDS, DS-N, DS-NDS, and N-NDS images, for the first and second observer on the ATL HDI-3000 and the ATL HDI-5000 scanner, respectively. The results of the Wilcoxon rank sum test for the visual perception evaluation were mostly significantly different (S) showing large intraobserver and interobserver variability for the different preprocessing procedures (NF-DS, NF-N, NF-NDS, DS-N, DS-NDS, N-NDS) for both scanners. Not significantly (NS) different values were obtained for both scanners, after normalization and speckle reduction filtering, showing that this improves the optical perception evaluation.

Table 4.2 presents the results of the statistical and texture features for the 80 images recorded from each image scanner. The upper part of Table 4.2 shows that the effect of speckle reduction filtering (DS) for both scanners was similar; that is, the mean and the median were preserved, the standard deviation was reduced, the skewness and the kurtosis were reduced, and the speckle index was reduced, where it is shown that the gray-value line profiles are smoother and less flattened. Furthermore, Table 4.2 shows that some statistical measures such as the skewness, kurtosis, and speckle index were better than the original (NF) and speckle reduction (D) images after normalization (N) for both scanners, and were even better after normalization and speckle reduction (NDS). However, the mean was increased for the N and NDS images for both scanners.

The bottom part of Table 4.2 shows that the entropy was increased and the contrast was reduced significantly in the cases of DS and NDS for both scanners. The entropy was slightly increased and the contrast was slightly reduced in the cases of N images for both scanners. The ASM was reduced for the DS images for both scanners and for the NDS images for the ATL HDI-5000 scanner.

We now look at the results of the Wilcoxon rank sum test for the statistical and texture features (see Tables 4.1 and 4.2), performed on the NF-DS, NF-N, NF-NDS, DS-N, DS-NDS, and N-NDS images on the ATL HDI-3000 scanner. No statistically significant difference was found between the most of the metrics when performing the nonparametric Wilcoxon rank sum test at $p < 0.05$, between the NF and DS, the NF and N, and NF and NDS features for both scanners. Statistically significant differences were obtained for the ASM, contrast, and entropy.

Table 4.2   Statistical and Texture Features Results of Mean Values for 80 Images Processed from Each Scanner for the Original (NF), Speckle Reduction (DS), Normalized (N), and Normalized Speckle Reduction (NDS) Images

Scanner	ATL HDI-3000				ATL HDI-5000			
Images	NF	DS	N	NDS	NF	DS	N	NDS
Statistical Features (SF)								
Mean	22.13	21.78	26.81	26.46	22.72	22.35	27.81	27.46
Median	3.07	4.53	3.56	5.07	3.73	5.23	4.59	6.07
Standard deviation	40.67	36.2	45.15	41.48	41.22	36.7	45.9	42.31
Skewness (σ^3)	2.88	2.49	2.23	2.00	2.84	2.45	2.17	1.94
Kurtosis (σ^4)	12.43	10.05	7.94	6.73	12.13	9.82	7.56	6.43
Speckle Index	0.29	0.27	0.25	0.24	0.28	0.27	0.24	0.23
SGLDM Range of Values								
Entropy	0.24	0.34	0.25	0.34	0.40	0.48	0.41	0.48
Contrast	667	309	664	303	618	302	595	287
ASM	0.36	0.35	0.38	0.37	0.37	0.33	0.39	0.35

Source: [4].

Furthermore, Table 4.2 shows that the entropy, which is a measure of the information content of the image, was higher for the ATL HDI-5000 in all cases. The ASM, which is a measure of the inhomogeneity of the image, is lower for the ATL HDI-5000 in the DS and NDS image cases. Furthermore, the entropy and the ASM were more influenced by speckle reduction than normalization because they reach their best values after speckle reduction filtering.

Table 4.3 illustrates the image quality evaluation metrics, for the 80 ultrasound images recorded from each image scanner, between the NF-DS, NF-N, NF-NDS, and N-NDS images. Best values were obtained for the NF-N images with lower RMSE, Err3, and Err4, higher SNR, and PSNR for both scanners. The GAE was 0.00 for all cases, which can be attributed to the fact that the information between the original and the processed images remains unchanged. Best values for Q and SSIN were obtained for the NF-N images for both scanners, whereas best values for SNR were obtained for the ATL HDI-3000 scanner on the NF-N images.

Table 4.3 shows that the effect of speckle reduction filtering was more obvious on the ATL HDI-3000 scanner, which shows that the ATL HDI-5000 scanner produces images with lower noise and distortion. Moreover, it was obvious that all quality metrics presented here are equally important for image quality evaluation. Specifically, for the most of the quality metrics, better measures were obtained between the NF-N, followed by the NF-NDS and N-NDS images, for both scanners. It is furthermore important to note that a higher PSNR (or equivalently, a lower RMSE) does not necessarily imply a higher subjective image quality, although they do provide some measure of relative quality.

Furthermore, the two experts visually evaluated 10 B-mode ultrasound images with different types of plaque [5] by delineating the plaque at the far wall of the carotid artery. The visual perception evaluation, and the delineations made by the two experts, showed that the plaque may be better identified on the ATL HDI-5000 scanner after normalization and speckle reduction (NDS), whereas the borders of the plaque and the surrounding tissue may be better visualized on the ATL HDI-5000 when compared with the ATL HDI-3000 scanner.

Table 4.3 Image Quality Evaluation Metrics Between the Original–Speckle Reduction (NF-DS), Original–Normalized (NF-N), Original–Normalized Speckle Reduction (NF-NDS) and the Normalized–Normalized Speckle Reduction (N-NDS) Images

Evaluation Metrics	ATL HDI-3000				ATL HDI-5000			
	NF-DS	NF-N	NF-NDS	N-NDS	NF-DS	NF-N	NF-NDS	N-NDS
MSE	1.4	1.3	2.0	1.3	1.2	0.3	1.9	1.3
RMSE	1.2	0.4	1.4	1.1	1.1	0.5	1.3	1.1
Err3	3.8	0.8	3.9	3.5	3.7	0.8	3.8	3.5
Err4	8.2	1.2	8.0	7.5	8.1	1.3	7.8	7.5
GAE	0	0	0	0	0	0	0	0
SNR	5.0	16.5	4.8	5.4	5.3	15.9	5.1	5.4
PSNR	48.0	59	45.6	44.6	47.4	58.5	46	44.6
Q	0.7	0.93	0.73	0.69	0.72	0.93	0.72	0.71
SSIN	0.9	0.95	0.92	0.83	0.94	0.95	0.91	0.83

Source: [4].

Table 4.4 summarizes the image quality evaluation results of this study, for the visual evaluation (Table 4.1), the statistical and texture analysis (Table 4.2), and the image quality evaluation metrics (Table 4.3). A double plus sign in Table 4.4 indicates very good performance, whereas a single plus sign indicates a good performance. Table 4.4 can be summarized as follows: (1) The NDS images were rated visually better on both scanners; (2) the NDS images showed better statistical and texture analysis results on both scanners, (3) the NF-N images on both scanners showed better image quality evaluation results, followed by the NF-DS on the ATL HDI-5000 scanner and the NF-DS on the HDI ATL-3000 scanner; and (4) the ATL HDI-5000 scanner images have considerable higher entropy than those of the ATL HDI-3000 and thus more information content. However, based on the visual evaluation by the two experts, both scanners were rated similarly.

4.6 Discussion

Normalization and speckle reduction filtering are very important preprocessing steps in the assessment of atherosclerosis in ultrasound imaging. We have therefore, investigated the usefulness of image quality evaluation in 80 ultrasound images of the carotid bifurcation, based on image quality metrics and visual perception after normalization and speckle reduction filtering using two different ultrasound scanners (ATL HDI-3000 and ATL HDI-5000). Specifically, the images were evaluated before and after speckle reduction, after normalization, and after normalization and speckle reduction filtering. The evaluation was based on visual evaluation by two experts, statistical and texture features, image normalization, speckle reduction, and on image quality evaluation metrics. Note that to the best of our knowledge, there are no other studies found in the literature for evaluating ultrasound image quality, based on speckle reduction filtering and normalization performed on carotid artery images, acquired by two different ultrasound scanners.

The main findings of this study can be summarized as follows:

1. The NDS images were rated visually better on both scanners.
2. The NDS images showed better statistical and texture analysis results on both scanners.
3. Better image quality evaluation results were obtained between the NF-N images for both scanners, followed by the NF-DS images for the ATL HDI-5000 scanner and the NF-DS on the HDI ATL-3000 scanner.

Table 4.4 Summary Findings of Image Quality Evaluation in Ultrasound Imaging of the Carotid Artery

Ultrasound Scanner	Visual Evaluation (Table 4.1)				Statistical and Texture Analysis (Table 4.2)				Image Quality Evaluation (Table 4.3)			
	NF	DS	N	NDS	NF	DS	N	NDS	NF-DS	NF-N	NF-NDS	N-NDS
ATL HDI-3000			++		+		++			++	+	
ATL HDI-5000			++		+		++		+	++		

Source: [4].

4. The ATL HDI-5000 scanner images have considerably higher entropy than the ATL HDI-3000 scanner and thus more information content. However, based on the visual evaluation by the two experts, both scanners were rated similarly.

It has been shown that normalization and speckle reduction produce better images. Normalization was also proposed in other studies using blood echogenicity as a reference and applied in carotid artery images [29]. References [5, 13] showed that normalization improves the image comparability by reducing the variability introduced by different gain settings, different operators, and different equipment. Note that the order in which these processes (normalization and speckle reduction filtering) are applied affects the final result. Based on unpublished results, we have observed that by applying speckle reduction filtering first and then normalization, distorted edges are produced. The preferred method, then, to obtain better results is to apply normalization first and then speckle reduction filtering.

Recent studies [4, 6, 7] have shown that the preprocessing of ultrasound images of the carotid artery with normalization and speckle reduction filtering improves the performance of the automated segmentation of the intima-media thickness [7] and plaque [6]. More specifically, [7] showed that a smaller variability in segmentation results was observed when performed on images after normalization and speckle reduction filtering, compared with the manual delineation results made by two medical experts. Furthermore, in another study [8], we have shown that speckle reduction filtering improves the percentage of correct classifications score of symptomatic and asymptomatic images of the carotid. Speckle reduction filtering was also investigated by other researchers on ultrasound images of liver and kidney [30] and on natural scenery [14], using an adaptive two-dimensional filter similar to the *lsmv* speckle reduction filter used in this study. In [14, 30], speckle reduction filtering was evaluated based only on visual perception evaluation made by the researches.

Verhoeven and Thijssen [31] applied mean and median filtering in simulated ultrasound images and in ultrasound images with blood vessels. The lesion SNR was used to quantify the detectability of lesions after filtering. Filtering was applied on images with fixed and adaptive size windows in order to investigate the influence of the filter window size. It was shown that the difference in performance between the filters was small but the choice of the correct window size was important. Kotropoulos and Pitas [32] applied adaptive speckle reduction filtering in a simulated tissue-mimicking phantom and liver ultrasound B-mode images. They showed that the proposed maximum likelihood estimator filter was superior to the mean filter.

Although in this study, speckle has been considered as noise, in other studies speckle, which was approximated by the Rayleigh distribution, was used to support automated segmentation. Specifically, in [17], an automated luminal contour segmentation method based on a statistical approach was introduced, whereas in [19] ultrasound intravascular images were segmented using knowledge-based methods. Furthermore, in [18] a semiautomatic segmentation method for intravascular ultrasound images, based on gray-scale statistics of the image, was proposed in which the lumen, IMT, and plaque were segmented in parallel by utilizing a fast marching model.

Some statistical measures, as shown in the upper part of Table 4.2, were better after normalization, whereas others, shown in the bottom part of Table 4.2, were better after speckle reduction. Table 4.2 also shows that the contrast was higher for the NF and N images on both scanners and was significantly different (S) after normalization and speckle reduction filtering. All other measures presented in Table 4.2 were comparable and showed that better values were obtained on the NDS images. Moreover, it was shown that the entropy that is a measure of the information content of the image [22] was higher for both scanners in the cases of the NDS and DS images. Significantly different entropy values were obtained mostly after normalization and speckle reduction filtering. Low-entropy images have low contrast and large areas of pixels with the same or similar gray-level values. An image that is perfectly flat will have a zero entropy. On the other hand, high-entropy images have high contrast and thus higher entropy values [25]. The ATL HDI-5000 scanner, therefore, produces images with higher information content.

The entropy was also used in other studies to classify the best liver ultrasound images [33], where it was shown that the experts rated images with higher entropy values better. In [21], entropy and other texture features were used to classify between symptomatic and asymptomatic carotid plaques for assessing the risk of stroke. It has also been shown [13] that asymptomatic plaques tend to be brighter, have higher entropy, and are more coarse, whereas symptomatic plaques tend to be darker, have lower entropy (i.e., the image intensity in neighboring pixels is more unequal), and are less coarse. Furthermore, note that texture analysis could also be performed on smaller areas of the carotid artery, such as the plaque, after segmentation [6, 7].

In previous studies [2, 3, 9–11, 26, 27], researchers evaluated image quality on natural scenery images using either only the visual perception by experts or some of the evaluation metrics presented in Table 4.3. In this study, MSE and RMSE values were in the range of 0.4 to 2.0, for all cases; Err3, Err4, SNR, PSNR, Q, and SSIN were better between the NF-N images for both scanners, showing that normalization increases the values of these measures. In [15], speckle reduction filtering was investigated on ultrasound images of the heart. The MSE values reported after speckle reduction for the adaptive weighted median filtering, wavelet shrinkage enhanced filter, wavelet shrinkage filter, and nonlinear coherence diffusion were 289, 271, 132, and 121, respectively. Loupas et al. [34] applied an adaptive weighted median filter for speckle reduction in ultrasound images of the liver and gallbladder and used the speckle index and the MSE for comparing the filter with a conventional mean filter. The filter improved the resolution of small structures in the ultrasound images. References [3, 26, 35] also documented that the MSE, RMSE, SNR, and PSNR measures are not objective for image quality evaluation and that they do not correspond to all aspects of the visual perception nor they correctly reflect artifacts [24, 27].

Recently the Q [3], and SSIN [26] measures for objective image quality evaluation have been proposed. The best values obtained in this study were $Q = 0.95$ and SSIN = 0.95 and were obtained for the NF-N images for both scanners. These results were followed with $Q = 0.73$, and SSIN = 0.92 in the case of NF-NDS for the HDI ATL-3000 scanner, and $Q = 0.72$ and SSIN = 0.94 in the case of NF-DS for the HDI ATL-5000 scanner. In [3], where natural scenery images were distorted by speckle noise, the values for Q reported were 0.4408, whereas the values for Q after contrast stretching were 0.9372.

Finally, in other studies [36, 37], a natural scene statistical model was proposed for assessing the quality of JPEG2000 compressed images blindly, without having the reference image available.

The methodology presented in this study may also be applicable in future studies to the evaluation of new ultrasound and telemedicine systems in order to compare their performance. It is also important to note that the methodology consists of a combination of subjective and objective measures that should be combined to arrive at a proper image quality evaluation result [27].

In conclusion, the results of this study showed that normalization and speckle reduction filtering are important processing steps favoring image quality. Additionally, the usefulness of the proposed methodology based on quality evaluation metrics combined with visual evaluation in ultrasound systems and in wireless telemedicine systems needs to be further investigated.

Acknowledgments

This work was partly funded through the Integrated System for the Support of the Diagnosis for the Risk of Stroke (IASIS) project of the Fifth Annual Program for the Financing of Research of the Research Promotion Foundation of Cyprus 2002–2005, as well as through the Integrated System for the Evaluation of Ultrasound Imaging of the Carotid Artery (TALOS) project of the Program for Research and Technological Development 2003–2005 of the Research Promotion Foundation of Cyprus.

References

[1]   McDicken, N., *Diagnostic Ultrasonics,* 3rd ed., New York: Churchill Livingstone, 1991.

[2]   Krupinski, E., et al., "The Medical Image Perception Society, Key Issues for Image Perception Research," *Radiology,* Vol. 209, 1998, pp. 611–612.

[3]   Wang, Z., and A. Bovik, "A Universal Quality Index," *IEEE Signal Proc. Letters,* Vol. 9, No. 3, 2002, pp. 81–84.

[4]   Loizou, C. P., et al., "Quantitative Quality Evaluation of Ultrasound Imaging in the Carotid Artery," *Med. Biol. Eng. Comput.,* Vol. 44, No. 5, 2006, pp. 414–426.

[5]   Elatrozy, T., et al., "The Effect of B-Mode Ultrasonic Image Standardization of the Echodensity of Symptomatic and Asymptomatic Carotid Bifurcation Plaque," *Int. Angiol.,* Vol. 7, No. 3, 1998, pp. 179–186.

[6]   Loizou, C. P., et al., "An Integrated System for the Segmentation of Atherosclerotic Carotid Plaque," *IEEE Trans. on Inf. Tech. in Biomedicine,* Vol. 11, No. 5, 2007, pp. 661–667.

[7]   Loizou, C. P., et al., "Snakes Based Segmentation of the Common Carotid Artery Intima Media," *Med. Biol. Eng. Comput.,* Vol. 45, No. 1, 2007, pp. 35–49.

[8]   Loizou, C. P., et al., "Comparative Evaluation of Despeckle Filtering in Ultrasound Imaging of the Carotid Artery," *IEEE Trans. on Ultrasonics Ferroelectrics and Frequency Control,* Vol. 52, No. 10, 2005, pp. 653–669.

[9] Ahumada, A., and C. Null, "Image Quality: A Multidimensional Problem," in *Digital Images and Human Vision,* A. B. Watson, (ed.), Cambridge, MA: Bradford Press, 1993, pp. 141–148.

[10] Fedorovskaya, E. A., H. De Ridder, and F. J. Blomaert, "Chroma Variations and Perceived Quality of Color Images and Natural Scenes," *Color Research and Application,* Vol. 22, No. 2, 1997, pp. 96–110.

[11] Deffner, G., "Evaluation of Display Image Quality: Experts vs. Nonexperts," *Symp. Society for Information and Display Digest,* Vol. 25, 1994, pp. 475–478.

[12] Philips Medical System Company, *Comparison of Image Clarity, SonoCT Real-Time Compound Imaging Versus Conventional 2-D Ultrasound Imaging,* ATL Ultrasound, Report, 2001.

[13] Kyriakou, E., et al., "Ultrasound Imaging in the Analysis of Carotid Plaque Morphology for the Assessment of Stroke," in *Plaque Imaging: Pixel to Molecular Level,* J. S. Suri et al., (eds.), Washington, D.C.: IOS Press, 2005, pp. 241–275.

[14] Lee, J. S. "Digital Image Enhancement and Noise Filtering by Use of Local Statistics," *IEEE Trans. on Pattern Anal. Mach. Intellig.,* Vol. 2, No. 2, 1980, pp. 165–168.

[15] Abd-Elmonien, K., A. B. Youssef, and Y. Kadah, "Real-Time Speckle Reduction and Coherence Enhancement in Ultrasound Imaging Via Nonlinear Anisotropic Diffusion," *IEEE Trans. on Biomed. Eng.,* Vol. 49, No. 9, 2002, pp. 997–1014.

[16] Dutt, V., "Statistical Analysis of Ultrasound Echo Envelope," Ph.D. Dissertation, Rochester, MN: Mayo Graduate School, 1995.

[17] Brusseau, E., et al., "Fully Automatic Luminal Contour Segmentation in Intracoronary Ultrasound Imaging—A Statistical Approach," *IEEE Trans. on Med. Imag.,* Vol. 23, No. 5, 2004, pp. 554–566.

[18] Cardinal, M. R., et al., "Intravascular Ultrasound Image Segmentation: A Fast-Marching Method," *Proc. MICCAI, LNCS 2879,* 2003, pp. 432–439.

[19] Olszewski, M. E., et al., "Multidimensional Segmentation of Coronary Intravascular Ultrasound Images Using Knowledge-Based Methods," *Proc. SPIE Medical Imaging: Image Processing,* Vol. 5747, 2005, pp. 496–504.

[20] Nicolaides, A., et al., "The Asymptomatic Carotid Stenosis and Risk of Stroke Study," *Int. Angiol.,* Vol. 22, No. 3, 2003, pp. 263–272.

[21] Christodoulou, C. I., et al., "Texture Based Classification of Atherosclerotic Carotid Plaques," *IEEE Trans. on Med. Imag.,* Vol. 22, No. 7, 2003, pp. 902–912.

[22] Haralick, R. M., K. Shanmugam, and I. Dinstein, "Texture Features for Image Classification," *IEEE Trans. on Systems, Man, and Cybernetics,* Vol. 3, 1973, pp. 610–621.

[23] Netravali, A. N., and B. G. Haskell, *Digital Pictures: Representation, Compression and Standards,* 2nd ed., New York: Plenum, 2000.

[24] Chen, T. J., et al., "A Novel Image Quality Index Using Moran I Statistics," *Phys. Med. Biol.,* Vol. 48, 2003, pp. 131–137.

[25] Gonzalez, R., and R. Woods, *Digital Image Processing,* 2nd ed., Upper Saddle River, NJ: Prentice-Hall, 2002, pp. 419–420.

[26] Wang, Z., et al., "Image Quality Assessment: From Error Measurement to Structural Similarity," *IEEE Trans. on Image Processing,* Vol. 13, No. 4, 2004, pp. 600–612.

[27] Winkler, S., "Vision Models and Quality Metrics for Image Processing Applications," Ph.D. Thesis, Lausanne, Switzerland: Ecole Polytechnique Federale de Lausanne, 2000.

[28] Sakrison, D., "On the Role of Observer and a Distortion Measure in Image Transmission," *IEEE Trans. on Communications,* Vol. 25, 1977, pp. 1251–1267.

[29] Wilhjelm, J. E., et al., "Quantitative Analysis of Ultrasound B-Mode Images of Carotid Atherosclerotic Plaque: Correlation with Visual Classification and Histological Examination," *IEEE Trans. on Med. Imag.,* Vol. 17, No. 6, 1998, pp. 910–922.

[30] Bamber, J. C., and C. Daft, "Adaptive Filtering for Reduction of Speckle in Ultrasonic Pulse-Echo Images," *Ultrasonics,* Vol. 24, 1986, pp. 41–44.

[31] Verhoeven, J. T. M., and J. M. Thijssen, "Improvement of Lesion Detectability by Speckle Reduction Filtering: A Quantitative Study," *Ultrasonic Imaging*, Vol. 15, 1993, pp. 181–204.

[32] Kotropoulos, C., and I. Pitas, "Optimum Nonlinear Signal Detection and Estimation in the Presence of Ultrasonic Speckle," *Ultrasonic Imaging*, Vol. 14, 1992, pp. 249–275.

[33] Wu, C. M., C. Y. Chang, and H. K. Sheng, "Texture Features for Classification of Ultrasonic Liver Images," *IEEE Trans. on Med. Imag.*, Vol. 11, No. 3, 1992, pp. 141–152.

[34] Loupas, T., W. N. McDicken, and P. L. Allan, "An Adaptive Weighted Median Filter for Speckle Suppression in Medical Ultrasonic Images," *IEEE Trans. on Circuits and Systems*, Vol. 36, 1989, pp. 129–135.

[35] Pommert, A., and K. Hoehne, "Evaluation of Image Quality," in *Medical Volume Visualization: The State of the Art*, T. Dohi and R. Kikinis, (eds.), *Medical Image Computing and Computer-Assisted Intervention Proc., Part II,* Lecture Notes in Computer Science 2489, Berlin: Springer-Verlag, 2002, pp. 598–605.

[36] Sheikh, H. R., A. C. Bovik, and I. Cormack, "No-Reference Quality Assessment Using Natural Scene Statistics: JPEG 2000," *IEEE Trans. on Image Processing*, Vol. 14, No. 11, November 2005, pp. 1918–1927.

[37] Sheikh, H. R., A. C. Bovik, and C. Veciana, "An Information Fidelity Criterion for Image Quality Assessment Using Natural Scene Statistics," *IEEE Trans. on Image Processing*, Vol. 14, No. 12, December 2005, pp. 2117–2128.

CHAPTER 5

User-Independent Plaque Segmentation and Accurate Intima-Media Thickness Measurement of Carotid Artery Wall Using Ultrasound

Filippo Molinari, Silvia Delsanto, Pierangela Giustetto, William Liboni, Sergio Badalamenti, and Jasjit Suri

Cardiovascular diseases (CVDs) are responsible for one-third of all global deaths. Among all CVDs, the ischemic stroke and the cardiac attack represent the most dangerous threat to health. A recent study by the World Health Organization revealed that by 2010 cerebrovascular and cardiac diseases would be responsible for 20% and 26% of all global deaths, respectively. Moreover, if nowadays CVDs are a specific problem of industrialized countries, by 2020 they will also strongly affect emerging and third world countries.

Interestingly, CVDs are a problem of both the sexes: Cerebrovascular accidents are prevalent in men under 65 years, but then, after 65 years of age, they tend to be represented more in women. Early prevention is believed to be the key to successfully lowering the impact of CVD on health.

Several studies provide evidence of the relationships between carotid artery wall status and CVDs [1]; specifically, an increased intima-media thickness (IMT) is correlated with an augmented risk of brain infarction or cardiac attack. Moreover, the presence of carotid plaques has been correlated not only to CVDs, but also to degenerative pathologies such as vascular dementia and Alzheimer's disease [2]. Hence, the assessment of carotid wall status is also essential for the early identification of risk conditions in asymptomatic patients.

A suitable clinical examination for the early detection of carotid wall pathologies has to be noninvasive, low cost, fast, highly repeatable, user independent, and safe for the subjects. Ultrasound scanning is a technique that possesses all these characteristics; in clinical practice, in fact, the status of the carotid artery is usually assessed by means of B-mode longitudinal images.

This chapter presents an innovative algorithm for the accurate segmentation of ultrasound images of the common tract of the carotid artery. Besides being completely user independent, this technique allows for accurate IMT measurements and for carotid plaque characterization.

5.1 Ultrasound Imaging of the Carotid Wall

5.1.1 Carotid Artery Anatomy and Pathology

Arteries and veins present a common structure constituted by three coaxial *tunicae,* or layers. Proceeding from the external tissue to the inner vessel, these three layers are named, respectively, *adventitia, media,* and *intima* [3].

The inner coat is constituted by a layer of pavement endothelium, with polygonal, oval, or fusiform cells, with very distinct round or oval nuclei, followed by a subendothelial layer, formed by delicate connective tissue with branched cells lying in the interspaces of the tissue, and finally an elastic or fenestrated layer, composed by a network of elastic fibers, named fenestrated membrane. This last part represents the main component of the intimal wall.

The middle coat is composed mainly of muscular tissue, and it is characterized by a transversal arrangement of fibers, as opposed to the longitudinal direction of those of the intimal layer. In smaller arteries, the muscular fibers are arranged in lamellae and arranged circularly around the vessel, whereas in larger arteries, such as the iliac, femoral, and carotid, layers of muscular fibers alternate with elastic fibers. The elastic component is even more present in the aorta and the innominate arteries, where bundles of white connective tissue have also been found. The muscle fiber cells constituting the media are around 50 μm and contain well-marked, rod-shaped nuclei.

The tunica adventitia is composed of bundles of connective tissue and of elastic fibers, except in the smallest arteries. The elastic tissue is located prevalently near the middle coat and is sometimes described as forming a special layer, the *tunica elastica externa* of Henle. This layer is especially evident in arteries of medium size.

The term *atherosclerosis* literally means "hardening of the arteries" and is derived from the Greek words *athera,* meaning "porridge," with reference to the foamy appearance of plaques under high-power ultralight microscopy, and *sclerosis,* meaning "hardening." Clinically this term indicates an ensemble of pathological conditions having the common factors of thickening and elasticity loss in the arterial walls. This pathology is the most common cause of death and morbidity in the Westernized world.

For more than 40 years, the mechanisms involved in the development of this pathology have been one of the most fiercely debated arguments in the medical arena. Scientists now generally agree that atherosclerosis is a chronic inflammatory process, triggered by a series of risk factors, the most notable of which is an elevated level of cholesterol in blood serum, apparently unique in its ability to induce atherosclerosis even in the absence of the other known risk factors. Many hypotheses have been formulated throughout the years, each emphasizing a different aspect of the atherogenetic process, at times presented as the major cause in the development of the pathology [4].

The progression of atherosclerosis may cause the appearance of plaques on the artery wall. In 1994–1995 the American Heart Association (AHA) proposed the well-known classification of plaques into six categories on the basis of their composition and morphology. Further studies revealed that plaque type is associated with the patient's clinical course [5, 6]. Hence, assessment of the status of the carotid wall is of paramount importance in clinical practice to detect early signs of atherosclerotic pathology, to follow up on the progression of existing lesions, and to select the proper therapy.

5.1.2   Principles and Validity of Ultrasound Analysis of the Carotid Artery

The B-mode ultrasound scan of the common tract of the carotid artery (CCA) is a widely adopted clinical exam for the assessment and diagnosis of several pathologies. In fact, in the presence of a disease, the CCA tract may be subjected to morphological changes (such as stenosis or plaques) that can perturb the normal blood circulation. In asymptomatic subjects it may represent a risk index for some severe pathologies such as acute stroke or cardiac infarction. It is well known that increased thickness of the CCA wall is correlated with a higher incidence of cardiovascular and cerebral acute events.

5.1.2.1   Normal Carotid Wall

The principles and the validity of the carotid analysis performed by means of ultrasound B-mode images was firstly described by Pignoli et al. [7]. Common features are observable in B-mode carotid images: The carotid artery wall appears as a double-layered structure made of two lines of different echogenicity. The first line of echoes is generated by the interface between the vessel lumen and the intima layer; the second line of echoes is generated by the interface between the media and the adventitia layers.

Figure 5.1 reveals an intrinsic limitation of the ultrasound carotid analysis: The intima layer is not distinguishable from the media layer. Considering the distal wall of the common carotid artery, note that the artery wall is made of a first layer of variable gray tones that depicts the progressive variation of the acoustic impedance in the media layer; then, a sharp-edged layer of high echogenicity is found at the media–adventitia interface.

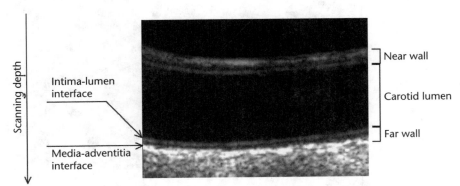

Figure 5.1   B-mode ultrasound image of a CCA tract (normal subject). The probe is positioned on the top of the figure and the direction of propagation of the ultrasound beam is downward. The far wall of the carotid artery is observable at the bottom of the figure. Specifically, the carotid tunica is characterized by two stripes that originate ultrasonic echoes: The first gray line is relative to the intimal and medial layers, underlined by the bright stripe corresponding to the adventitial layer. Note that in this kind of imaging, the intimal layer cannot be distinguished by the medial one. (*From:* [8]. © 2007 IEEE. Reprinted with permisson.)

The exploitation of these simple observations led to the first segmentation algorithms proposed in the literature. These techniques, which will be fully analyzed in next section, were mainly devoted to computer-assisted measurement of the IMT. In fact, during a clinical examination, the operator usually measures the IMT directly on the console of the ultrasound device by freezing a B-mode image and placing two markers (Figure 5.2): the first one corresponding to the lumen–intima interface, and the second corresponding to the media–adventitia interface. The device automatically measures the absolute value of the geometric distance of the two markers, providing the IMT measurement (given in millimeters or inches).

The dependence on the operator of such a measure is clear; previous studies reported high interoperator and intraoperator variability when measuring IMT. In 1991 a study by Wendelhag and colleagues [9] reported a coefficient of variation for intraobserver and inter observer variability in the measurement of the mean IMT of the common carotid artery of approximately 10%. Hence, the possibility of both reducing the time employed by the operator in determining the IMT through repeated measurements on the wall and augmenting the reproducibility of the measured values through image processing algorithms appears to be an attractive solution.

5.1.2.2   Ultrasound Analysis of Plaques

The basis of many currently used classification schemes for carotid plaques analyzed by means of ultrasonography is the Gray-Weale scale [10]. According to Gray-Weale, plaques are subdivided into four classes:

- Type I plaques are predominantly echolucent, that is, hypoechoic, with a thin echogenic (fibrous) cap.

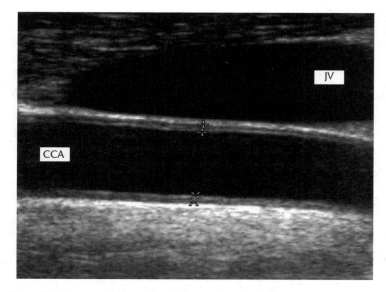

Figure 5.2   B-mode ultrasound image of a CCA tract (normal subject). To estimate the IMT, the operator, via the ultrasound device console, uses the markers placed on the near and far wall. It is evident that such a measure is punctual and, moreover, user dependent.

- Type II plaques are substantially echolucent, with small areas of echolucency (less than 25%).
- Type III plaques are predominantly echogenic with small areas of echolucency (less than 25%).
- Type IV plaques are uniformly echogenic.

Pathological correlation of plaques evaluated following this scheme reveals an association between intraplaque hemorrhage and/or ulceration with types I and II and fibrous tissue with types III and IV.

In the Tromso study, plaques were graded on the basis of their echogenicity on a 1 to 4 scale, with 1 signifying echolucent plaques. Plaques were also classified as heterogeneous if more than 20% of the plaque area corresponded to an echogenicity of at least two grades different from the rest of the plaque. Interobserver and intraobserver rates were good on plaque echodensity ($k = 0.80$ and $k = 0.79$, respectively), less so on plaque heterogeneity ($k = 0.71$ and $k = 0.54$). Hence, a major problem of these analyses is the need for quantitative measures [11]. In fact, computer-based image processing is expected to further improve the robustness of classification, but introduces new issues related to instrumentation-related variability, due to the type of equipment and acquisition parameters such as probe frequency and time gain compensation, which may alter the gray-level distribution on which the classification is based.

Ultrasound plaque segmentation may be used to:

- *Measure plaque dimensions with respect to the vessel diameter.* Even though this such measure is not accurate enough to calculate the degree of stenosis and even though plaque dimension itself is insufficient to establish the need for surgical treatment, nevertheless plaque dimension is an important clinical parameter to assess the carotid wall degeneration.
- *Perform a gray-scale classification of the plaque.* Several approaches have been proposed for the computer-based measurement of texture features, such as the gray-level dependence matrices, the gray-level difference statistics, and first-order statistics [11, 12].
- *Measure border irregularities.* Aside from echolucency and heterogeneity, other ultrasonographic characteristics have been identified as typical of rupture-prone plaques [13]. For instance, irregular borders, which might be histologically related to ulceration, have been suggested as a possible criterion for establishing increased cardiovascular risk associated with the imaged plaque.

Hence, the ultrasound analysis of the CCA wall is of great importance in clinical practice, because it represents a low-cost, fast, reliable and quite repeatable tool for assessing a subject's cardiovascular risk. Objective and quantitative measurements are needed to support diagnosis.

5.2   Algorithms for the Segmentation of Carotid Artery Ultrasound Images

The problem of carotid artery image segmentation, as will be more extensively clarified in the following, is composed of two conceptually different steps. In the first step, identification of the region of interest (ROI) is performed; that is, the region of the image corresponding to the carotid artery, the object of the segmentation, is identified. In the second step, the intimal and medial layers of the carotid artery wall are distinguished from the carotid lumen, from pixels belonging to the adventitial wall, and from pixels belonging to surrounding tissues.

5.2.1   Different Approaches for ROI Extraction

This first step is of paramount importance because it defines the portion of the image in which the carotid wall is contained. The extraction of a proper ROI: (1) enables the selection of parameters for the segmentation phase and (2) facilitates the segmentation procedure, avoiding further misclassification of pixels with characteristics similar to those of the vessel wall, but belonging to other structures.

In previously proposed algorithms, this step is almost always performed with a certain level of user interaction.

In the approach proposed by Liguori et al. [14], the user had to draw a rectangle on the image in which the carotid wall was contained. The authors developed a graphical user interface that could also be used for some preprocessing of the image (i.e., smoothing, gain calibration, data storage). Similarly, a ROI containing the carotid wall had to be selected by the human operator in the algorithm by Gutierrez et al. [15].

Cheng et al. [16] proposed an automated algorithm for the intimal and adventitial layers detection of the CCA based on snakes. In their solution, segmentation was performed by means of an active contour that required initialization. Hence, the user had to provide the algorithm with the coordinates of two points located just above the intima and in the carotid lumen. This initialization also restricted the portion of the image to be segmented and allowed for the avoiding of other confounding borders that could attract the snake (i.e., the jugular vein). A similar initialization is required for the gradient-based segmentation method described by Touboul et al. [17]. Regarding Cheng's technique, note that it is capable of segmenting only the distal border of the CCA.

All previous methods have the major drawback of requiring a certain degree of user interaction. In fact, once the starting points or rectangles are selected, segmentation will be performed only within those boundaries. This may be a major limitation in the presence of plaque if the ROI is not accurately selected, since a portion of the lesion may be ignored by the segmentation procedure. Moreover, consistency of results in function of the selected ROI is questionable: As an example, the selection of different starting points for the algorithm in [16] would lead to different profiles of the intimal and adventitial layers.

The major advantage of human-based ROI selection is insensitivity to superimposed noise. One of the major problems when segmenting B-mode ultrasound vascular images is the presence of noise, usually given by blood backscattering. An example is illustrated in Figure 5.3. International guidelines and clinical practice indicate that the gain of the echographic channels should be reduced as much as possible to avoid excessive noise. However, in some patients, lowering the gain makes the carotid wall tunicae less visible. When the operator delimits the ROI, the problem of noise is lessened, because the expert operator always draws the ROI in the portion of the image where he can more easily distinguish the carotid wall; we could say that human ROI selection usually corresponds to the portion of the ultrasound image where the SNR is higher.

We developed a completely user-independent algorithm for layer extraction of the CCA wall from ultrasound images. Hence, our technique allows for the automatic detection of the ROI. Section 5.3 describes the automatic procedure for ROI detection as well as the overall structure of the algorithm. Sections 5.4 to 5.7 show the performances on a large image database.

A major advantage of our approach is that its results are completely independent of the user. This ensures that:

- Results are consistent since there is no way to have different starting conditions on the same image;
- Segmentation is performed throughout the entire length of the image with no restrictions;
- The procedure is suitable for the automatic processing of large database sets, since there is no need for human interaction with the computer program.

5.2.2   Approaches to Image Segmentation

Once the ROI has been determined in the first step, the pixels belonging to the intima-media layer must be identified. Because the pixels belonging to the intimal and medial layers are separated from pixels belonging to the lumen and the

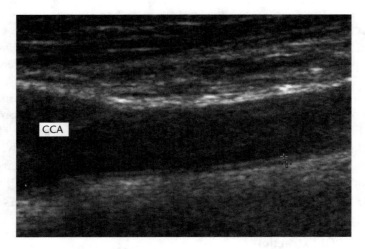

Figure 5.3   B-mode ultrasound image of a CCA tract (normal subject). During this examination, excessive gain of the echographic channels determined blood backscattering. The CCA lumen is not black and homogeneous, but it is textured. The SNR of such an image is poor and its segmentation may be problematic.

adventitia by a relatively simple interface (especially in normal cases), boundary detection techniques (as opposed to other classification techniques such as clustering) have been the methods of choice for performing the intima–lumen segmentation.

Various techniques have been proposed in the literature for this purpose. The simpler approach is the gradient-based technique [14, 18], which is sometimes combined with a model-based technique [18]. Carotid segmentation had also been performed by means of active contours [16] or by means of dynamic programming [18, 19].

Some of these proposed algorithms achieved quite good results. Cheng et al. [16] reported an average segmentation error lower than a pixel on a database of 32 images. The core of their algorithm is a snake driven by the Macleod operator as an external force. An interesting feature of this approach is that geometric and topological considerations about carotid morphology are taken into account to improve segmentation: A gradient filter is used to enhance the edge having a black region above and a white region below it (i.e., the adventitia layer). This simple consideration exploits geometrical characteristics into pixel fuzziness that is useful for segmentation.

Finally, Gutierrez et al. [15] used a deformable model proposed by Lobregt and Viergever [20]. Their technique is based on a geometrical deformable model made of points that can shift toward the image discontinuities under the effect of external dynamic forces. This implementation is very close to a snake, with the advantage that its formulation of the internal and external forces is simpler. This technique proved effective in segmenting the carotid artery wall, despite a sensible overestimation of the IMT.

Among all of the described algorithms, a thorough characterization of the performances of the techniques in terms of plaques is missing. In the following, we will show that our technique proves effective in estimating IMT and in segmenting fibrous and type II echolucent plaques.

5.3 Structure of the CULEX Algorithm

We propose a technique for the completely user-independent layer extraction (CULEX) of the CCA wall from ultrasound images. As discussed earlier, our algorithm consists of two distinct parts: (1) automatic ROI extraction, and (2) segmentation of the pixels in the ROI by means of a fuzzy C-means approach.

Figure 5.4 shows a block diagram of the CULEX structure. The meaning of the blocks is fully described in the following sections. Conceptually, the algorithm first automatically determines the region in which the carotid walls are located, then performs a fuzzy classification of the pixels belonging to the ROI. The output of the fuzzy classifier is then postprocessed in order to extract the layers contour.

5.3.1 ROI Extraction

ROI extraction is an iterative algorithm made of six steps. To correctly identify this region, morphological considerations are crucial: The basis of our ROI identifica-

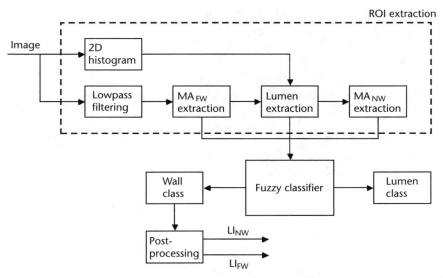

Figure 5.4   Schematic block diagram of the CULEX algorithm. The algorithm is made up of two distinct parts. The ROI extraction phase (dashed rectangle) deals with local and punctual measures on the input image; its outputs are the indication of the CCA lumen, the profile of the far wall media–adventitia (MA_{FW}) and the profile of the near wall media–adventitia (MA_{NW}). The second phase consists of a fuzzy classifier that separates lumen pixels from wall pixels and produces the final contours of the far and near wall lumen–intima interfaces (LI_{FW} and LI_{NW}, respectively).

tion procedure is the observation that the vessel lumen is characterized by pixels with low intensity and relatively low variance (low ultrasound echoes of the blood cells and relative homogeneity of the medium), surrounded by high-intensity pixels belonging to the carotid walls. Hence, the basic idea is to search for high-intensity pixels possibly belonging to the distal adventitial wall, starting from the deepest imaged layers of the scan. Then, considering the image column-wise and proceeding from the bottom of the image to the top, the lumen vessel and the near adventitial wall are extracted.

To perform this identification, we compute a bidimensional histogram of the image. For each pixel in the image we consider a 10×10 square neighborhood and then derive two measures: the mean intensity and the variance of the group of pixels. All of the mean values and variances of the neighborhoods are grouped into classes in a bidimensional histogram. It can be observed that the bidimensional distribution of these two variables for the pixels belonging to the lumen is condensed in the lowest classes. Figure 5.5 (right panel) reports an example of the distribution of the neighborhood mean and variance for the CCA image (left panel) of a healthy subject; the black pixels represent lumen pixels, whereas the gray pixels are all the other pixels in the image. Similar results are reported in [21].

The histogram allowed us to differentiate between lumen pixels and pixels belonging to the tissues. The extraction of the adventitial layers was performed considering the intensity profile of each column. Specifically, the peak intensities that can be observed in the intensity profile of each column derive from the adventitia layers, which are the most echogenic surfaces in the image. Figure 5.6 shows an example of

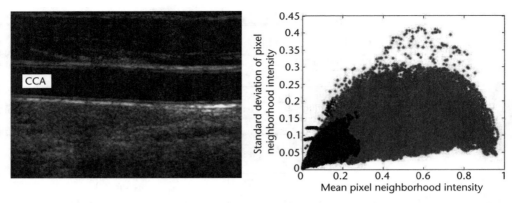

Figure 5.5 Distribution of mean and standard deviation (right panel) of a 10 × 10 square neighborhood for the pixels belonging to a B-mode ultrasound image of a normal CCA (left panel). The black pixels represent the points in the vessel lumen, whereas the gray pixels are all the other pixels in the image. The lumen pixels are characterized by low values for both the neighborhood mean and standard deviation. This neat discrimination of the pixels located in the lumen of the CCA is used by the algorithm for automatic ROI selection (i.e., to ensure the vessel lumen has been correctly located). (*From:* [8]. © 2007 IEEE. Reprinted with permission.)

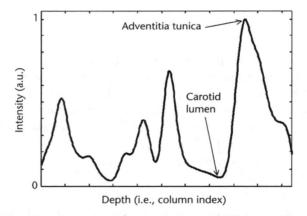

Figure 5.6 Intensity profile relative to a column of a B-mode CCA image. The abscissa represents depth (i.e., column index of the image) increasing from left to right; the vertical axis represents pixel brightness. The deepest peak in the intensity profile is the adventitia tunica; the previous minimum is the vessel lumen. This intensity profile is used to automatically determine the ROI (both the adventitia border, then the carotid lumen). (*From:* [8]. © 2007 IEEE. Reprinted with permission.)

an intensity profile for a CCA image: The deepest peak corresponds to the far adventitial wall, the carotid lumen is characterized by a minimum value of intensity, and the near adventitial layer corresponds to the peak intensity closer to skin surface. The combination of punctual measures (i.e., depth and intensity of a pixel) with local measures (i.e., average intensity of the neighborhood and standard deviation) allowed for an effective determination of the ROI.

In detail, the fundamental steps of ROI identification are as follows:

1. Lowpass filtering of the image (Gaussian filter of the 50th order, STD equal to 10) in order to increase the SNR.

2. Determination of the far adventitial wall from the intensity profile of the image considered column-wise. Specifically, for each column of the image, we identify the pixels possibly belonging to the adventitia as those local maxima whose intensity is in the 90th percentile of the intensity distribution (see Figure 5.6 as an example).

3. Individuation of the carotid lumen as the minimum point on the intensity profile whose mean intensity and variance of the neighborhood belong to the lowest classes of the bidimensional histogram. The minimum is searched descending the intensity profile from the maxima found in step 2. When the minimum is reached, the corresponding point is taken as a marker of the carotid lumen and the starting maximum as a marker of the adventitia.

4. Identification of the successive maximum as the adventitia layer of the near wall is accomplished.

5. Iteration of steps 2, 3, and 4 is done for all columns of the image.

6. The near wall media–adventitia interface is determined by step 4. To extract the far wall media–adventitia interface, a gradient-based contour extraction is performed and successively refined by applying a snake.

We adopted a simple snake to adjust border detection. Specifically, considering the curvilinear representation of a parametric model:

$$s : v(s) = \left[x(s), y(s) \right]$$
$$s \in [0,1]$$

the energy functional of the snake, which tends to modify the curvilinear representation in the bidimensional plane, at a given profile may be expressed as follows:

$$E = \int_{0}^{1} \frac{1}{2} \alpha |v'(s)|^2 + \gamma E_{ext}\left(v(s)\right) \, ds \qquad (5.1)$$

Details, implementations, and extensions of this algorithm can be found in [21]. We used a value of α equal to 0.1 and a value of γ equal to 0.01.

As an example, Figure 5.7 represents an ultrasound CCA scan with ROI overlapped: The white continuous line labeled ROI_L represents the vessel lumen, the white continuous line at the bottom of the image corresponds to the far wall media–adventitia layer (MA_{FW}), and the white continuous line at the top of the image is relative to the near wall media–adventitia layer (MA_{NW}).

The ROI extends from the near wall media–adventitia layer (MA_{NW}) to the far wall media–adventitia layer (MA_{FW}). The carotid lumen is taken as a check of the feasibility of the segmentation: In some cases ROI extraction fails (a detailed description of the robustness of the CULEX algorithm is provided in Section 2.5) because lumen cannot be correctly identified. The columns in which one or more of the three components of the ROI is missing are excluded by the segmentation process. This choice allowed us to develop an algorithm that is less susceptible to spa-

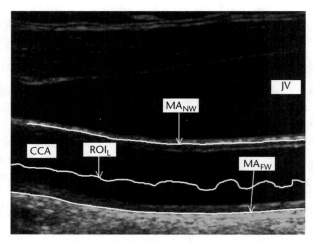

Figure 5.7 Example of ROI extraction performed by CULEX on a CCA image of a healthy subject. The ROI is determined by the pixels comprised between the near wall media–adventitia interface (MA_{NW}) and the far wall media–adventitia interface (MA_{FW}). The white line in the center of the image is the carotid lumen (ROI_L) that we use as a check.

tial variations of SNR, because if noise corrupts a portion of the image, making ROI detection impossible, the remaining part can still be correctly processed.

5.3.2 Segmentation of the Lumen and Carotid Wall

As previously discussed, we chose to perform image segmentation by means of a fuzzy c-means approach. Only the pixels comprised between the near and far media–adventitia walls are classified. We used the pixel intensity as feature and allowed two classes: one relative to the carotid wall and the other to the lumen. The class memberships were in the range of [0,1], with 0 meaning no membership and 1 meaning full membership in the specific cluster. Decisions were made using a threshold value equal to 0.1: Pixels with a wall-class membership of greater than 0.1 are considered to belong to the wall.

At this stage, two final postprocessing steps are required in order to come to the final segmentation:

1. Successive dilation and erosion operations (disk structure element) on the binary image obtained as output of the (defuzzified) clustering process and filling of the holes in the regions belonging both to the lumen and the walls [23];

2. Contour refinement by means of a snake [as in (5.1) with $\alpha = 0.1$ and $\gamma = 0.01$]. Near and far wall media–adventitia interfaces (indicated by MA_{NW} and MA_{FW}, respectively) and near and far wall lumen–intima interfaces (indicated by LI_{NW} and LI_{FW}, respectively) are available.

Figure 5.8 shows an example of the resulting automatic segmentation performed by CULEX on four CCA ultrasound images of healthy individuals.

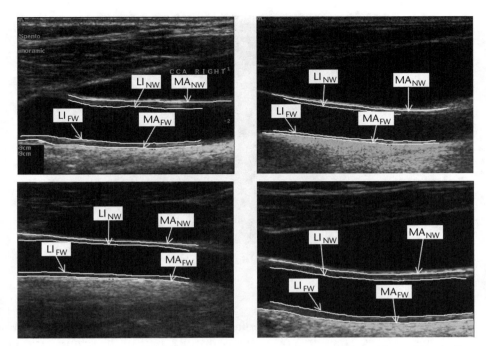

Figure 5.8   CULEX segmentation of four CCA images relative to healthy subjects. MA_{NW} indicates the near wall media–adventitia interface, MA_{FW} indicates the far wall media–adventitia interface, LI_{NW} indicates the near wall lumen–intima interface, and LI_{FW} indicates the far wall lumen–intima interface. The four images are quite different in terms of image resolution, level of noise, and presence of markers or written parts on the image itself. CULEX provides suitable segmentations in all of the images; hence, it also proves effective on previously recorded images that have already been processed.

The images in Figure 5.8 are different in terms of superimposed noise, spatial resolution, and presence of signs, but CULEX segmentation is still correct. This technique also proved effective on recorded images that had been previously processed by an operator who added markers, written parts, measures, and so on.

Figure 5.9 represents a zoomed vision of the CCA of a healthy subject. This figure allows us to evaluate the quality of CULEX segmentation.

5.3.3   Notes About the CULEX Segmentation Strategy

The basic idea of the CULEX algorithm is the preservation of pixel fuzziness, which is used for both automatic ROI extraction and for segmentation. Depending on the purpose, pixel features (i.e., intensity value, average intensity of a regional neighborhood, standard deviation of the neighborhood and depth) are combined in different fashions to allow for proper classification/clustering. For instance, to proficiently detect the ROI in which the CCA walls are comprised, CULEX needs to take into account at least three pixel features; whereas to perform fuzzy c-means classification, we observed that pixel intensity alone gives good results, provided the ROI has been selected properly.

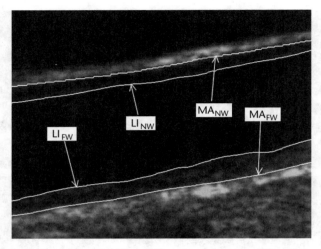

Figure 5.9 Expanded vision of the CULEX segmentation of a tract of the CCA.

This tight dependence of the two phases (ROI selection and subsequent wall segmentation) is enforced in the media–adventitia segmentation process. In fact, even though the two steps are conceptually distinct, part of the segmentation procedure begins at the end of the ROI extraction phase: The media–adventitia contours are extracted in step 6 of the ROI selection procedure. The subsequent segmentation does not affect the media–adventitia layer extraction. In other words, we found the gradient-based segmentation process to be the most suitable technique for evidencing the border between the media and the adventitia tunicae. The application of a further step of fuzzy clustering did not improve segmentation. Moreover, by restricting the pixel classification procedure to the ROI comprised between the two adventitial walls, we could develop an algorithm that proved suitable for both IMT measurement and plaque imaging and segmentation.

Due to the differing echogenicity of the tissues, the same observation does not apply to the lumen–intima segmentation. To accurately trace the LI_{NW} and LI_{FW} borders, tissue must be separated from vessel lumen by a fuzzy clustering. This second step, in fact, is the one that allows us to accurately adhere to the carotid wall in normal scans and to follow a plaque profile in pathology.

Preliminary tests we made (not reported in this chapter) showed that the classification of all pixels of the image without any user interaction is extremely difficult. The first reason is that in many ultrasound CCA images the jugular vein (JV) is present. In that case, for a fuzzy classifier it is impossible to separate pixels belonging to the carotid lumen from pixels belonging to the JV lumen. Moreover, JV lumen is often more homogeneous and characterized by lower backscattering than CCA lumen and this could bias the results of an intensity-based fuzzy classifier. Also, the relative position of the JV with respect to the CCA is unpredictable, because it depends on anatomy as well as on the insonation performed by the user.

The second reason is that tissue texture could easily be confounded with plaque texture. In fact, in many cases, the distribution of the gray tones in tissues surrounding the CCA is quite similar to that of an echolucent plaque. Therefore, without a previous ROI selection step, classification would be extremely complicated.

5.4   CULEX Characterization and Image Databases

As in many other cases in the medical imaging arena, the ground truth is unavailable. Comparison with a manually obtained segmentation is possible, of course, but interrater agreement, especially in this particular application, may be low, so that the evaluation of the algorithm performance may be strongly biased on the basis of who performed the manual segmentation. A partial solution may be to make a comparison to the mean of many manual segmentations performed by several experts. However, this does not lend information on the quality of the segmentation versus the interrater variability, which is in fact the best indication of which degree of precision it is possible to achieve.

In the following sections, we will present the results of the validation following two different approaches, whose common goal is to compare the algorithm's performance with the performance of human experts.

5.4.1   Percent Statistics Algorithm

The Percent Statistics algorithm for the evaluation of contour extraction was first proposed by Chalana and Kim [24] and later improved by Alberola-Lòpez et al. [25]. Its objective is to determine whether "the computer-generated boundaries differ from the manually outlined boundaries as much as the manually outlined boundaries differ from one another." The basic idea is that if the computer-generated boundary (CGB) behaves like the expert-outlined boundaries (EOB), it must have the same probability of falling within the interobserver range as the manual segmentations. Defining, for each contour C_i, the distance D_{ij} $(i \neq j)$ from any other contour C_j, and the maximum distance D_m between any two contours:

$$D_m = \max_{i,j}(D_{ij}) \text{ for } i \neq j$$

A contour falls in the interobserver range if the distances separating it from the other contours are all less than D_m. Two different distance metrics have been suggested [24, 25], the Hausdorff distance and the average distance between the curves. We considered the average distance.

Assuming then that the EOBs are independent and identically distributed and that the CGB behaves like the EOBs, the probability of the computer-generated contour falling into the interobserver range is equal to:

$$P(C_0 \in IR) = \frac{n-1}{n+1}$$

where C_0 is the CGB, $n + 1$ is the total number of contours (n EOBs + 1 CGB), and $n - 1$ is the number of "internal" contours (i.e., the total number of contours minus the two contours who are distant D_m). Denoting then as X_j the event "the computer-generated boundary lies within the interobserver range for the jth image," for j ranging from 1 to N, we have that X_j is a Bernoulli distributed variable, with parameter $p = (n - 1)/(n + 1)$, $q = 1 - p$.

Let us now consider the random variable $Z = \sum_{j=1}^{N} X_j / N$. Because we are trying to determine whether the CGB falls outside the interobserver range more frequently than the EOBs, we seek the one-sided confidence interval for variable Z, that is, the ε value for which $P(p - Z > \varepsilon) = \alpha$, which is $1 - \alpha$ the confidence level of the interval. For N sufficiently large, we can apply the De Moivre-Laplace approximation and assume that Z is normally distributed, with mean p and standard variance pq/N. Then, $\varepsilon = \sqrt{\dfrac{pq}{N}} z_{1-\alpha}$, where $z_{1-\alpha}$ is the value of a normal standard distribution leaving an area equal to $1 - \alpha$ to its right. The acceptance region for this test is the region in which Z has a value greater than $p - \varepsilon$.

5.4.2 Confidence Interval for the Average Segmentation

In this approach, we consider for each column the points corresponding to the EOBs. The basic assumption is that the operator-traced points follow a Gaussian distribution, whose mean corresponds to the true value of the lumen–intima and media–adventitia interfaces.

We then verify on each column whether or not the CGB falls within the confidence interval for the mean. Because the variance of the distribution is unknown, the t-Student distribution is used. Thus, we require:

$$\overline{\text{EOB}}_j - t_{\alpha/2, n-1} \frac{s}{\sqrt{N}} < \text{CGB}_j < \overline{\text{EOB}}_j + t_{\alpha/2, n-1} \frac{2}{\sqrt{N}}$$

where $\overline{\text{EOB}}_j$ is the sample mean of the EOB's ordinate values for the jth column, CGB_j is the CGB ordinate value for the jth column, s is the sample variance, N is the sample size (i.e., the number of operators), and $t_{\alpha/2, n-1}$ is the value for which $H(t_{\alpha/2, n-1}) = \alpha/2$, where $H(\cdot)$ is the distribution function of a t-Student variable, with $n - 1$ degrees of freedom. The percentage of times that the CGB falls within the confidence interval for the mean is then computed for each image.

5.4.3 Segmentation Error

An estimation of the CULEX segmentation error can be obtained by comparing the algorithm's lumen–intima and media–adventitia borders to the average borders traced by the human operators. Considering the ith image in the database, we defined the segmentation error for each column of the image as follows:

$$e_B^i(c) = \left| B_{\text{CULEX}}^i(c) - B_{\text{OP}}^i(c) \right| \tag{5.2}$$

where

- B_{CULEX}^i is the border contour generated by CULEX.
- B_{OP}^i is the average border segmented by the four human operators.

- c is an index spanning through the columns of the image.

We evaluated the segmentation error for all the four borders in the image: LI_{NW}, MA_{NW}, LI_{FW}, and MA_{FW}.

5.4.4 Image Database and Ultrasound Equipment

To test our algorithm, we used two different databases: The first one consists of 200 images relative to 50 healthy individuals (age range = 35 to 89 years; mean ± STD = 63 ± 7.5 years) without previous history of cardiovascular or cerebrovascular diseases, the second one is made of 100 images relative to 20 patients (age range = 41 to 89 years; mean ± STD = 56 ± 4.6 years). Thirty images were relative to intima–media thickening (IMT values ranging from 0.09 to 0.59 cm), 30 to homogeneous echogenic (stable) carotid plaques, and 40 to inhomogeneous echolucent type II (unstable) plaques.

All of the images were relative to the common tract of the carotid artery, insonated according to internationally recognized standards [1]. Patients were randomly selected among those who performed ultrasound examinations in the Neurology Division of the Presidio Sanitario Gradenigo of Torino, Italy. All of them signed an informed consent before being enrolled in this study.

A neurologist, a cardiologist, and two radiology technicians independently segmented all the images in our database by means of a graphical user interface (GUI) we developed. The manually traced profiles were used for the performance evaluation of CULEX. The GUI was developed in a MATLAB (The MathWorks, Natick, Massachusetts) framework and an example of the GUI's appearance is shown in Figure 5.10. This GUI allows for segmenting of the images and for reviewing saved tracings. New operators can be added automatically. The near and far wall sections are kept separate in order to allow the user to focus on a specific target. The images are stored in a structure and it is possible to complete segmentation in different sessions.

All of the tests were conducted on a G4 dual-processor, 2.5-GHz PowerMac, equipped with 4 GB of RAM memory and a 30-inch flat screen.

For acquisition of the images, we used two different devices: a Philips HDI ATL-5000 and a General Electric LOGIQ 500. The devices were equipped with 10-MHz (Philips) and 9-MHz (GE) linear probes. All of the images were transmitted by the US device to a host computer in DICOM format, discretized on 8 bits, and represented by means of a grayscale linear mapping of the pixel intensity.

In the following section, we will present the results in terms of the segmentation capabilities of the CULEX algorithm. To provide a full and comprehensive presentation of the algorithm's performances in the various scenarios, we separated the results and the discussions relative to ROI extraction, IMT measurement, and plaque imaging and segmentation.

As discussed in the previous section, because the ground truth is unavailable, we relied on manually traced contours to assess the quality of the computer-generated segmentations.

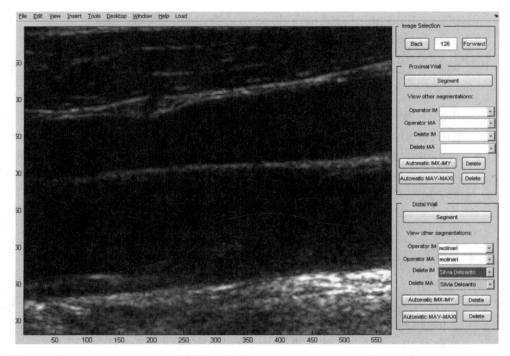

Figure 5.10 GUI developed in the MATLAB environment to evaluate CULEX performance and allow operators to segment the images. The GUI also allows for reviewing of the results: The image databases are stored in a structure together with the operators' manual segmentations. The GUI also allows new segmentations to be added and previous tracings can be reviewed.

5.5 ROI Detection and Robustness to Noise

CULEX correctly extracted the ROI in 295 images out of 300. All 200 images of the first database were correctly processed, whereas in 5 images of the second database (i.e., the one relative to patients), the ROI was incorrectly chosen due to image artifacts. The evaluation of the correctness of the ROI extraction was done by visual inspection of the images.

At the current stage of development, an error in the ROI selection precludes the correct segmentation of the carotid wall and cannot be recovered.

Generally, different noise sources and artifacts characterize ultrasound images. Some of these sources are caused by instrumentation (i.e., speckle noise [26]), some result from the biological environment (i.e., artifacts produced by air, bones, high density deposits), and some derive from incorrect use of the device (i.e., excessive blood backscattering, low contrast, details out of focus). Our tests revealed that blood backscattering represents the major threat to a suitable ROI extraction. In fact, an excessive level of gray tones dispersed into the vessel lumen may produce two effects: (1) the gray tone distribution is considered part of a partially echolucent plaque, and (2) gradient-based segmentation (i.e., step 6 of the CULEX algorithm) fails in determining the adventitia layer due to poor contrast of the wall with the lumen. Figure 5.11 (left panel) represents an image in which the ROI was incorrectly chosen due to excessive blood backscattering.

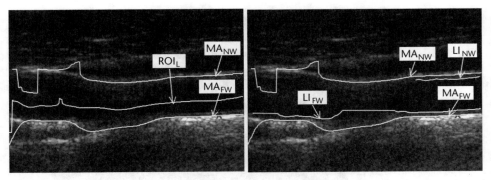

Figure 5.11 Segmentation of an image with high blood backscattering. The left image shows the ROI extraction: The lumen (ROI$_L$) and MA_{FW} are correctly individuated, but MA_{NW} extraction is problematic due to backscattering. This profile falls into the lumen vessel (leftmost part of the image) and then explodes beyond the adventitia tunica. Segmentation in this region is not feasible. The right image shows the segmentation performed by CULEX in the remaining portion of the image. Note that even if part of the image is corrupted by noise, the unaffected part can nevertheless be correctly segmented.

Note that if lumen and MA_{FW} are correctly detected, the adventitia layer of the near wall is not extracted properly. Specifically, the leftmost part of the image shows the MA_{NW} profile falling into the vessel lumen, whereas immediately after it exceeds the artery boundaries. Both of these effects are due to insufficient contrast between the adventitia and carotid lumen. Moreover, in the central portion, this image shows a wide zone in which backscatter is even higher; obviously, in that region ROI extraction is unreliable given the loss of detail in the image itself. Figure 5.11 (right panel) shows the segmentation performed by CULEX. Note how the portion of the image that is unaffected by blood backscattering is correctly segmented.

Another major problem for most segmentation algorithms is the presence of ultrasound artifacts. In CCA clinical examination, one of the most important artifacts is the presence of shadow cones. Hard tissues, usually calcium deposits or hard plaques, originate in these regions of poor image quality. A previous version of our algorithm [27] was unable to deal with such artifacts. In this version, ROI selection is also used to restrict image segmentation: if in some parts of the image ROI is incomplete (i.e., one among the lumen, MA_{FW}, and MA_{NW} cannot be identified), segmentation in those specific portions of the image does not take place. Hence, the remaining portion of the image that is unaffected by image artifacts can be correctly processed. An example is shown in Figure 5.12, where the large white arrow indicates a calcium deposit present in the near wall media tunica that projects a shadow cone onto the image. CULEX provides suitable segmentation of the remaining part of the image.

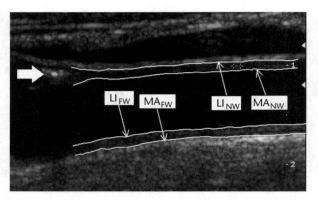

Figure 5.12 B-mode image of a CCA showing a calcium deposit in the near wall media tunica. The large white arrow indicates the calcium and the shadow cone the deposit projects onto the image. Segmentation performed by CULEX is, however, correct in the portion of the image unaffected by the artifact.

5.6 Segmentation of the Lumen–Intima and Media–Adventitia Interfaces and IMT Measurement

The evaluation of the CULEX performances in terms of segmentation accuracy were characterized using the methods defined in Section 5.4. Results herein presented are relative to the image database containing images of normal carotid walls.

We asked four experts to manually segment the images. However, due to different support for the segmentations, we decided to evaluate the segmentation errors at 50 points per image. Hence, results are relative to a total amount of 10,000 points (50 points × 200 images).

5.6.1 Absolute Segmentation Errors and IMT Measurements

The segmentation errors defined in (5.2) were computed on each image of the database for the lumen–intima and media–adventitia interfaces of the near and far wall. Figure 5.13 depicts the distribution of the segmentation error for the four interfaces. The abscissa reports the absolute error value, expressed in pixels, and the vertical axis reports the percent of time a specific error value occurred in the 10,000 evaluated points. Note that the segmentation error is always very low, except for the MA_{NW} that presents a greater error.

The boxplot in Figure 5.14 reports the distribution of the segmentation errors for the interfaces of the far wall. The crosses represent the outliers, estimated using a confidence level equal to 95%. It can be seen that the average segmentation error on both the interfaces is lower than a pixel. Figure 5.15 reports the boxplot for the near wall: The LI_{NW} interface is segmented with a very low average error, in this case lower than a pixel, whereas the MA_{NW} is segmented with an average equal to approximately 2 pixels.

The second column of Table 5.1 reports in a synthetic way the segmentation performances in terms of average segmentation error and standard deviation.

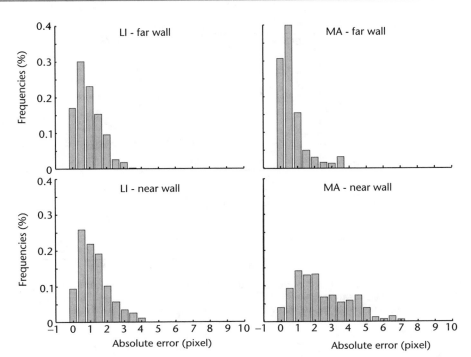

Figure 5.13   Distribution of the segmentation errors estimated on 10,000 points (50 points × 200 images). The histograms are relative to the LI_{FW} (upper left), to the MA_{FW} (upper right), to the MA_{NW} (lower right), and to the LI_{NW} (lower left). The abscissa reports the absolute value of the segmentation error [defined in (5.1)] and the bars represent the percentage of times that specific error occurs. A summary of the mean errors is shown in the second column of Table 5.1.

Table 5.1   Performance of the CULEX Technique in Terms of Segmentation Error (Second Column), of the Percent Statistics Test (Third Column), and of the Confidence Interval (Fourth Column)

Interface	Segmentation Error (Pixel) Mean ± STD	Percent Statistics Z Score	Confidence Interval: Percentage Points Falling into the 95% Confidence Interval (Mean ± STD)
LI_{NW}	1.2/0.9	0.514*	92/14
MA_{NW}	2.1/1.9	0.300	52/28
LI_{FW}	0.9/0.8	0.570*	81/18
MA_{FW}	0.8/0.7	0.800*	85/21

* For the percent statistics test, the asterisk indicates that the test has been passed (i.e., it is possible to accept the hypothesis that CULEX behaves like a human operator in segmenting that specific interface).

Figure 5.16 reports the relative error committed in the IMT measurement. Data are relative only to the distal border of the images, because several studies reported how it could be clinically misleading and difficult to estimate IMT from the proximal artery wall. The average error on the IMT measurement is equal to 9% ± 9%, which corresponds, in our image database, to approximately 1 pixel.

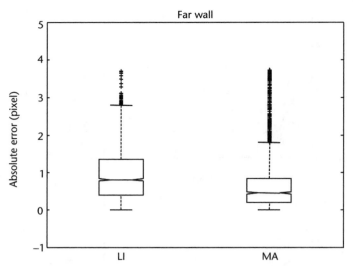

Figure 5.14 Boxplot of the segmentation errors for the LI (left box) and MA (right box) interfaces of the far wall. Mean errors and standard deviations are comparable. The crosses represent the outliers (calculated by considering a confidence level equal to 95%).

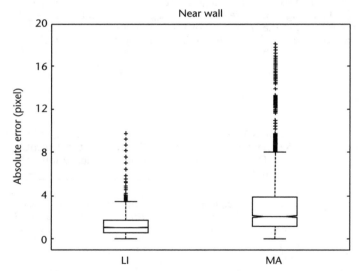

Figure 5.15 Boxplot of the segmentation errors for the LI (left box) and MA (right box) interfaces of the near wall. Note that the algorithm commits a greater error in the MA segmentation. The crosses represent the outliers (calculated by considering a confidence level equal to 95%).

Comparing the performance of CULEX to previously developed algorithms, we found that the segmentation error is in line with other operator-dependent techniques [15, 16] and is slightly better than other segmentation approaches [13]. However, a direct comparison of human-dependent techniques with CULEX is not straightforward, because there may be a correlation of the segmentation error with the manual ROI selection. In fact, the presence of an operator-dependent error in the

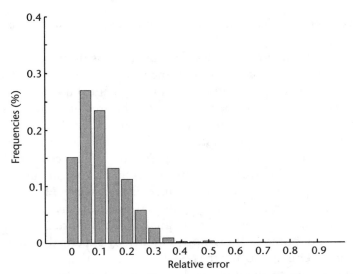

Figure 5.16   Distribution of the absolute percentage error on the IMT estimation obtained on 10,000 points. The abscissa reports the IMT measurement error in terms of percentage; the bars represent the percentage of times that specific error occurs. The average error is equal to 9% ± 9% (mean ± STD).

computer-generated boundaries (e.g., in the case of nonoptimal selection of the ROI in partially automated approaches) introduces a further element of variability, which complicates the performance evaluation. More importantly still, a well-known fact is that the characteristics of the image database may be fundamental in determining the quality of the computer-generated contour. In some cases, such as in [14], this is due to the fact that the algorithm was developed on the basis of specific hypotheses, for example, the fact that the CCA remains horizontal in the viewed image. More generally, the segmentation performances are always related to the image quality and the segmentation error per se does not take this factor into account. In this sense, a comparison with the variability of the segmentations performed by the operators is fundamental, because it yields important information on the quality of the guess of the contour made by the operators, which is strongly dependent on factors such as image noise.

Our technique is thus interesting in that CULEX allows a completely user-independent extraction of the lumen–intima and media–adventitia profiles with a segmentation error (compared to the average segmentation of four expert operators) that is acceptable in clinical practice, without restrictions on the orientation of the CCA in the image.

5.6.2   Percent Statistics

For the Percent Statistics test, choosing $\alpha = 0.05$, we find that the $p - \varepsilon$ has a value of 0.513. The Z value for the far wall lumen–intima interface was equal to 0.57, whereas the one for the media–adventitia was equal to 0.8. Hence, we may say that Z falls within the 95% confidence interval (CI), that is, that CULEX statistically behaves like a human operator in segmenting the distal carotid wall in our image database.

The corresponding values for the near wall were equal to 0.514 for the lumen–intima interface and 0.3 for the media–adventitia interface. This latter interface is the only one that did not pass the statistical test. Results are in line with the measurements of the segmentation error: the greater difficulties can be found in the outlining of the near wall media–adventitia border. The third column of Table 5.1 reports the results for this test.

To our knowledge, none of the previously referenced algorithms for the carotid artery segmentation were tested by means of the Percent Statistics algorithm; hence, a direct comparison of the performances is impossible. However, we believe this test is important because in many practical applications (i.e., image reconstruction, image fusion, carotid wall texture analysis) it is more important to know that the CGBs are not different from those obtained by experts than to have a low segmentation error. Our algorithm passed this statistical test, even though p value was extremely close to the limit of the acceptance region. Clearly, we are working to increase the number of human operators segmenting our sample database, since this would enforce the strength of the statistical test.

5.6.3   Confidence Interval for the Mean Segmentation

Figure 5.17 represents the distribution of the results for the confidence interval. The CGB was tested against the average segmentation of the four humans. The bars represent the percentage of times the CGB falls within the confidence interval ($\alpha = 0.05$) for the average LI profiles (left graphs, far wall at the top, near wall at the bottom) and MA profiles (right graphs, far wall at the top, near wall at the bottom). The number of tested points was 10,000 (50 points per image). The probability of falling into the 95% CI for the average segmentation was always quite high, and no significant differences were found among LI_{NW}, LI_{FW}, and MA_{FW} in terms of CULEX segmentation. Once again, however, the MA_{NW} interface was more problematic; Figure 5.17 shows, in the bottom right histogram, that in 40% of the cases, 40% or fewer of the pixels fell into the CI. In fact, due to the characteristics of the ultrasound images, the near wall media–adventitia layer is often difficult to track, even for a trained expert. At present, our algorithm is capable of performing a segmentation that is reliable, but it overestimates the real position of the MA_{NW} border.

Finally, Figure 5.18 shows an example of segmentation with the CIs overlapped. Note that, in this image, nearly the 100% of the points of the CGBs are within the 95% CI of the average manual segmentations.

This test is a further step toward ensuring that CULEX performs reasonably like a human expert when segmenting the CCA walls.

5.7   Plaque Imaging

The availability of a user-independent technique for the segmentation of carotid wall plaques is of paramount importance in clinical practice as well as in the engineering field. In fact, most of the time, the segmentation of a plaque is a preliminary

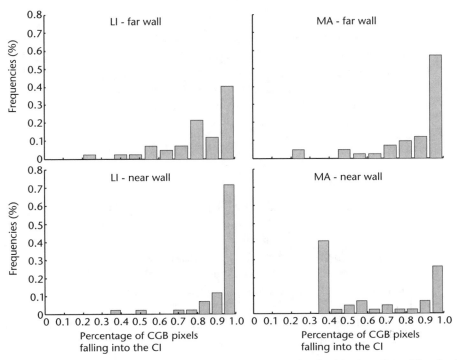

Figure 5.17   Distribution of the pixels falling into the 95% CI for the four interfaces. The abscissa represents the percentage of pixels falling into the 95% CI in a given image; the bars report the percentage frequencies over 200 images. The mean percentage values were 92% for the LI_{NW}, 81% for the LI_{FW}, and 85% for the MA_{FW} interface. The MA_{NW} interface presents an average value equal to 52%, confirming that this specific border is the most difficult to track by means of our technique.

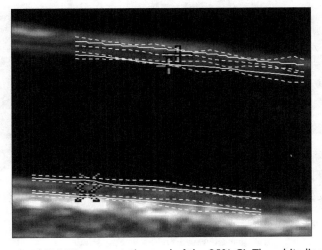

Figure 5.18   Example of CULEX segmentation and of the 95% CI. The white lines represent the borders traced by CULEX; the dashed white lines represent the 95% CI. In this example, nearly 100% of the pixels fall into the CI.

step for the application of some other analysis technique, ranging from 3-D morphological reconstruction to plaque classification. CULEX proved suitable for segmenting both fibrous and echolucent type II plaques. We tested CULEX on our second database, consisting of 100 images of plaques and thickened carotid walls.

When in presence of increased IMT, CULEX provided performances that were statistically equal to the case of normal carotid wall. ROI selection was effective on all 30 images in the database, with segmentation errors comparable to the normal case.

5.7.1   Stable Plaques

Figure 5.19 depicts the CULEX segmentation of an image containing fibrous plaques. The left panel of the upper row depicts the original image; the right panel depicts the CULEX segmentations. The image is relative to a subject showing an important plaque located on the proximal wall (as indicated by the large white arrows); the far wall is comparable to a normal wall. This image indicates that CULEX correctly characterizes plaque on the near wall and segment of the distal wall.

We would like to remark once again that this segmentation is completely user independent, and the four profiles are extracted in a single shot, without the need to

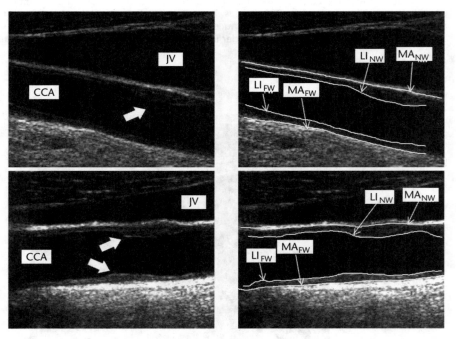

Figure 5.19 CULEX segmentation of stable plaques. The upper row shows the case of a fibrous plaque located on the near CCA wall; the lower row shows a CCA with a distal and a proximal plaque. Left images are the raw ultrasound B-mode images. Large white arrows point to plaques. The right images report CULEX segmentation of the plaques, where we can see that this technique allows for excellent characterization of the plaque profile.

tune the algorithm or select parameters. Specifically, in this image, CULEX was capable of segmenting contemporarily a plaque and a normal carotid wall.

The left panel of the lower row in Figure 5.19 shows two fibrous plaques: one on the near wall and one on the far wall. Again, CULEX was able to characterize plaques (right image).

To determine how CULEX performed on the plaque database, we calculated the percentage of misclassified pixels in relation to the total number of pixels corresponding to plaque (i.e., an error expressed as percentage of misclassification over the plaque area). For the 30 images constituting our database, we found an error equal to 8% ± 5% (mean ± STD).

We also compared the maximum stenosis diameter of the CCA lumen as detected by CULEX and compared it to the one derived from human tracings; we found a relative error lower than 1% (STD = 1%).

In this case too, a direct comparison of the CULEX performances with previous proposed techniques is not possible. Some approaches are not suitable for plaque characterization [14, 15], whereas other algorithms fail to segment proximal wall malformations [16]. Besides being completely user independent, our approach allows for direct extraction of a plaque contour, independent of its location.

5.7.2   Unstable Plaques (Echolucent Type II)

Figure 5.20 illustrates the CULEX segmentation of two echolucent plaques located on the distal CCA wall. The left panel shows the raw images, whereas the right panel depicts the segmentations. Note that echolucent type II plaques can also be correctly characterized by our technique. Of the 40 images in our database, the segmentation error was equal to 12% ± 4% (mean ± STD). The difference between maximum the stenotic diameter was, in this case, about 10% (STD = 3%).

The development of an algorithm capable of also segmenting echolucent plaques poses a problem of SNR. Specifically, if the image is too noisy or characterized by excessive blood backscattering, the lumen–intima interface segmentation may become difficult. Some pixels belonging to the lumen, in fact, could be classified as plaque due to the fact that they represent light-gray spots in the black of the vessel lumen. Similarly, pixels belonging to an echolucent plaque may be classified as lumen if the SNR is poor. A partial solution could be to filter the image, but such an operation enhances the signal power, but decreases the image detail and contrast, resulting in greater bias in the automatically segmented profiles. Hence, to correctly segment echolucent plaques, the image must be of good quality.

Again, no other technique that we are aware of allows for the segmentation of echolucent plaques in a user-independent fashion. Most of the previously described techniques are not capable of distinguishing echolucent plaques from image noise [14–16]. Hence, despite a segmentation error that is of the order of one-tenth of the plaque area, our approach allows for the individuation and characterization of echolucent plaques without being instructed by the user. It must be considered, also, that with this structure it is possible to segment a CCA ultrasound image in a completely blind fashion: CULEX proved suitable for normal scans as well as for pathological carotid images.

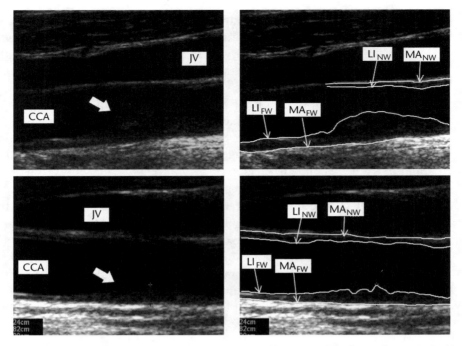

Figure 5.20   CULEX segmentation of unstable plaques. The upper row shows the case of a big echolucent type II plaque located on the near CCA wall; the lower row shows a plaque image in which the operator fixed four markers for the delimitation of the plaque area. The images in the left panel are the raw ultrasound B-mode images. Large white arrows point to plaques. The images on the right depict CULEX segmentation of the plaques, in which we can see that this technique allows for excellent characterization of the plaque profile. Specifically, the percentage of misclassified pixels with respect to the plaque area is lower than 9% in these examples.

5.8   Discussion and Conclusions

The CULEX algorithm can be proficiently used for the segmentation of ultrasound images of the common carotid artery wall. This algorithm has the advantage of being completely user independent; that is, the raw US image in DICOM format can be processed without any interaction with the operator. Moreover, the CULEX algorithm is able to characterize both echogenic and echolucent plaques in the carotid wall. Pixel fuzziness and user independence are preserved and the obtained IMT design is extremely close to the gold-standard tracing.

On normal carotids, the characterization of the algorithm showed that the contour traced by the CULEX is statistically comparable to that of a human trained operator. Moreover, the average error the algorithm commits in segmenting the lumen–intima and the media–adventitia interfaces is lower than 1 pixel for the far wall, equal to 1.2 pixels for the lumen–intima interface of the near wall, and equal to 2.1 pixels for the media–adventitia layer of the near wall.

When estimating the IMT, the average error is equal to 9%: note that this performance is in line with that obtained by means of user-dependent techniques.

On stable plaques and on increased IMT images, the CULEX algorithm performed a segmentation substantially equal to that obtained on normal images. When processing echolucent plaques, the segmentation error increased slightly because of the weak difference in gray tones between pixels belonging to the wall and pixels belonging to the lumen.

This algorithm is a step forward in the completely automatic detection of the LI and MA interfaces in an ultrasound B-mode carotid scan, suitable for healthy as well as pathological subjects.

One of the major advantages of this technique, in fact, is that it can also be used in presence of pathology. The algorithm proved effective in automatically detecting and segmenting echolucent type II plaques, echogenic plaques, and carotid walls with increased IMT. Despite a small bias in the estimation of the plaque dimensions, we found the performance of our algorithm to be extremely useful for the assessment of the carotid media layer, as a computer aid to the early diagnosis of wall degeneration in asymptomatic patients.

This system may be extremely helpful when segmentation is just a preliminary step for further processing such as coregistration with other imaging techniques, 3-D reconstruction, and carotid wall assessment. As an example, this automatic algorithm could be used for plaque identification before performing plaque texture analysis, especially when the effects of several image processing steps need to be validated.

The results are very promising and the system is being integrated into clinical setups for automatic pathological carotid wall analysis.

References

[1]   Touboul, J. P., et al., "Carotid Intima-Media Thickness, Plaques, and Framingham Risk Score as Independent Determinants of Stroke Risk," *Stroke*, Vol. 36, No. 8, 2005, pp. 1741–1745.

[2]   Watanabe, T., et al., "Small Dense Low-Density Lipoprotein and Carotid Atherosclerosis in Relation to Vascular Dementia," *Metabolism*, Vol. 53, No. 4, 2004, pp. 476–482.

[3]   Gray, H., *Gray's Anatomy: Anatomy of the Human Body*, Philadelphia, PA: Lea & Febiger, 1918.

[4]   Steinberg, D., "An Interpretive History of the Cholesterol Controversy: Part I," *J. Lipid Research*, Vol. 45, 2004, pp. 1583–1593.

[5]   el-Barghouti, N., et al., "The Relative Effect of Carotid Plaque Heterogeneity and Echogenicity on Ipsilateral Cerebral Infarction and Symptoms of Cerebrovascular Disease," *Int. Angiol.*, Vol. 15, 1996, pp. 300–306.

[6]   Sabetai, M. M., et al., "Carotid Plaque Echogenicity and Types of Silent CT-Brain Infarcts. Is There an Association in Patients with Asymptomatic Carotid Stenosis?" *Int. Angiol.*, Vol. 20, 2001, pp. 51–57.

[7]   Pignoli, P., et al., "Intimal Plus Medial Thickness of the Arterial Wall: A Direct Measurement with Ultrasound Imaging," *Circulation*, Vol. 74, 1986, pp. 1399–1406.

[8]   Molinari, F., et al., "Characterization of a Completely User-Independent Algorithm for Carotid Artery Segmentation in 2-D Ultrasound Images," *IEEE Trans. on Instrum. & Meas.*, Vol. 56, No. 4, 2007.

[9]   Wendelhag, I., et al., "Ultrasound Measurement of Wall Thickness in the Carotid Artery: Fundamental Principles and Description of a Computerized Analysing System," *Clin. Physiol.*, Vol. 11, 1991, pp. 565–577.

[10] Gray-Weale, A. C., et al., "Carotid Artery Atheroma: Comparison of Preoperative B-Mode Ultrasound Appearance with Carotid Endarterectomy Specimen Pathology," *J. Cardiovasc. Surg.*, Vol. 29, 1988, pp. 676–681.

[11] Suri, J. S., et al., *Plaque Imaging: Pixels to Molecular Levels,* Amsterdam: IOS Press, 2005.

[12] Christodoulos, C. I., et al., "Texture-Based Classification of Atherosclerotic Carotid Plaques," *IEEE Trans. on Medical Imaging*, Vol. 22, No. 7, 2003, pp. 902–912.

[13] Widder, B., et al., "Morphological Characterization of Carotid Artery Stenoses by Ultrasound Duplex Scanning," *Ultrasound Med. Biol.*, Vol. 16, No. 4, 1990, pp. 349–354.

[14] Liguori, C., A. Paolillo, and A. Pietrosanto, "An Automatic Measurement System for the Evaluation of Carotid Intima-Media Thickness," *IEEE Trans. on Instr. and Meas.*, Vol. 50, 2001, pp. 1684–1691.

[15] Gutierrez, M. A., et al., "Automatic Measurement of Carotid Diameter and Wall Thickness in Ultrasound Images," *Computers in Cardiology*, Vol. 29, 2002, pp. 359–362.

[16] Cheng, D., et al., "Using Snakes to Detect the Intimal and Adventitial Layers of the Common Carotid Artery Wall in Sonographic Images," *Computer Methods and Programs in Biomedicine*, Vol. 67, 2002, pp. 27–37.

[17] Touboul, P. J., et al., "Use of Monitoring Software to Improve the Measurement of Carotid Wall Thickness by B-Mode Imaging," *J. Hypertens.*, Vol. 10, No. 5 (Suppl.), July 1992, pp. S37–S41.

[18] Gustavsson, T., et al., "Implementation and Comparison of Four Different Boundary Detection Algorithms for Quantitative Ultrasonic Measurements of the Human Carotid Artery," *IEEE Computers in Cardiology*, Vol. 24, 1997, pp. 69–72.

[19] Liang, Q., et al., "A Multiscale Dynamic Programming Procedure for Boundary Detection in Ultrasonic Artery Images," *IEEE Trans. on Medical Imaging*, Vol. 19, 2000, pp. 359–369.

[20] Lobregt, S., and M. Viergever, "A Discrete Dynamic Contour Model," *IEEE Trans. on Medical Imaging*, Vol. 14, 1995, pp. 12–24.

[21] Delsanto, S., et al., "CULEX—Completely User-Independent Layers EXtraction: Ultrasonic Carotid Artery Images Segmentation," *27th Annual International Conference of the IEEE EMBS*, Shanghai, China, September 2005.

[22] Xu, C., and J. L. Prince, "Snakes, Shapes, and Gradient Vector Flow," *IEEE Trans. on Image Processing*, Vol. 7, 1998, pp. 359–369.

[23] Soille, P., *Morphological Image Analysis: Principles and Applications,* Berlin: Springer-Verlag, 1999, pp. 173–174.

[24] Chalana, V., and Y. Kim, "A Methodology for Evaluation of Boundary Detection Algorithms on Medical Images," *IEEE Trans. on Medical Imaging*, Vol. 16, 1997, pp. 642–652.

[25] Alberola-Lòpez, C., M. Martin-Fernandez, and J. Riuz-Alzola, "Comments on: A Methodology for Evaluation of Boundary Detection Algorithms on Medical Images," *IEEE Trans. on Medical Imaging*, Vol. 23, 2004, pp. 658–659.

[26] Bruckhardt, C. B., "Speckle in Ultrasound B-Mode Scans," *IEEE Trans. on Sonics Ultrason.*, Vol. SU-25, No. 1, 1978, pp. 1–6.

[27] Delsanto, S., et al., "Characterization of a Completely User-Independent Algorithm for the Segmentation of Carotid Artery Ultrasound Images," *IEEE Instrumentation and Measurement Technology Conference (IMTC)*, Sorrento, Italy, April 2006.

PART III

2-D/3-D Breast Ultrasound Imaging

CHAPTER 6

Breast Lesion Classification Using 3-D Ultrasound

Ruey-Feng Chang, Chii-Jen Chen, Kau-Yih Chiou, Woo Kyung Moon, Dar-Ren Chen, and Jasjit S. Suri

6.1 Introduction

Breast cancer is a progressive disease, and the probability of developing invasive breast cancer in women is not the same at different age intervals [1]. According to recent studies [2, 3], early detection of breast cancer can reduce morbidity and mortality rates. In the past, mammography has been the most widely used screening modality, but it has a low false-negative rate [4]. False-negatives can be attributed to the existing technological limitations of mammography, human error, and quality assurance failures. Many researchers have found that only 5.1% of malignancies (193 of 3,753) were detected in women with an abnormal or benign finding on the mammogram [5]. Recently, it has been shown that the rates of false-negatives can be reduced by using a different modality, such as ultrasound [6].

Because ultrasound is widely available and relatively inexpensive to perform, it has become a valuable diagnostic adjunct to mammography [7–9]. Zonderland et al. [10] reported a sensitivity of 91% and specificity of 98% for ultrasound compared to mammography (83% and 97%, respectively) based on their experiment. Although different ultrasound operators can produce different US images, ultrasound is still preferred due to its convenience, noninvasiveness, and low cost. To reduce the effect of operator dependence in ultrasound, several computerized methods have been proposed to help the classification between benign and malignant tumors [11–18].

The general idea of computer-aided diagnosis (CAD) has been developed to assist physicians in diagnosing ultrasonic abnormalities suspicious for breast cancer. With the assistance of CAD, physicians can efficiently detect lesions, such as spiculations, and discriminate more accurately between begin and malignant tumors [11, 16, 19–25]. Sawaki et al. [11] proposed a CAD system that uses fuzzy inference for breast sonography and adopted six different criteria to classify lesions: lesion shape, border, edge shadows, internal echoes, posterior echoes, and halo. However, their system's accuracy, sensitivity, and specificity were only 60.3%, 82.1%, and 42.9%, respectively.

Chen et al. [20] proposed an autocorrelation coefficient to analyze the texture information. However, there was a fundamental weakness with texture-based strategies. The settings of US machine parameters had to be fixed for acquiring ultrasound images. If the ultrasound parameter setting subsequently changed, CAD performance became very unstable [21]. Moreover, a CAD system trained with images from one US machine needs to be trained again for a different US machine because the new machine will have different image resolution and image quality. Hence, Chen et al. [21] proposed nearly setting-independent features based on shape information. Their system was very robust and powerful: The statistical data using a receiver operating characteristic (ROC) curve were all greater than 0.95. Chang et al. [22] also used six shape features in their analysis: factor, roundness, aspect ratio, convexity, solidity, and extent.

Note that all of the mentioned strategies were used on 2-D ultrasound images. Recently, 3-D US [23] has shown promising signs of overcoming the limitations of traditional 2-D US and allowing physicians to view the anatomy in three dimensions interactively, instead of assembling the sectional images in their minds. Chen et al. [16, 24, 25] proposed a 3-D US texture features autocorrelation matrix and run difference matrix. However, as pointed earlier, the preceding texture-based CAD approach is highly machine dependent and utilizes these machine settings all of the time.

Shape is also an important visual feature for an image. Shape representation seeks primarily effective and important shape features based on either shape boundary information or a boundary joined with interior points [26]. Thus, our chapter adapts a methodology based on geometric deformable models such as a level set, which has the ability to extract the topology of shapes of breast tumors. In the level set paradigm, tracking of the evolution of contours and surfaces is solved using numerical methods. The efficacy of this scheme has been demonstrated with numerical experiments on some synthesized images and some low-contrast medical images [27, 28]. Recently, the medical shape extraction approach in medical images has been dominated by the level set framework of Suri et al. [29–31].

In this chapter, we propose a CAD system based on a combination of two stages: capturing the topology of a shape, thus preserving its irregularity, and then statistically quantifying regular and irregular shapes accurately. The chapter examines the use of CAD for improving the diagnostic accuracy of breast lesion detection in 3-D breast ultrasound. Further, it helps to characterize breast lesions into benign and malignant lesions, which is clinically useful information for breast radiologists. Our CAD algorithm for detection of breast lesions uses a level set framework and our classification strategy uses a perceptual shape analysis. In addition, beam angle statistics [32, 33] is utilized for shape quantification of lesions, because this approach handles the concavities and convexities of lesions.

The database consists of 134 pathology-proven 3-D breast ultrasound cases including 70 benign and 64 malignant breast tumors. All of them are applied to our proposed CAD method to analyze the classification of breast lesions. For the similarity measurement on the 3-D acquisition method, the similarity values for benign and malignant lesions are $1,733 \pm 759$ and $4,081 \pm 1,683$, respectively. These values are statistically significant because the p-value is smaller than 0.001 (by Student's t-test). To compare system performance, the proposed framework is also applied to

2-D ultrasound images, called the 2-D acquisition method. In this method, each 2-D image is obtained by selecting the middle slice from 3-D data. The system accuracy, sensitivity, specificity, positive predictive value, and negative predictive value are 85.07% (114 of 134), 87.50% (56 of 64), 82.86% (58 of 70), 82.35% (56 of 68), and 87.88% (58 of 66), respectively, for the 2-D acquisition method; 83.58% (112 of 134), 93.75% (60 of 64), 74.29% (52 of 70), 76.92% (60 of 78), and 92.86% (52 of 56), respectively, for the 3-D acquisition method. The ROC area indexes are 0.915 and 0.934, respectively, for 2-D and 3-D acquisition methods. In our experiments, the proposed CAD system is suitable for both the 2-D and 3-D acquisition methods.

Section 6.2 reviews the level set segmentation framework. The proposed architecture for the CAD system is described in Section 6.3. In Section 6.4, the experimental protocol will be used to evaluate the performance of the CAD system. Some special issues will be discussed in Section 6.5. Finally, a conclusion is drawn and future works discussed in Section 6.6.

6.2 Level Set Segmentation Method

The level set segmentation technique is capable of capturing the topology of shapes in medical images [27, 34]. This technique supports recovery of arbitrarily complex shapes, such as corners and protrusions, and automatically handles topological changes such as merging and splitting. Recently, a large number of implementations for computer vision and medical imaging based on this basic concept have been published [22, 23, 27–29, 35–37].

6.2.1 Basic Level Set Theory

As a starting point and motivation for the level set approach, consider a closed curve moving in a plane. Let $C(0)$ be a smooth and closed initial curve in Euclidean plane $\Re^2$, and let $C(t)$ be the family of curves that is generated by the movement of initial curve $C(0)$ along the direction of its normal vector after time t. The main idea of the level set approach is to represent the front $C(t)$ as the level set $\{\phi = 0\}$ of a function ϕ. Thus, given a moving closed hypersurface $C(t)$, that is, $C(t = 0):[0, \infty) \rightarrow \Re^N$, we hope to produce a formulation for the motion of a hypersurface propagating along its normal direction with speed F, where F can be a function of various arguments, such as curvature or normal direction. Let $\phi(\mathbf{x}, t = 0)$, where $\mathbf{x}$ is a point in $\Re^N$, be defined by $\phi(\mathbf{x}, t = 0) = \pm d$ is the signed distance from $\mathbf{x}$ to $C(t = 0)$, and a plus (minus) sign is chosen if the point $\mathbf{x}$ is outside (inside) the initial front.

The goal is to produce an equation for the evolving function $\phi(\mathbf{x}, t)$, which contains the embedded motion of $C(t)$ as the level set $\{\phi = 0\}$. Let $x(t)$, $t \in [0, \infty)$ be the path of a point on the propagating front, that is, $\mathbf{x}(t = 0)$ is a point on the initial front $C(t = 0)$. Because the evolving function ϕ is always zero on the propagating front $C(t)$, $\phi(\mathbf{x}, x(t), t) = 0$. By the chain rule [27],

$$\phi_t + \sum_{i=1}^{N} \phi_{x_i} x_{i_t} = 0 \tag{6.1}$$

where x_{i_t} is the ith component of $\mathbf{x}$ at time t, and N is the number of components. Because

$$\sum_{i=1}^{N} \phi_{x_i} x_{i_t} = \left(\phi_{x_1}, \phi_{x_2}, \phi_{x_3}, \ldots, \phi_{x_N}\right)\left(x_{1_t}, x_{2_t}, x_{3_t}, \ldots, x_{N_t}\right)$$
$$= F\left(\mathbf{x}(t)\right)|\nabla\phi| \tag{6.2}$$

where $F(\mathbf{x}(t))$ is the speed function with the front for all components of $\mathbf{x}$ at time t. Combining (6.1) and (6.2), the final curve evolution equation is given as:

$$\frac{\partial\phi}{\partial t} = F(\kappa)|\nabla\phi| \tag{6.3}$$

where ϕ is the level set function and $F(\kappa)$ is the speed with which the front (or zero-level curve) propagates. This fundamental equation describes the time evolution of the level set function ϕ in such a way that the zero-level curve of this evolving function is always identified with the propagating interface [29]. The preceding equation is also called a Eulerian Hamilton-Jacobi equation [29, 34]. Equation (6.3) for 2-D and 3-D cases can be generalized as follows:

$$\frac{\partial\phi}{\partial t} = F_\kappa(x, y)|\nabla\phi| \text{ and}$$
$$\frac{\partial\phi}{\partial t} = F_\kappa(x, y, z)|\nabla\phi| \tag{6.4}$$

where (x, y) and (x, y, z) are the points on the propagation directions, and $F_\kappa(x, y)$ and $F_\kappa(x, y, z)$ are curvature-dependent speed functions in 2-D and 3-D, respectively [29]. The evolution of the front in the normal direction with speed F is the zero-level curve at time t of the function $\phi(x, y, t)$, as shown in Figure 6.1.

6.2.2 Summary of Level Set Theory

Level set theory can be used for image segmentation by using image-based features such as mean intensity, gradient, and edges in the governing differential equation. In a typical approach, a contour is initialized by a user and is then evolved until it fits the topology of an anatomic structure in the image [27].

There are challenges associated with using the level set framework by itself on the raw images. This is because the propagation of the fronts can sometimes lead to bleeding of the fronts due to noise structures present in the images [38]. One way to handle this is by removing the noisy structures and then enhancing the structures to be segmented. The combined effect of these two processes will assist the level set front propagations to recognize the edges of the structures, thereby bringing the speed close to zero at the structure edges. Keeping this in mind, we have developed our breast lesion identification chain. Once successful identification of breast lesions has occurred, our mission is to quantify the segmented shape and then classify it into breast lesion type to yield an image with clinical utility. Thus, combining such steps for breast radiologists is like developing a system for CAD.

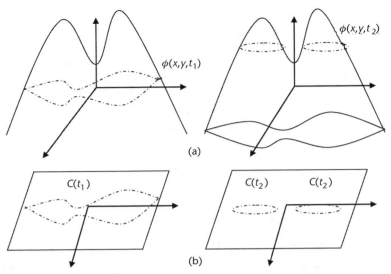

Figure 6.1   Level set methodology and curve evolution. (a) The evolving level set function and (b) the zero-level set value of the surface.

6.3   Proposed Architecture for the CAD System

The architecture for the breast CAD system is shown in Figure 6.2. The input to the system consists of 3-D breast slices acquired using a 3-D acquisition method. The main component of the system architecture consists of segmentation of the breast lesions, the dependent engine that finds the shapes of different breast lesions. Besides the main engine, there are components in the architecture that provide stability to the segmentation engine. This is the preprocessing subsystem for removing the system noise or acquisition noise, hence a diffusion filter, named after the method it uses. The final component of the system is the shape analysis subsystem. This subsystem accepts the silhouettes of the breast lesions and helps in quantifying the shapes and classifying the shapes. Because this subsystem uses statistics derived from the contours and their angular relationships, one block in Figure 6.2 is labeled "Beam Angle Statistics." In the following sections, we will present the blocks of the system architecture in detail.

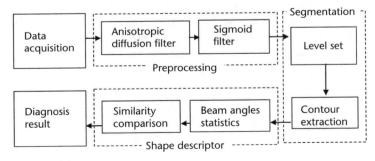

Figure 6.2   System architecture.

6.3.1 Data Acquisition

For the study in this chapter, the 3-D ultrasound breast images were acquired using a Voluson 730 (General Electric, USA) scanner with a Voluson small part transducer S-VNW5-10. The transducer with a linear array of crystal elements has the following specifications: 5- to 10-MHz scan frequency, 40-mm scan width, and 20° to 30° sweep angles. A set of 2-D ultrasound images was obtained by the slow tilt movement of a sectorial mechanical transducer. After completing the scanning process (about 10 seconds per case), the 3-D US image was obtained by reconstructing these 2-D US images. This scanning process of acquiring the successive 2-D US images at regular angular intervals arranges the set of 2-D image planes in a fanlike geometry.

6.3.2 Preprocessing Subsystem Architecture

6.3.2.1 Anisotropic Diffusion Filtering

To improve image quality, the speckle noise in ultrasound images must be reduced. Although conventional lowpass filtering and linear diffusion can be used to reduce speckle, the edge information may be blurred simultaneously. The anisotropic diffusion filtering cannot only reduce the speckle noise on the ultrasound image, but can also preserve the edge information [39]. It also satisfies the fundamental requirements discussed earlier. Consider the anisotropic diffusion equation

$$I_t = \mathrm{div}\big(c(x,y,t)\nabla I\big) \tag{6.5}$$

where div is the divergence operator, $c(x, y, t) = f(\|\nabla I(x, y, t)\|)$ is the diffusion coefficient, (x, y) is spatial position, and t is an artificial time parameter. The anisotropic diffusion equation uses the diffusion coefficient that is selected locally as a function of the magnitude gradient of the intensity function in the image for estimating the image structure. This function is selected to satisfy $f(z) \to 0$ when $z \to \infty$ so that diffusion processing is stopped across the edges. Therefore, the speckle noise within a homogeneous region can be blurred and the region boundary can be preserved.

6.3.2.2 Image Enhancement Via Sigmoid Mapping

In a breast ultrasound image, different tissues have different intensities, and the intensity values of malignant and benign tumors tend to be mostly low [40]. For accurate segmentation of the tumor region, we enhance the tumor boundary by mapping the intensities based on a sigmoid intensity transform. The sigmoid transform is a nonlinear mapping that maps a specific intensity range into a new intensity range with a very smooth and continuous transition. This mapping can focus attention on a specific set of intensity values and progressively decrease the values outside the range. We will focus on the tumor regions and fade out the other regions. The following equation represents the sigmoid intensity transformation:

$$\tilde{I} = (I_{max} - I_{min}) \frac{1}{\left(1 + e^{-\left(\frac{1-\beta}{\alpha}\right)}\right)} + I_{min} \tag{6.6}$$

where I and $\tilde{I}$ are the intensity of the input and output pixels; I_{max} and I_{min} are the maximum and minimum values of the input image; α is the width of the input intensity range; and β defines the intensity around which the range is centered. The values of $(\alpha, \beta) = (8, 60)$ for the parameters are often adopted in the experiments. However, the parameters are not the same in each case; they can be slightly adjusted. The results for different lesion cases obtained via an anisotropic diffusion filter and sigmoid function are shown in Figure 6.3. Note that this enhancement acts as a clamper or stopper when using the level set framework.

6.3.3   Segmentation Subsystem Based on the Level Set Framework

In the level set framework, the fundamental Eikonal equation, (6.3), describes the time evolution of the level set function Φ in such a way that the zero-level curve of this evolving function is always identified with the propagating interface [29]. The speed function here has two components, the gradient speed and curvature speed function. The gradient speed function helps the deformable level set to clamp near the irregularity of the breast lesion. Thus, the gradient information plays an important role in making sure that the best boundaries are determined, thus preserving the true lesion of the breast.

To improve the performance of the level set, the gradient speed function is padded with two additive steps, which are actually hidden in the process. First, partial differential equation (PDE) diffusion is adopted for smoothing the images and preserving the edges of the breast lesions [31]. Second, the enhancement of the lesions is done by using the sigmoid function, which acts like a stopper in the level set framework. Note carefully that the sigmoid function is more like a region-based stopper because the sigmoid function really enhances those edges which are truly corre-

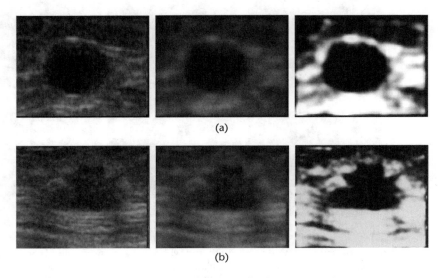

(a)

(b)

Figure 6.3   The preprocessing results of image enhancement using anisotropic diffusion filtering and sigmoid filtering. (a) and (b) are the benign and malignant cases, respectively. In each case, the first image is original input, the second image has been processed with an anisotropic diffusion filter ($t = 5$), and the last image has been enhanced by sigmoid filtering ($\alpha = 8$, $\beta = 60$).

sponding to the breast lesion in the ultrasound image. The stopper is the critical function for making the level set framework work and preventing bleeding of the propagating front.

Note that ultrasound images have high speckle and noise in them. Thus level set work would be likely to give misleading, irregular results if it were to be run on the raw images or images without enhancement processing. Note that the implementations are done using the PDE for surface propagation and evolution [29, 31]. The implementations are done in the narrow band and the initial curve is placed inside the breast lesion. Because the capture range of the level set is large, it immediately clamps to the lesion shape and its irregularity. The irregularity of the lesion is preserved in the level set framework as the topology of the lesion is extracted. The output of the level set segmentation engine leads to the extracted boundary and its (x, y) locations of the breast lesion.

We ran our system-based segmentation strategy with and without the preprocessing block to understand the importance of our sigmoid stopper to the level set framework. In the first scenario, we ran the software architecture without the preprocessing block, that is, without the so-called "stopper," as shown in Figure 6.4 (middle columns). Then we ran our complete system consisting of the PDE diffusion–based noise smoother and sigmoid-based stopper followed by level set–based propagation, as seen in Figure 6.4 (last column). Note carefully, that due to large speckle noise in the ultrasound breast slices, the results from the first scenario lead not only to inaccurate results but are likely to bleed over more iterations. In contrast, with the smoother and stopper, system performance is not only robust, but accurate.

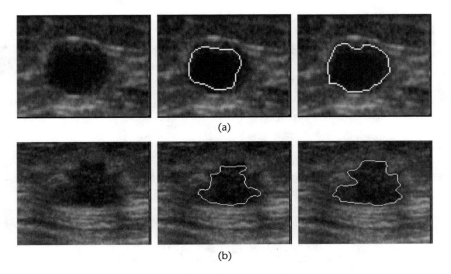

(a)

(b)

Figure 6.4 The segmentation results with and without the preprocessing block in the CAD architecture. (a) and (b) are the benign and malignant cases, respectively. In each case, the first image is the raw image; the second image has had the level set without the PDE smoother and sigmoid enhancer run on it; and the last image has been processed by the proposed level set-based framework with the PDE smoother and sigmoid enhancer.

6.3.4   Beam Angle Statistics

The classification of lesions is the main goal of this research. The classification criteria are based on the shapes of the breast lesion, which in turn depends on how well we statistically compute the measure of irregularity of the breast lesions. Thus, we need a measure that can define the description of the silhouette of the breast lesions. One such method is to use a shape description, given the silhouette, represented as a discrete set of contour points. One must ascertain that this descriptor has properties such as invariance to scale, rotation and translation, and a low sensitivity to noise. Once such descriptor was developed previously [32, 33] using the statistics taken around the points, but represented by the neighborhood angles of the ordered set of points. The authors termed this strategy *beam angle statistics* (BAS) because the angles computed were formed with the help of segments resembling beams. The idea of BAS originates from the concept of curvature of the contour. Due to the nature of breast lesions, however, the fluctuations of the contours are so rough that the curvature method for BAS computation is not justified. We thus implemented a constrained method, where BAS was computed on a limited set of contour points around the Kth point on the contour. We call this method *K-curvature BAS statistics*.

Before we define the K-curvature method, we first define mathematically the curvature function, represented by $k(t)$, where t represents the arc length parameterization. This curvature function virtually represents how much a curve bends at each point at the shape boundary and is mathematically defined by the variant of the slope $s(t)$ along the arc length t. If $(x(t), y(t))$ represents the curve $k(t)$, then the slope (angle) of the curve is represented as:

$$s(t) = \tan^{-1} \frac{dy(t)}{dx(t)} \tag{6.7}$$

The curvature function is nothing but the derivative of the slope, which can be mathematically given as follows [41]:

$$k(t) = \frac{ds(t)}{dt} = \frac{x'(t)y''(t) - y'(t)x''(t)}{\left(x'(t)^2 + y'(t)^2 \right)^{3/2}} \tag{6.8}$$

where x' is the first-order differentiation and x'' is the second-order differentiation; the same definitions are applied to y' and y''.

Because the fluctuations on the boundary have an extreme influence on the curvature function, we introduce smoothing to prevent noise in the slope function by the K-slope method [32, 33]. The K-slope at a boundary point is defined by the slope of a line that points with its Kth right neighbor or left neighbor. Hence, the K-curvature at a boundary point is defined by the difference between the K-slope values of its Kth right neighbor and left neighbor. However, a good value for parameter K is difficult to choose. A single K-value would not capture an accurate curvature for all varieties of shapes. This problem can be solved when the K-curvature function is considered to be the output of a stochastic process [32, 33].

Let the shape boundary presented by the ordered set of points be given as $B = \{f(1), ..., f(N_b)\}$, where

$$f(i) = \left(x(i), y(i)\right) \quad i = 1, ..., N_b \tag{6.9}$$

where N_b is the number of boundary points. The beams are the lines connecting the reference point with the remains of the points on the boundary, as shown in Figure 6.5(a). For each point $f(i)$, the beams of $f(i)$ are defined by the set of vectors

$$B\left(f(i)\right) = \left\{V_{i+j}, V_{i-j}\right\} \quad \text{for } i = 1, ..., N_b, \quad j = 1, ..., N_b/2 \tag{6.10}$$

where V_{i+j} and V_{i-j} are the forward and backward vectors of $f(i)$ connecting to $f(i+j)$ and $f(i-j)$. The slope of each beam is computed as

$$S_{V_{i+l}} = \tan^{-1} \frac{\Delta y_{i+l}}{\Delta x_{i+l}}, \quad l = \pm 1, ..., N_b/2 \tag{6.11}$$

Therefore, the beam angle $\theta_K(i)$ of point $f(i)$ between the forward and backward vectors is calculated as:

$$\theta_K(i) = \left(s_{V_{i-l}} - s_{V_{i+l}}\right) \tag{6.12}$$

An example of the beam angle when l equals 4 is shown in Figure 6.5(b).

The angle between each couple of beams is taken as the random variable at each boundary point. For this purpose, BAS may offer a compact descriptor for a shape representation by the mth moment $\theta^m(i)$ of the random variable $\theta_K(i)$, which is defined as:

$$E\left[\theta^m(i)\right] = \sum_K \theta_K^m P_K\left(\theta_K(i)\right) \quad m = 0, 1, 2, 3, ... \tag{6.13}$$

where E is the expected value, and $P_K(\theta_K(i))$ is estimated by the histogram of $\theta_K(i)$ at each boundary point $f(i)$. Then, each boundary point $f(i)$ is represented as a vector $\Gamma(i)$, whose elements are the moments of the beam angles as:

$$\Gamma(i) = \left\lfloor E\left[\theta^1(i)\right], E\left[\theta^2(i)\right], E\left[\theta^3(i)\right], ... \right\rfloor \tag{6.14}$$

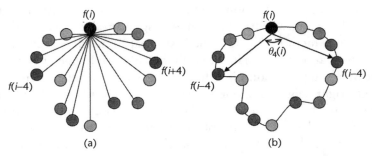

(a)   (b)

Figure 6.5   (a) The beams of point $f(i)$ and (b) the beam angle of 4-curvature at point $f(i)$.

Thus, a contour can be represented by plotting the moments of the $\theta_K(i)$'s for all boundary points.

The different order moments of BAS for a shape boundary are illustrated in Figure 6.6. The boundary of a breast lesion after level set segmentation is shown in Figure 6.6(a), and we can observe that the variation of the boundary in the dotted-line range is huge. The corresponding BAS for different order moments are also shown in dotted-line range in Figure 6.6(b–d). Generally, the amplitude of a curve will become bigger in a higher-order moment of BAS, so that a higher-order moment can be sufficient for representing most variation of shape boundary. In this example, the variations of the lesion's boundary can be clearly observed and analyzed on the fifth-order moment of BAS. Therefore, the fifth-order moment of BAS is used in our experiments, and the other reason for using a higher-order moment will be shown in the following section.

6.3.5   Similarity Between Shapes: A Classification Strategy

For breast ultrasound images, there are several distinguishable properties between benign masses and malignant masses [42]. Most benign masses are generally well circumscribed, compact, and roughly elliptical; malignant masses usually have an irregular semblance, blurred boundary, and are surrounded by a radiating pattern of linear spicules. However, the irregular situations must still be considered because some lesions may have the opposite characteristics.

To solve this problem, the idea is to use a best-fit ellipse to cover the segmented lesion. For most benign tumors with round and well-circumscribed features, the segmented lesion could be covered by its corresponding best-fit ellipse. If the difference between the segmented tumor and its best-fit ellipse is small, the tumor should be benign undoubtedly, and vice versa.

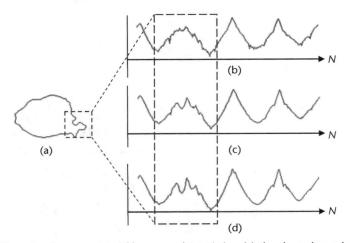

Figure 6.6   Different order moments of beam angle statistics: (a) the shape boundary, (b) first-order moment, (c) third-order moment, and (d) fifth-order moment. The variations of the lesion boundary can be clearly observed and analyzed on higher-order moment of BAS. Note: *N* is the number of boundary points.

6.3.5.1 Best-Fit Ellipse Computation

Consider a shape represented by a region R containing N_r pixels. Let (m, n) be a pixel in R, and the center $(\overline{m}, \overline{n})$ of mass is defined as

$$(\overline{m}, \overline{n}) = \left(\left(\frac{1}{N_r} \sum_{(m,n) \in R} \sum m \right), \left(\frac{1}{N_r} \sum_{(m,n) \in R} \sum n \right) \right) \tag{6.15}$$

The (p, q)-order central moments are

$$\mu_{p,q} = \sum_{(m,n) \in R} \sum (m - \overline{m})^p \cdot (n - \overline{n})^q \tag{6.16}$$

The preceding formulas for center of gravity of regions and the higher-order moments will be very useful ahead when computing the best-fit ellipse. Let a and b be the lengths of semimajor and semiminor axes of the best-fit ellipse, as shown in Figure 6.7(a). The least and the greatest moments of inertia for an ellipse are defined as [43]:

$$l_{\max} = \frac{\pi}{4} a^3 b, \; l_{\min} = \frac{\pi}{4} a b^3 \tag{6.17}$$

For a contour with orientation angle θ, the preceding formula can be computed as:

$$l'_{\max} = \sum_{(m,n) \in R} \sum \left((n - \overline{n}) \sin \theta + (m - \overline{m}) \cos \theta \right)^2 \text{ and}$$

$$l'_{\min} = \sum_{(m,n) \in R} \sum \left((n - \overline{n}) \cos \theta - (m - \overline{m}) \sin \theta \right)^2 \tag{6.18}$$

where θ is defined as the angle of axis of the least moment of inertia. It can be obtained as:

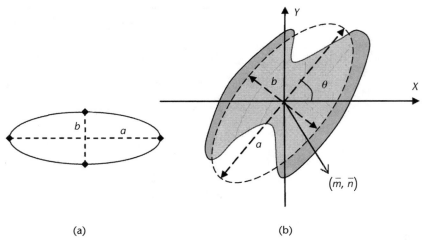

(a)   (b)

Figure 6.7 (a) The a represents the semimajor axis and b the semiminor axis. (b) The best-fit ellipse for the contour with orientation angle θ.

$$\theta = \frac{1}{2} \tan^{-1} \left(\frac{2\mu_{1,1}}{\mu_{2,0} - \mu_{0,2}} \right) \tag{6.19}$$

where $\mu_{i,j}$ is defined in (6.16). Let $l_{max} = l'_{max}, l_{min} = l'_{min}$. Then the lengths of the semimajor $\hat{a}$ and semiminor axes $\hat{b}$ are estimated as:

$$\hat{a} = \left(\frac{4}{\pi} \right)^{\frac{1}{4}} \left[\frac{(l'_{max})^3}{l'_{min}} \right]^{\frac{1}{8}} \text{ and } \hat{b} = \left(\frac{4}{\pi} \right)^{\frac{1}{4}} \left[\frac{(l'_{min})^3}{l'_{max}} \right]^{\frac{1}{8}} \tag{6.20}$$

Thus, the best-fit ellipse based on (6.20) can be obtained, as shown in Figure 6.7(b). For the similarity comparison, the tumorous contour and its corresponding best-fit ellipse are represented by BAS. At that point, we then have a similarity measure for these two BAS variations.

6.3.5.2   Similarity Measurement Between a True Lesion and a Fitted Lesion

The similarity measurement can be achieved by the optimal correspondent subsequence (OCS) algorithm [44], which can return a similarity value (S_m).

Consider C_1 and C_2 to be two contours with P_1 and P_2 pixels, and $C_1(i)$ and $C_2(j)$ are one of the pixels on C_1 and C_2. To compare the difference from C_1 and C_2, let $f(i)$ be a mapping function from C_1 to C_2. The pixel $C_1(i)$ is mapped to the point $C_2(f(i))$; the distance between $C_1(i)$ and $C_2(f(i))$ is defined as dist($C_1(i), C_2(f(j))$), which is the sum of square of the moment difference in Euclidean distance between $C_1(i)$ and $C_2(f(i))$. The similarity measure can be solved by minimizing the recurrent relation as:

$$D(i,j) = \min \left\{ \begin{array}{l} \left[D(i-1, j-1) + dist\left(C_1(i), C_2\left(f(i) \right) \right) \right] \\ \left[D(i-1, j) + t \right], \left[D(i, j-1) + t \right] \end{array} \right\} \tag{6.21}$$

where $D(I, j)$ is the cumulative distance function, and t is a constant. In this chapter, $D(0, 0)$ equals 0 and t is 10. The pseudoalgorithm is as follows:

```
Set D(0, 0) = 0
Set D(i, 0) = i * t   (for i = 1 … P₁)
Set D(0, j) = j * t   (for j = 1 … P₂)
For i = 1 to P₁
    For j = 1 to P₂
        T₁ = D(i-1, j-1) + dist(C₁(i), C₂(f(j)))
        T₂ = D(i-1, j) + t
        T₃ = D(i, j-1) + t
        D(i, j)= min{T₁, T₂, T₃}
    End For j
End For i
Return Sₘ = D(P₁, P₂)
```

The other reason for using the higher-order moment of BAS can be described now. In Figure 6.8(a), the corresponding best-fit ellipse of the tumor is shown as a dashed-line ellipse and the different order of moments for their BAS variations are

also shown in Figure 6.8(b–d). When the higher-order moment is adopted, the variations between the tumor and best-fit ellipse can apparently be discriminated, especially for malignant lesion cases. The choice of the higher-order moment can avoid
the miscarriage of a similarity measurement for the ambiguous cases. However, considering the time complexity of the system, the fifth-order moment of BAS is used in
the similarity measurement. The efficiency of this decision for moment order can be
proven in our experiments.

6.4   Experimental Protocol, Data Analysis, and Performance Evaluation

In the experiments, 134 pathology-proven cases, including 70 benign and 64 malignant breast tumors, were used to verify the classified accuracy of the proposed system. All cases were first verified by a radiologist (one of the authors). The goal of the
experiments was to confirm that our proposed CAD system can automatically classify lesions as well as a physician can. To compute the similarity effectively between
the tumor and its best-fit ellipse, the fifth-order moment of BAS was used in the
experiments. In the statistical analyses, Student's t-test and the ROC curve are used
to validate the performance of the CAD system.

6.4.1   Similarity Measure Analysis

The end product of 3-D US breast acquisition consists of 2-D US image slices represented as orthogonal views. The CAD architecture accepts those slices that have
breast lesions or tumors in them. This is manually driven by the trained
sonographer. The whole concept of classification of breast lesions lies in accurately
determining the similarity measure between the segmented tumor and best-fit

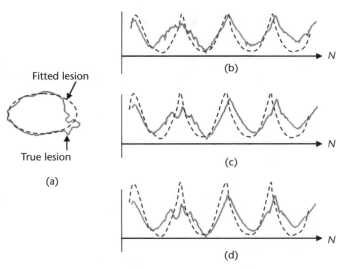

Figure 6.8   Different order moments for BAS between the true shape boundary (solid line) and
corresponding best-fit ellipse (dashed line): (a) the shape boundary, (b) first-order moment, (c)
third-order moment, and (d) fifth-order moment. The higher-order moment of BAS has the more
obvious difference.

ellipse. The best-fit ellipse was computed using (6.20) based on moment features of the breast lesion shape. We ran the protocol and computed the similarity measure (see Section 6.3.5) and the results are shown in Figures 6.9 and 6.10. The chart shows that benign tumors have lower similarity values (S_m) compared to malignant tumors. This is a clinical decision maker for the CAD architecture to classify the breast lesions. In most lesion cases, the differences between S_m can be efficaciously used to distinguish and classify whether a tumor is malignant or not. The means of the similarity values for 134 pathology cases are listed in Table 6.1.

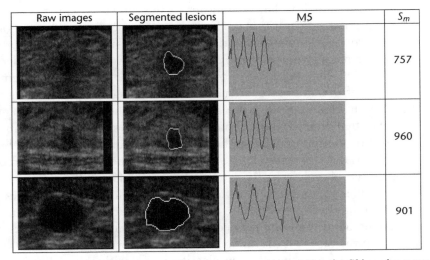

Figure 6.9 The similarity measurements for benign cases. Note: M5 is the fifth-order moment of BAS, and S_m is the similarity value between the true breast tumor and the best-fit ellipse of this breast tumor. The last case is a misdiagnosis due to the higher similarity value.

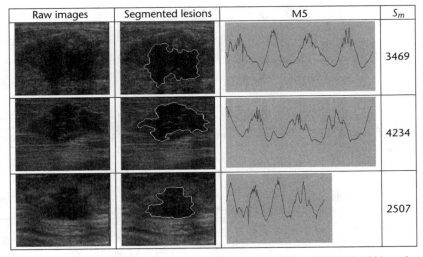

Figure 6.10 The similarity measurements for malignant cases. Note: M5 is the fifth-order moment of BAS, and S_m is the similarity value between the true breast tumor and the best-fit ellipse of this breast tumor. The last case is a misdiagnosis due to the smaller similarity value.

6.4.2 Performance Evaluation

To compare the performance, the proposed framework was applied to 2-D US images, called the 2-D acquisition method. In this method, each 2-D image is obtained by selecting the middle slice from 3-D data. Comparing the 2-D and 3-D acquisition methods, the similarity values with best-fit ellipse for benign and malignant breast tumors are listed in Table 6.1. The means of the similarity values for benign and malignant cases in the 2-D acquisition method are all smaller than for the 3-D acquisition method. That is because only one slice is computed for each 2-D method; the reliability of the 2-D method is inferior to the 3-D method. However, through the statistical data, the means of the similarity values for benign cases on both 2-D and 3-D acquisition methods are all smaller than malignant ones. That is to say, the proposed similarity algorithm is efficacious and robust for both 2-D and 3-D acquisition methods, because the corresponding p-values are all smaller than 0.001.

The ROC curve is also used to estimate the proposed CAD system, and the performance of similarity measurements can be evaluated by examining the ROC area index, A_z. In Figure 6.11, the A_z value (0.934) for the 3-D acquisition method is a little higher than that for the 2-D method. That is to say, the performance of the CAD architecture cannot only be applied to the 3-D acquisition method, but also can be suitable for the 2-D method. The preceding argument can be proven by the p-value between two A_z values by the z-test, which is 0.249 (>0.05).

In Tables 6.2 and 6.3, the different thresholds for similarity values (S_m) and their corresponding best-fit ellipse are evaluated to decide whether tumors are benign or malignant. Some items are defined here: TP is the number of positives correctly classified as positive; TN is the number of negatives correctly classified as negative; FP is the number of negatives falsely classified as positive; and FN is the number of positives falsely classified as negative. Five performance indices including accuracy, sensitivity, specificity, positive predictive value (PPV), and negative predictive value (NPV) are defined as follows:

$$\text{Accuracy} = (\text{TP} + \text{TN})/(\text{TP} + \text{TN} + \text{FP} + \text{FN})$$
$$\text{Sensitivity} = \text{TP}/(\text{TP} + \text{FN})$$
$$\text{Specificity} = \text{TN}/(\text{TN} + \text{FP}) \tag{6.22}$$
$$\text{PPV} = \text{TP}/(\text{TP} + \text{FP})$$
$$\text{NPV} = \text{TN}/(\text{TN} + \text{FN})$$

Furthermore, the different thresholds on Tables 6.2 and 6.3 provide the physicians with various choices for physicians to examine and achieve the capability of automation.

Table 6.1 The Means of Similarity Values for 134 Pathology-Proven Cases on 2-D and 3-D Acquisition Methods (by Student's t-Test)

S_m	2-D		3-D	
	Benign	Malignant	Benign	Malignant
Mean	$1{,}279 \pm 527$	$2{,}598 \pm 951$	$1{,}733 \pm 759$	$4{,}081 \pm 1{,}683$
p-value	<0.001		<0.001	

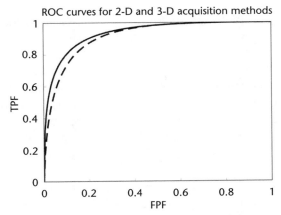

Figure 6.11   The ROC curves for the 2-D and 3-D acquisition methods using BAS. The *p*-value is 0.249 (>0.05) and the CAD for the 3-D acquisition method shows a slight improvement compared to the 2-D acquisition method.

Table 6.2   Results (in %) of Performance Using Different Thresholds to Decide Whether a Tumor Is Benign or Malignant for the 2-D Acquisition Method

Threshold	(TP, TN)	Accuracy	Sensitivity	Specificity	PPV	NPV
1,200	(62, 36)	73.13	96.88	51.43	64.58	94.74
1,400	(60, 52)	83.58	93.75	74.29	76.92	92.86
1,600	(56, 58)	85.07	87.50	82.86	82.35	87.88
1,800	(47, 62)	81.34	73.44	88.57	85.45	78.48
2,000	(41, 65)	79.10	64.06	92.86	89.13	73.86

Table 6.3   Results (in %) of Performance Using Different Thresholds to Decide Whether a Tumor Is Benign or Malignant for the 3-D Acquisition Method

Threshold	(TP, TN)	Accuracy	Sensitivity	Specificity	PPV	NPV
1,600	(63, 44)	79.85	98.44	62.86	70.79	97.78
1,800	(62, 49)	82.84	96.88	70.00	74.70	96.08
2,000	(60, 52)	83.58	93.75	74.29	76.92	92.86
2,200	(57, 56)	84.33	89.06	80.00	80.28	88.89
2,400	(56, 60)	86.57	87.50	85.71	84.85	88.24

6.5   Discussion

6.5.1   Discussion About the Dual Level Set

Segmentation is one of the critical components required to extract information about true irregularities of breast lesions. In our laboratory, we have tried several segmentation strategies, ranging from deformable models to regional models. Though the level set does optimize in the sense that it has the ability to capture the irregularity of the lesions, in some cases even the level set is powerless. These are misdiagnosis cases; that is, those cases in which a benign case can be considered malignant and vice versa.

These types of misdiagnosis are a result of broken islands or patches of lesions that are not well connected to each other in ultrasound images. Such broken islands can result from the partial volume effect during the ultrasound image acquisition process (discussion of which is beyond the scope of this chapter). Another reason for such broken patches is that sometimes lesions are not concentrated at one location like a mass, but are instead scattered around in multiple masses with a fair number of gaps in between. Here our level set assumes that the lesion is more concentrated as a mass and hence shows a uniform intensity in ultrasound images. Thus, when the level set is made to run under the speed functions governed by the PDE smoother and sigmoid enhancer, it does capture it accurately in a majority of the cases, resulting in high sensitivity. But when the level set is made to run in multiple concentrated masses, it fails because of its inability to find the same uniform propagating field while propagating.

One way to handle this situation is to use a dual level set. This is not discussed in this chapter, but we are working on this strategy to handle multiple mass regions in ultrasound images. Suri and his team have developed a prototype, but it needs more testing before it is ready for clinical usage. One way to improve the performance of the system is to exclude multiple island cases from our database. This is not a very good idea, but we have identified it to improve further our CAD architecture. We anticipate on working on this in the near future, but the current system shows very promising results on central massed regions with large noise patterns.

6.5.2 Discussion About the Similarity Measure

The similarity measure is a measure of the similarity of two shapes, but it is expressed in a reverse engineering or indirect way. Here we use a base shape, which is like a standard shape, and we compare how much deviation exists between the true shape and the base shape. This deviation is—in a least-squares sense—a strategy to compute how much the error between the true shape and the base shape is. The larger the least-square error or deviation, the worse the irregularity of the true shape. It also signifies that the similarity is at its lowest. Thus, if the deviation between the true shape and least-square shape is low, then it means that the nature of the lesions is less irregular (or more regular) and hence of the benign type. In contrast, if the deviation between the true shape and least-square shape is large, then it means that the nature of the lesions are more irregular (or less regular) and hence of the malignant type. The main advantage of this strategy is that it is simple and straightforward to compute the best-fit ellipse.

6.5.3 Discussion About Threshold Selection

Another important part that affects the overall performance of the CAD architecture is the selection of the threshold. Too low or too high a threshold is not a very acceptable choice. We observed that for the 2-D acquisition method, the threshold value of 1,600 was a reasonable choice. On the other hand, for the 3-D acquisition method, threshold values of 2,000, 2,200, and 2,400 were acceptable and well adapted for our data sets. Note that here the threshold has acted as a tuner to the system and can be tuned in the CAD architecture depending on the data acquisition method (2-D or 3-D) and the choice of the ultrasound machine vendor (since the physics involved for reconstruction may change).

Regardless of which machine vendor is chosen, the threshold can be a priori trained for the system. There is no doubt that we are clearly able to establish the empirical strategy for threshold computation, but we must bear in mind that external factors such as the vendor's hardware and the image database collected from that acquisition can affect the value of the threshold. Again, such changes are beyond the scope of this research. In the authors' laboratory, however, we are experimenting with different machine types as one aspect of future research work.

6.6   Conclusion and Future Works

The contribution of this work lies in the development of a tool for analyzing breast lesions given 3-D slices acquired using 3-D methodology. The novelty of this tool lies in the way we automatically detect and classify breast lesions. The detection is done by capturing the irregularity of breast lesions using a topology-capturing tool, such as a level set. Another novelty of the system lies in the way in which the geometric deformable model uses enhancement as a stopping criterion. To make the system robust, the level set method is implemented on noise-free images, which are done in our preprocessing subsystem framework. Once the irregularity of breast lesions is detected, the shape of the lesions can be used as a criterion for classification of breast disease. Embedding this decision-making tool in the detection process results in a good diagnostic tool. We believe that developing such a tool for early breast cancer detection is where the novelty of the system lies.

We also emphasize that our CAD architecture is well suited for both 2-D and 3-D acquisition data sets. We demonstrated a system accuracy and sensitivity of 84% and 94%, respectively. Our system shows very promising results, but we will admit that the segmentation strategy could be improved as could quantification of shapes by using a better shape descriptor. We are involved in researching a dual level set and incorporation of spatial information of neighboring slices for more robust segmentation. We also intend to make this system fully automatic, by setting the machine a priori.

In the future, we will devote research to improving the performance of the CAD architecture. For example, the proposed shape descriptor can be extended into the 3-D form to present more complex shape representation. Moreover, physicians will not need to point out the tumor information any more, and this fully automatic CAD system will assist them in all of their analyses of breast lesions.

References

[1]   American Cancer Society, *Cancer Facts and Figures 2004*, 2004.

[2]   Leucht, W., and D. Leucht, *Teaching Atlas of Breast Ultrasound*, New York: Thieme Medical, 2000, pp. 24–38.

[3]   Cox, B., "Variation in the Effectiveness of Breast Screening by Year of Follow-Up," *J. Natl. Cancer Inst. Monogr.*, No. 22, 1997, pp. 69–72.

[4]   Ganott, M. A., et al., "Analysis of False-Negative Cancer Cases Identified with a Mammography Audit," *Breast J.*, Vol. 5, No. 3, May 1999, pp. 166–175.

[5]   American Cancer Society, *Cancer Prevention & Early Detection Facts & Figures 2004*, 2004.

[6]   Crystal, P., et al., "Using Sonography to Screen Women with Mammographically Dense Breasts," *AJR Am. J. Roentgenol.*, Vol. 181, No. 1, July 2003, pp. 177–182.

[7]   Jackson, V. P., "The Role of US in Breast Imaging," *Radiology*, Vol. 177, No. 2, November 1990, pp. 305–311,

[8]   Bassett, L. W., et al., "Usefulness of Mammography and Sonography in Women Less Than 35 Years of Age," *Radiology*, Vol. 180, No. 3, September 1991, pp. 831–835.

[9]   Stavros, A. T., et al., "Solid Breast Nodules: Use of Sonography to Distinguish Between Benign and Malignant Lesions," *Radiology*, Vol. 196, No. 1, July 1995, pp. 123–134.

[10] Zonderland, H. M., et al., "Diagnosis of Breast Cancer: Contribution of US as an Adjunct to Mammography," *Radiology*, Vol. 213, No. 2, November 1999, pp. 413–422.

[11] Sawaki, A., et al., "Breast Ultrasonography: Diagnostic Efficacy of a Computer-Aided Diagnostic System Using Fuzzy Inference," *Radiat. Med.*, Vol. 17, No. 1, January 1999, pp. 41–45.

[12] Chang, R. F., et al., "Improvement in Breast Tumor Discrimination by Support Vector Machines and Speckle-Emphasis Texture Analysis," *Ultrasound Med. Biol.*, Vol. 29, No. 5, May 2003, pp. 679–686.

[13] Chen, D. R., et al., "Use of the Bootstrap Technique with Small Training Sets for Computer-Aided Diagnosis in Breast Ultrasound," *Ultrasound Med. Biol.*, Vol. 28, No. 7, July 2002, pp. 897–902.

[14] Chang, R. F., et al., "Computer-Aided Diagnosis for Surgical Office-Based Breast Ultrasound," *Arch. Surg.*, Vol. 135, No. 6, June 2000, pp. 696–699.

[15] Kuo, W. J., et al., "Computer-Aided Diagnosis of Breast Tumors with Different US Systems," *Acad. Radiol.*, Vol. 9, No. 7, July 2002, pp. 793–799.

[16] Chen, D. R., et al., "Computer-Aided Diagnosis for 3-Dimensional Breast Ultrasonography," *Arch. Surg.*, Vol. 138, No. 3, March 2003, pp. 296–302.

[17] Horsch, K., et al., "Computerized Diagnosis of Breast Lesions on Ultrasound," *Med. Phys.*, Vol. 29, No. 2, February 2002, pp. 157–164.

[18] Drukker, K., et al., "Computerized Lesion Detection on Breast Ultrasound," *Med. Phys.*, Vol. 29, No. 7, July 2002, pp. 1438–1446.

[19] Horsch, K., et al., "Performance of Computer-Aided Diagnosis in the Interpretation of Lesions on Breast Sonography," *Academic Radiology*, Vol. 11, No. 3, March 2004, pp. 272–280.

[20] Chen, D. R., R. F. Chang, and Y. L. Huang, "Computer-Aided Diagnosis Applied to US of Solid Breast Nodules by Using Neural Networks," *Radiology*, Vol. 213, No. 2, November 1999, pp. 407–412.

[21] Chen, C. M., et al., "Breast Lesions on Sonograms: Computer-Aided Diagnosis with Nearly Setting-Independent Features and Artificial Neural Networks," *Radiology*, Vol. 226, No. 2, February 2003, pp. 504–514.

[22] Chang, R. F., et al., "Automatic Ultrasound Segmentation and Morphology Based Diagnosis of Solid Breast Tumors," *Breast Cancer Res. Treat.*, Vol. 89, No. 2, January 2005, pp. 179–185.

[23] Chang, R. F., et al., "Computer-Aided Diagnosis for 2-D/3-D Breast Ultrasound," in *Recent Advances in Breast Imaging, Mammography, and Computer-Aided Diagnosis of Breast Cancer*, J. S. Suri and R. M. Rangayyan, (eds.), Bellingham, WA: SPIE Press, 2006.

[24] Chen, W. M., et al., "Breast Cancer Diagnosis Using Three-Dimensional Ultrasound and Pixel Relation Analysis," *Ultrasound Med. Biol.*, Vol. 29, No. 7, July 2003, pp. 1027–1035.

[25]   Chen, W. M., et al., "3-D Ultrasound Texture Classification Using Run Difference Matrix," *Ultrasound Med. Biol.,* Vol. 31, No. 6, June 2005, pp. 763–770.

[26]   Zhang, D. S., and G. J. Lu, "Review of Shape Representation and Description Techniques," *Pattern Recognition,* Vol. 37, No. 1, January 2004, pp. 1–19.

[27]   Sethian, J. A., *Level Set Methods and Fast Marching Methods: Evolving Interfaces in Computational Geometry, Fluid Mechanics, Computer Vision, and Materials Science,* 2nd ed., Cambridge, U.K.: Cambridge University Press, 1999.

[28]   Malladi, R., J. A. Sethian, and B. C. Vemuri, "Shape Modeling with Front Propagation: A Level Set Approach," *IEEE Trans. on Pattern Analysis and Machine Intelligence,* Vol. 17, No. 2, 1995, pp. 158–175.

[29]   Suri, J. S., et al., "Shape Recovery Algorithms Using Level Sets in 2-D/3-D Medical Imagery: A State-of-the-Art Review," *IEEE Trans. on Information Technol. Biomed.,* Vol. 6, No. 1, March 2002, pp. 8–28.

[30]   Suri, J. S., S. Singh, and L. Reden, "Fusion of Region and Boundary/Surface-Based Computer Vision and Pattern Recognition Techniques for 2-D and 3-D MR Cerebral Cortical Segmentation (Part II): A State-of-the-Art Review," *Pattern Analysis and Applications,* Vol. 5, No. 1, 2002, pp. 77–98.

[31]   Suri, J. S., et al., "Modeling Segmentation Via Geometric Deformable Regularizers, PDE and Level Sets in Still/Motion Imagery: A Revisit," *Int. J. Image and Graphics,* Vol. 1, No. 4, December 2001, pp. 681–734.

[32]   Arica, N., and F. T. Y. Vural, "BAS: A Perceptual Shape Descriptor Based on the Beam Angle Statistics," *Pattern Recognition Letters,* Vol. 24, No. 9–10, June 2003, pp. 1627–1639.

[33]   Arica, N., and F. T. Yarman-Vural, "A Compact Shape Descriptor Based on the Beam Angle Statistics," *Proc. Image and Video Retrieval,* Vol. 2728, 2003, pp. 152–162.

[34]   Osher, S., and J. Sethian, "Fronts Propagating with Curvature-Dependent Speed: Algorithms Based on Hamilton-Jacobi Equations," *J. Comput. Phys.,* Vol. 79, No. 1, 1988, pp. 12–49.

[35]   Moon, W. K., et al., "Solid Breast Masses: Classification with Computer-Aided Analysis of Continuous US Images Obtained with Probe Compression," *Radiology,* Vol. 236, No. 2, August 2005, pp. 458–464.

[36]   Sussman, M., and E. Fatemi, "An Efficient, Interface-Preserving Level Set Redistancing Algorithm and Its Application to Interfacial Incompressible Fluid Flow," *Siam Journal on Scientific Computing,* Vol. 20, No. 4, April 1999, pp. 1165–1191.

[37]   Suri, J. S., "Fast MR Brain Segmentation using Regional Level Sets," *International Journal of Engineering in Medicine and Biology,* Vol. 20, No. 4, July 2001, pp. 84–95.

[38]   Suri, J. S., "Leaking Prevention in Fast Level Sets Using Fuzzy Models: An Application in MR Brain," *Proc. IEEE-EMBS Int. Conf. Information Technology Applications in Biomedicine,* 2000, pp. 220–225.

[39]   Perona, P., and J. Malik, "Scale-Space and Edge Detection Using Anisotropic Diffusion," *IEEE Trans. on Patt. Anal. Mach. Intell.,* Vol. 12, No. 7, July 1990, pp. 629–639.

[40]   Rapp, C. L., "Sonography of the Breast," *Proc. Soc. Diagnostic Medical Sonography's 17th Annual Conf.,* Dallas, TX, 2003, pp. 57–67.

[41]   Costa, L. D. F., and R. M. Casar, Jr., *Shape Analysis and Classification: Theory and Practice,* Boca Raton, FL: CRC Press, 2000.

[42]   Rangayyan, R. M., et al., "Measures of Acutance and Shape for Classification of Breast Tumors," *IEEE Trans. on Med. Imaging,* Vol. 16, No. 6, 1997, pp. 799–810.

[43]   Jain, A. K., *Fundamentals of Digital Image Processing,* Upper Saddle River, NJ: Prentice-Hall, 1989.

[44]   Wang, Y. P., and T. Pavlidis, "Optimal Correspondence of String Subsequences," *IEEE Trans. on Pattern Anal. Mach. Intell.,* Vol. 12, No. 11, 1990, pp. 1080–1087.

CHAPTER 7

3-D Breast Ultrasound Strain Imaging

Ruey-Feng Chang, Chii-Jen Chen, Jasjit S. Suri, and Wei-Ren Lai

7.1 Introduction

Breast cancer is a common disease that influences the health of women; in fact, it ranks second among cancer deaths in women. To reduce the morbidity and mortality associated with breast cancer, early detection becomes a very important job. If the cancers could be diagnosed through regular breast cancer examinations at an earlier stage than is currently possible, the survival rate within 5 years would increase to about 95% [1].

During the past two decades, the ultrasound technique has become an indispensable imaging modality because it is inexpensive for detection and diagnosis of breast lesions. The main superiority of ultrasound lies in its noninvasiveness and ability to efficiently capture the tissue properties [2, 3]. Unlike X-ray mammography and MRI, ultrasound does not use ionizing radiation; therefore, it has become a very adjunct tool for real-time interactive visualization. In the past, conventional 2-D ultrasound has been adapted for both diagnosis and visual-based classification of lesions [4, 5]. The diagnosis of breast pathology was highly dependent on how the breast radiologists adapted the use of the transducer in 2-D conventional ultrasound. However, this is a very subjective task, because the ability to acquire breast images is based entirely on the knowledge and experience of breast radiologists. Mentally transforming the conventional 2-D images into 3-D images would affect the accuracy of diagnosis. Recently, 3-D ultrasound has emerged as a promising method for breast imaging [6–10].

With the advancement of ultrasound acquisition and the digital nature of 3-D volume acquisition, computer-aided diagnosis has become even more powerful, thereby improving the workflow of breast imaging and reducing the physician's workload [11–17]. The most important clinical question today in breast imaging lies in its ability to discriminate benign versus malignant lesions based on US image features such as shape, margin, and texture [18–21].

To answer the question about the characterization of malignancy types, researchers recently proposed strategies that involved the nature of tissue property. An example is the evaluation of tissue change or deformity at different pressure levels. Such an evaluation is important because a hard tissue lesion would behave differently from a soft tissue lesion at different pressure conditions. This property and assumption would yield different shapes or volumes for benign and malignant lesions. Such analysis would come under the field of elastography, which would help in tissue characterization. Elastography is nothing but computation of the elastic modulus of breast tissue [22–27]. In elastography, we compare ultrasound radio-frequency (RF) waveforms obtained before and after a slight pressure from the ultrasound probe. During pressure, soft tissues tend to flatten more than hard tissues. The use of this characteristic can be very useful for a CAD system.

This research uses CAD tools to collect lesion features in strain imaging framework, and then adapt a classification strategy based on a support vector machine (SVM) to distinguish benign versus malignant lesions. The novelty of our architecture lies in an application of strain feature extraction that is based on the level set strategy. Note clearly that our contributions are aimed at development of a CAD system that takes advantage of the elastic nature of the breast tissues. Thus, the integration of our strain feature extractor, in precompression and postcompression pressure modes followed by a classifier, provides an automatic system that has clinical utility for breast radiologists. Our strain features are computed after image alignment of precompressed and postcompressed volumes. Because a change in pressure results in deformation of lesions, our goal is to quantify the deformation in the form of change in the lesion volume, change in the lesion shape, shift of the breast lesion due to pressure, and finally the contour measurements, all within the level set framework. The main advantage of our approach is its simplicity in measurement of deformation and its accuracy. This comes from the ability of the level set to capture the topology of benign and malignant lesions reasonably well.

This research described in this chapter extended the conventional method to 3-D ultrasound imaging. The proposed architecture consisted of tumor shape analysis before and after the compression. This deformation was quantified using the level set strategy. Four different strain features were computed: volume difference (V_d), shape difference (S_d), contour difference (C_d), and shift distance (M_d). All features were the bases for classification of breast lesions using support vector machines. The experimental protocol was tested on a data set of 100 tumors with 60 of them being benign and 40 malignant.

We demonstrated an accuracy rate for the SVM classifier using 3-D strain features of 91% and a corresponding ROC area index A_z of 0.8855. Our 3-D system has sensitivity and specificity values of 90% and 92.67%, respectively. We also ran our protocol on ultrasound images acquired using a 2-D conventional framework and compared our performance index using the 3-D method. This comparative protocol showed an improvement of 16% in accuracy, 15% in sensitivity, and 15% in specificity.

The proposed architecture for the 3-D strain imaging system is introduced in Section 7.2. The 3-D strain feature extraction system is presented in Section 7.3. The results and data analysis of our experimental protocol are discussed in Section 7.4, and some special issues are discussed in Section 7.5. Finally, a conclusion is drawn and future works discussed in Section 7.6.

7.2   Review of the 2-D Continuous Strain Imaging System

In our previous work, we proposed a novel elasticity analysis technique that was based on the level set method [28]. The 2-D US image data were acquired with a Voluson 530 (Kretz Technik, Austria) scanner. On a Voluson 530 scanner, continuous 2-D images can be recorded at 15 images per second. The compression is completed in 4 seconds and 60 images are obtained.

In this CAD system, the level set segmentation plays a major role that will influence the results of elasticity analysis. Because the ultrasonic image suffers from noise, in order to remove the noise in the ultrasound images and to improve the segmentation performance, the proposed preprocessing procedures of anisotropic diffusion filtering and the stick method are also used. The flowchart for the proposed CAD system is shown in Figure 7.1.

After the segmentation of tumor contours from the continuous ultrasound images, four features were defined for labeling of tumors: contour difference (C_d), shift distance (M_d), area difference (A_d), and solidity (S_o). The first three are related to the strain, and the fourth is for shape. The four features are briefly introduced as follows:

1. *Contour difference* (C_d): This value is used to evaluate changes in lesion shape between two images.
2. *Shift distance* (M_d): This value uses the motion estimation technique to find the motion of two image contours.
3. *Area difference* (A_d): This feature was used to compare tumor areas between two images.
4. *Solidity* (S_o): Shape values can be used to distinguish between benign and malignant tumors. Benign tumors usually have smooth shapes, whereas malignant tumors are irregular. The normalized solidity value [29] was used in this study. It is defined as

$$\text{sol} = \frac{\text{Area}_C - \text{Area}_T}{\dfrac{1}{N}\sum_{i=1}^{N}\text{Area}_{C_i} - \text{Area}_{T_i}} \tag{7.1}$$

where N is the total number of cases, Area_C is the area of the convex hull of a tumor, and Area_T is the area of a tumor, as shown in Figure 7.2.

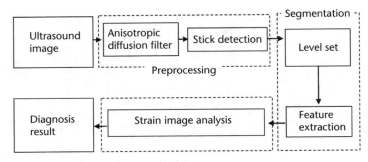

Figure 7.1   The 2-D breast lesion characterization system.

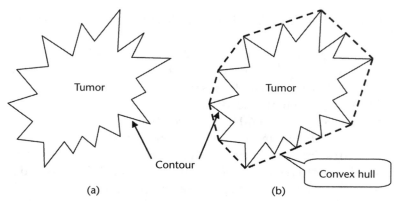

Figure 7.2 Drawings depicting (a) the contour of a tumor and (b) the convex hull of a tumor. The outer polygon is the convex hull of this tumor. The convex hull is the smallest convex set containing the tumor.

In the experiments, we used all 100 pathology-proven cases, including 60 benign breast tumors and 40 malignant ones, to evaluate the classification accuracy of the proposed method. The segmentation results in different lesion cases using the level set framework are shown in Figures 7.3 and 7.4. The mean values of all features (C_d, M_d, A_d, and S_o) for malignant tumors were $3.52\% \pm 2.12$, 2.62 ± 1.31, $1.08\% \pm 0.85$, and 1.70 ± 1.85, respectively; whereas those for benign tumors were $9.72\% \pm 4.54$, 5.04 ± 2.79, $3.17\% \pm 2.86$, and 0.53 ± 0.63, respectively. All of these statistical data were evaluated by Student's t-test, and the p-values (<0.0001) can prove the validity of these statistical data. The accuracy, sensitivity, specificity, PPV, and NPV values for this CAD system are 87%, 85%, 88%, 82% and 89%, respectively. These sufficient results proved that all performances of the proposed CAD system are significant.

7.3 Proposed Architecture of the 3-D Strain Imaging System

In this chapter, the strain imaging system is extended into 3-D ultrasound because 2-D ultrasound can only obtain one cross section of a tumor and the tumor may slip outside when the compression force is slanted. Three-dimensional ultrasound can

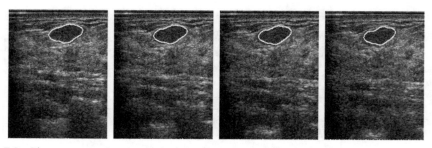

Figure 7.3 The segmentation results for a benign case for different image slices (numbers: 1, 17, 33, and 49) with $C_d = 16.90\%$, $M_d = 7.161$, $A_d = 1.71\%$, and $S_o = 0.1876$.

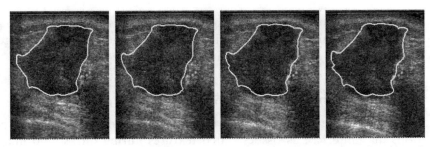

Figure 7.4   The segmentation results for a malignant case for different image slices (numbers: 1, 17, 33 and 49) with C_d = 1.20%, M_d = 1.241, A_d = 0.85%, and S_o = 3.6919.

scan the entire tumor and its probe is wider; therefore, the drawbacks of the 2-D US process can be avoided.

The 3-D strain imaging architecture is shown in Figure 7.5. In this strain imaging system, the preprocessor is used to remove speckle and enhance the breast lesions. The main purpose of this enhancement is to help the level set develop strong stopper capability during the deformation process. Then, the segmentor applies in segmenting the breast lesions from the ultrasound breast slices. We use the level set framework in the segmentation process. Note that the segmentor and preprocessor are cascaded and the outputs of these two processes are segmented breast lesions for precompressed data and postcompressed data, respectively. The output of the cascaded process goes into the feature extraction process, which yields four different sets of strain features. It is important to understand that it is the combination of both precompressed and postcompressed data sets that helps in computing the strain features. The strain features are then fed into the SVM-based classifier, which helps with the classification of the breast lesions into benign and malignant classes.

7.3.1   Data Acquisition

In this study, the 3-D ultrasound breast images were acquired using a Voluson 730 scanner with a Voluson small part transducer S-VNW5-10. The transducer with a linear array of crystal elements has the following specifications: 5- to 10-MHz scan frequency, 4-cm scan width, and 20° to 30° sweep angles. A set of 2-D US images was obtained by the slow tilt movement of a sector-based mechanical transducer. After completing the scanning process (on an average of 10 seconds per case), the

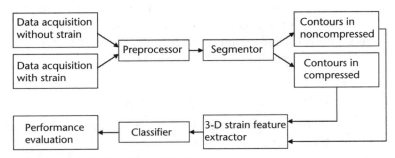

Figure 7.5   3-D breast strain imaging architecture.

3-D US image can be obtained by reconstructing these 2-D US images. This scanning process of acquiring the successive 2-D US images at regular angular intervals arranges the set of 2-D image planes in a fanlike geometry.

7.3.2   Image Enhancement Via Sigmoid Mapping

Ultrasound images are dominated by speckle phenomenon or noise. These perturbations may interfere as tissue-related textures during ultrasound image processing [30, 31]. This interference can affect the lesion boundary estimation processes. Thus, we need to enhance the boundary lesions and improve the performance of segmentation [32].

In a breast ultrasound image, different tissues have different intensities, and the intensity values of malignant and benign tumors mostly tend to be low [33]. For accurate segmentation of the tumor region, we enhance the tumor boundary by mapping the intensities based on a sigmoid intensity transform. The sigmoid transform is a nonlinear mapping that maps a specific intensity range into a new intensity range with a very smooth and continuous transition. This mapping can focus on a specific set of intensity values and progressively decrease the values outside the chosen range. We will focus on the tumor regions and fade out the other regions. The following equation represents the sigmoid intensity transformation:

$$\tilde{I} = (I_{max} - I_{min}) \frac{1}{\left(1 + e^{-\left(\frac{I-\beta}{\alpha}\right)}\right)} + I_{min} \qquad (7.2)$$

where I and $\tilde{I}$ are the intensity of the input and output pixels; I_{max} and I_{min} are the maximum and minimum values of the input image; α is the width of the input intensity range; and β defines the intensity around which the range is centered. The values of $(\alpha, \beta) = (10, 25)$ for the parameters are often adopted in the experiments. However, the parameters are not the same in each case; they can be slightly adjusted. The results for different lesion cases subjected to the sigmoid function are shown in Figure 7.6.

7.3.3   Segmentation Subsystem Based on the Level Set Framework

The segmentation technique of the level set is capable of capturing the topology of shapes in medical images [34, 35]. This technique supports recovery of arbitrarily complex shapes, such as corners and protrusions, and automatically handles topological changes such as merging and splitting. Recently, research on large numbers of implementations for computer vision and medical imaging based on this basic concept have been published [12, 28, 29, 35–39].

7.3.3.1   Basic Theory of the Level Set

As a starting point and motivation for the level set approach, consider a closed curve moving in a plane. Let $\phi(0)$ be a smooth and closed initial curve in a Euclidean plane $\Re^2$, and let $\phi(t)$ be the family of curves that is generated by the movement of initial curve $\phi(0)$ along the direction of its normal vector after time t. The main idea of the

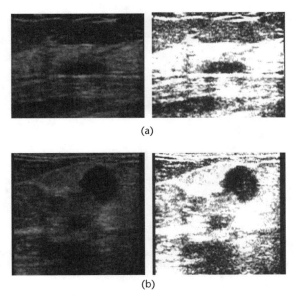

(a)

(b)

Figure 7.6   Results from sigmoid filtering on (a) benign and (b) malignant cases. The images in the left panel are raw images; the images on the right are enhanced by sigmoid mapping with parameters $\alpha = 10$, $\beta = 25$.

level set approach is to represent the front $\phi(t)$ as the level set $\{\Gamma = 0\}$ of a function Γ. Thus, given a moving closed hypersurface $\phi(t)$, that is, $\phi(t = 0)$: $[0, \infty) \rightarrow \Re^N$, we hope to produce a formulation for the motion of a hypersurface propagating along its normal direction with speed F, where F can be a function of various arguments, such as curvature or normal direction. The level set methodology and curve evolution are shown in Figure 7.7. Arrows in Figure 7.7(a) show the propagating direction of the front. During a period of time t_1, the front $\phi(t = t_1)$ and hypersurface $\Gamma(x, y, t = t_1)$ are shown in Figure 7.7(b).

In this way, the actual object contour can be obtained using the level set method. The details of the level set segmentation methodology are described next.

Let $\Gamma(\mathbf{x}, t = 0)$, where $\mathbf{x}$ is a point in $\Re^N$, be defined by $\Gamma(\mathbf{x}, t = 0) = \pm d$, where d is the signed distance from x to $\phi(t = 0)$, and a plus (minus) sign is chosen if the point $\mathbf{x}$ is outside (inside) the initial front. The goal is to produce an equation for the evolving function $\Gamma(\mathbf{x}, t)$, which contains the embedded motion of $\phi(t)$ as the level set $\{\Gamma = 0\}$. Let $\mathbf{x}(t)$, for $t \in [0, \infty)$ be the path of a point on the propagating front; that is, $\mathbf{x}(t = 0)$ is a point on the initial front $\phi(t = 0)$. Because the evolving function Γ is always 0 on the propagating front $\phi(t)$, then

$$\Gamma\big(\mathbf{x}(t), t\big) = 0 \tag{7.3}$$

By the chain rule [35],

$$\Gamma_t + \sum_{i=1}^{N} \Gamma_{x_i} x_{i_t} = 0 \tag{7.4}$$

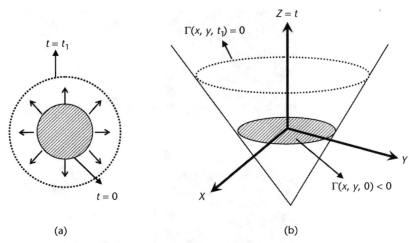

Figure 7.7 The level set methodology and curve evolution. (a) Arrows show the propagating direction of the front. (b) The level set hypersurface.

where x_{i_t} is the ith component of x at time t, and N is the number of components. Since

$$\sum_{i=1}^{N} \Gamma_{x_i} x_{i_t} = \left(\Gamma_{x_1}, \Gamma_{x_2}, \Gamma_{x_3}, \ldots, \Gamma_{x_N}\right) \cdot \left(x_{1_t}, x_{2_t}, x_{3_t}, \ldots, x_{N_t}\right) \tag{7.5}$$

$$= F\left(\mathbf{x}(t)\right)|\nabla \Gamma|$$

where $F(\mathbf{x}(t))$ is the speed function with the front for all components of $\mathbf{x}$ at time t. Combining (7.4) and (7.5), the final curve evolution equation is given as:

$$\frac{\partial \Gamma}{\partial t} = F(\kappa)|\nabla \Gamma| \tag{7.6}$$

where Γ is the level set function and $F(\kappa)$ is the speed with which the front (or zero-level curve) propagates. This fundamental equation describes the time evolution of the level set function Γ in such a way that the zero-level curve of this evolving function is always identified with the propagating interface [38]. The preceding equation is also called a Eulerian Hamilton-Jacobi equation [34, 38]. Equation (7.6) for 2-D and 3-D cases can be generalized as follows:

$$\frac{\partial \Gamma}{\partial t} = F_\kappa(x, y)|\nabla \Gamma| \quad \text{and}$$

$$\frac{\partial \Gamma}{\partial t} = F_\kappa(x, y, z)|\nabla \Gamma| \tag{7.7}$$

where (x, y) and (x, y, z) are the points on the propagation directions, and $F_\kappa(x, y)$ and $F_\kappa(x, y, z)$ are curvature-dependent speed functions in two and three dimensions, respectively [38].

The level set can be used for image segmentation by using image-based features such as mean intensity, gradient, and edges in the governing differential equation. In a typical approach, a contour is initialized by a user and then evolved until it fits the topology of an anatomic structure in the image [35]. There are challenges to using the level set framework by itself on the raw images. This is because the propagation of the fronts can sometimes lead to bleeding of the fronts due to noise structures present in the images [40]. One way to handle this is by enhancing the structures to be segmented. This effect will assist the level set front propagations to recognize the edges of the structures, thereby bringing the speed close to zero at the structure edges. Keeping this in mind, we have developed our breast lesion identification chain. The segmentation examples with and without the preprocessing block by using the level set method are illustrated in Figure 7.8. These conspicuous results also prove the performance of our proposed CAD architecture.

7.3.3.2   3-D Strain Imaging and Segmentation

In this system, 3-D ultrasound is used to generate two datasets with and without the pressure of a probe to provide additional information for tissue characterization of stiffness. The aim is to differentiate between different tissues based on their elasticity. For each tumor, a normal 3-D ultrasound image dataset is first obtained, and then the physician applies slight pressure to the probe to obtain the second 3-D ultrasound strain image dataset. Then, the level set method is used to extract the contour of each dataset, and the contours of the precompressed and postcompressed ultrasound images are aligned first and compared by using the proposed strain features. In Figures 7.9 and 7.10, we illustrate the segmentation results of different lesion cases for precompressed and postcompressed ultrasound images using the proposed level set framework. According to previous examples, the shape

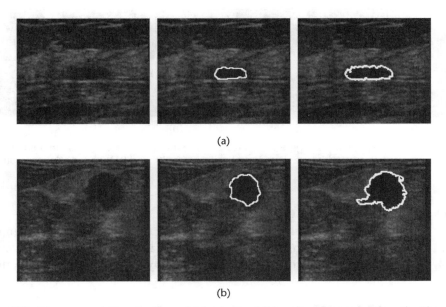

(a)

(b)

Figure 7.8   The segmentation results on (a) benign and (b) malignant cases with and without the preprocessing block of the 3-D strain imaging architecture. In each case, the first image is the raw image; the middle column is obtained by running the level set without the sigmoid enhancer (preprocessing block), and the last column shows the results using the entire architecture.

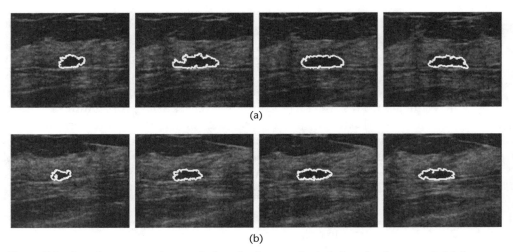

(a)

(b)

Figure 7.9 Sample segmentation results for a benign case in strain imaging framework: (a) the precompressed image slices (numbers: 57, 66, 70, and 79), and (b) the postcompressed image slices.

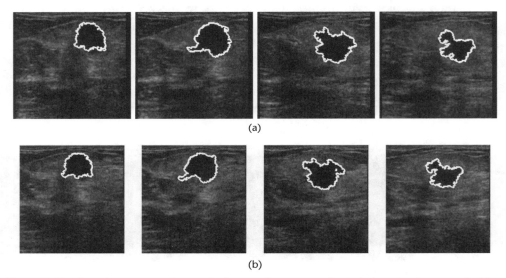

(a)

(b)

Figure 7.10 Sample segmentation results for a malignant case in strain imaging framework: (a) the precompressed image slices (numbers: 64, 68, 80, and 91), and (b) the postcompressed image slices.

variation and translation for malignant cases between precompressed and postcompressed are obviously smaller than benign ones. Hence, we will present some useful features for the lesion strain analysis in the following section.

7.3.4 Strain Features

As discussed earlier, the elasticity of tissue can be utilized as a framework for classification of breast lesions into two different classes: benign cases and malignant cases. This classification can be achieved by a powerful classifier, SVM, which can utilize the deformation features of the two kinds of breast lesions. Thus, our immediate goal should be to develop a strategy by which we can quantify tissue deformations in compression framework. The most straightforward and direct way would be to compute the change in quantified values in precompressed and postcompressed modes. We thus approach this in four different ways, utilizing features of volume, enveloped cuboids enveloping the 3-D tumors, silhouette change, and movement of lesions due to compression. These features are the ones mentioned earlier in the chapter: volume difference (V_d), shape difference (S_d), contour difference (C_d), and shift distance (M_d).

7.3.4.1   Normalized Volume Difference

Because the volume of the precompressed lesion changes as a result of the probe pressure, the purpose of this feature is to compare volumes between the precompressed and postcompressed lesions. If the lesion is hard, then its volume in the postcompressed images should be similar to that of the precompressed images. Of course, the volume difference of a large lesion will be larger than that of a small lesion. Hence, the volume difference should be normalized by the lesion volume. The volume difference feature is mathematically expressed as:

$$V_d = \frac{\left| \text{Vol}_{\text{pre}} - \text{Vol}_{\text{post}} \right|}{\text{Vol}_{\text{pre}}} \tag{7.8}$$

where Vol_{pre} and Vol_{post} are the lesion volumes in the precompressed and postcompressed datasets, respectively. The volume of the lesion is mathematically expressed as:

$$\text{Vol} = \sum_{R_i \in \text{Lesion}} \text{Area}(R_i) \tag{7.9}$$

where R_i is the lesion region in the image slice i, and $\text{Area}(R_i)$ is the number of pixels in region R_i. For example, the tumor area is 15 pixels in Figure 7.11(c), and the tumor area is 13 pixels in Figure 7.11(d). Then we calculate the volumes of a tumor by summing up the tumor area sizes of every image slice. If the volume of the tumor is 210 pixels in Figure 7.11(a) and the volume of the tumor is 197 pixels in Figure 7.11(b), then the number of difference pixels of these two tumors in Figures 7.11(a) and (b) is 13 pixels, and the volume difference is 13/210 = 0.0619.

7.3.4.2   Normalized Shape Difference

When a tumor is compressed, its height will become smaller but its width and length will become larger. If a tumor is soft, then the differences of height, width, and length will be larger than those of a hard tumor. In a hard tumor, the height, width, and length will remain at values similar to the precompressed values. Actually, the ratios of width–height (WH) and length–height (LH) can be used to represent the simple shape information. Hence, the shape difference of the precompressed and postcompressed tumors can be defined by these two ratios:

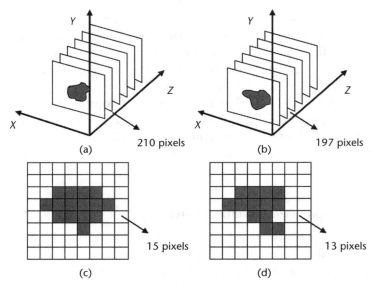

(a)   210 pixels   (b)   197 pixels

(c)   15 pixels   (d)   13 pixels

Figure 7.11 (a) Precompressed images. (b) Postcompressed images. (c) An image slice of the precompressed lesion. (d) An image slice of the postcompressed lesion.

$$S_d = \max\left(\frac{WH_{post} - WH_{pre}}{WH_{pre}}, \frac{LH_{post} - LH_{pre}}{LH_{pre}}\right) \qquad (7.10)$$

where WH_{pre} and WH_{post} are the width–height ratios for precompressed and postcompressed tumor, respectively; LH_{pre} and LH_{post} are the length–height ratios for precompressed and postcompressed tumor, respectively. To compute the length, height, and width of a tumor efficiently, we use a cube to cover the 3-D breast tumor, as shown in Figure 7.12. Thus, the information of length, height, and width can be obtained directly form this cube.

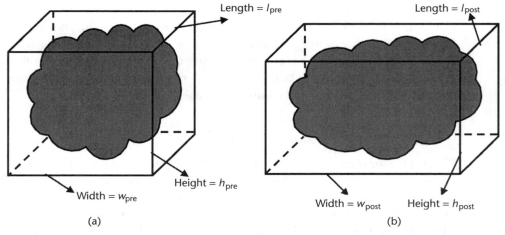

(a)   (b)

Length = l_{pre} Length = l_{post}

Width = w_{pre} Height = h_{pre} Width = w_{post} Height = h_{post}

Figure 7.12 Dimensions of a tumor in the strain imaging framework. (a) Precompressed tumor dimensions are given as l_{pre}, w_{pre}, and h_{pre}. (b) Postcompressed tumor dimension are given as l_{post}, w_{post}, and h_{post}. The filled shapes are represented as tumors in the breast.

7.3.4.3   Normalized Contour Difference

This feature is used to compare the contours between the precompressed and postcompressed lesions. The initial contours of precompressed and postcompressed lesions can be obtained by using the level set segmentation framework. These contours should be registered first because of the movement caused by the pressure of the probe. To simplify the alignment between two contours, a reference point should first be defined. A pure spatial alignment technique is used in this chapter. The centers of gravity are used as the basic reference points for the alignment of two lesion contours. For each pixel (x, y, z) inside the lesion, the center of gravity $(\bar{x}, \bar{y}, \bar{z})$ can be defined as

$$(\bar{x}, \bar{y}, \bar{z}) = \left(\sum_{(x,y,z) \in R} \frac{x}{N}, \sum_{(x,y,z) \in R} \frac{y}{N}, \sum_{(x,y,z) \in R} \frac{z}{N} \right) \tag{7.11}$$

where N is the total number of pixels in the tumor region R. Then, the contour difference is defined as:

$$C_d = \sum_i \frac{A_d\left(R_i, R_i^{\text{reg}}\right)}{\text{Vol}_{\text{pre}}} \tag{7.12}$$

where R_i is the lesion region in ith precompressed image slice; R_i^{reg} is the lesion region in ith registered postcompressed image slice; A_d is the number of different pixels in these two registered regions; and Vol_{pre} is the lesion volume in the precompressed dataset. The value as a number carries a quantification significance compared to other similar scenarios.

For example, in Figure 7.13, image slice B in the postcompressed dataset is registered to image slice A in the precompressed dataset. The center of gravity in slice A is at the $(5,5)$ position, and the center of gravity in slice B is at the $(5,3)$ position. Before alignment of these two image slices, the area in slice A is 12 pixels [see Figure 7.13(a)], and the area in slice B is 11 pixels [see Figure 7.13(b)]. In the alignment process, we consider only the spatial transition of two reference points; therefore, the number of difference pixels is 5 while the alignment of reference points is com-

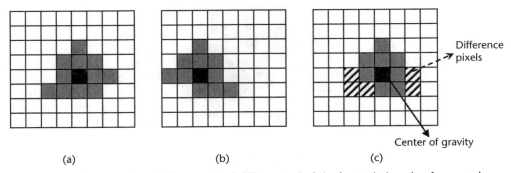

(a)   (b)   (c)

Figure 7.13 Computation of alignment and difference pixels in the strain imaging framework. Image slice B in the postcompressed dataset is aligned with image slice A in the precompressed dataset. (a) Precompressed slice A, (b) postcompressed slice B, and (c) alignment of these two images and determination of the contour difference.

pleted, as shown in Figure 7.13(c). Suppose that the total volume of the precompressed tumor is 210 and the total number of different pixels between two aligned tumors is 64. In such a case, the contour difference is 64/210 = 0.3048.

7.3.4.4 Shift Distance

The shift distance uses the concept of the motion estimation in video compression [41] to find the motion of the contours. At first, the centers of gravity of two lesions have to be aligned as done for the computation of the contour difference feature. For speed reasons, we compute the motion vectors for the contour pixels rather than for each pixel in the image. The contour of the current image slice will be equally divided into eight sections and eight points can be obtained. These points are so-called "cutting points" because they radially cut the breast lesion contour into different sections. It is this cutting point that needs to have a motion vector.

The motion estimation is block based, and a small block containing each cutting point will find the most similar block in the postcompressed image slice. To reduce the number of searched blocks in the postcompressed image slice, a search window could be defined to limit the searching region. The motion estimation technique [42] is used to find the motion vector of these eight blocks for each image slice. For a block at (m, n), if the most similar block in the postcompressed image slice is at $(m + i, n + j)$, then the motion vector for the block (m, n) is (i, j), as shown in Figure 7.14, and the distance between these two blocks is defined as dis $= \sqrt{i^2 + j^2}$. Thus, shift distance of the breast lesion can be defined as

$$M_d = \underset{A}{\mathrm{avg}}\left(\underset{k \in G}{\max}(\mathrm{dis}_k)\right) \tag{7.13}$$

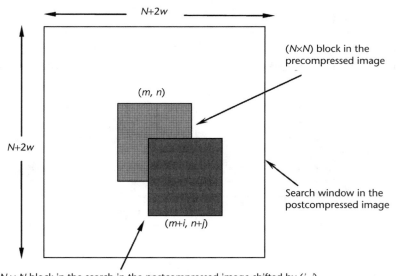

$N \times N$ block in the search in the postcompressed image shifted by (i, j)

Figure 7.14 Demonstration of the matching process. The current block and the most similar block in a search window of the postcompressed sliced image.

where G represented eight blocks in each image slice and A represented all slices in the postcompressed 3-D dataset. Only eight blocks on the contour need to find their most similar blocks. The maximum of these eight distances will be assigned as the shift distance of this slice. Then, the shift distance of this sequence will be the average of the shift distances of all slices in the sequence. Once the four features are determined, our goal is to use a classifier that can distinguish the benign and malignant breast lesions.

7.4   Experimental Protocol, Data Analysis, and Results

7.4.1   2-D and 3-D Experimental Protocols

Two sets of protocols were used for experimentation with the datasets. Protocol A used the 3-D ultrasound acquisition method, whereas protocol B used conventional 2-D images. With the 2-D strain imaging method, only one precompressed and one postcompressed image slice were used for feature analysis. These slices were obtained from the middle slice of images that contained tumor information. Four strain features—contour difference, shift distance, volume difference, and shape difference—are adopted for all lesion cases. In our experiments, we used 100 pathology-proven cases for the 2-D and 3-D strain method, including 60 benign breast tumors and 40 malignant ones, to evaluate the classification accuracy of the proposed method.

Three data analysis methods—ROC curves, Student's t-test, and the Chi-square test, are used to analyze these four features. An SVM [42–44] is also used to combine all four features to classify whether a tumor is malignant or not. The major reason to adopt an SVM classifier is that the SVM was found to be better than the Bayesian classifier, which is a classic linear classification method, for continuous ultrasound strain images [28]. This research is our previous study for continuous ultrasound strain images; the statistical significance have been proven especially in accuracy and specificity (87.0% versus 77.0%, 88.3% versus 70.0%, and p-value <0.01). According to a study by Chang et al. [45], the classification ability of the SVM at gray-scale ultrasound was equal to that of a neural network model, whereas the SVM had a much shorter training time (1 versus 189 seconds). Because of its performance, the SVM has become an effective tool for pattern recognition, machine learning, and data mining.

7.4.2   Segmentation Results

The level set method is first used to extract the contours of the 100 pathology-proven cases. Then these contours are used for the 2-D and 3-D strain image analysis to obtain four strain features. According to the experiments, most benign lesions usually have larger values for the four strain features than the malignant lesions. Unfortunately, some misdiagnoses may result due to the effects of speckle noise and tissue-related texture, which cannot be reduced by the preprocessing. This effect will result in an incomplete lesion contour and directly affect the accuracy of the strain feature analysis. In this situation, some feature values of benign cases are smaller compared to those of malignant cases. These cases would be misdiagnosed by the proposed CAD system. In Figures 7.9 and 7.10, we show segmentation results for both precompressed and postcompressed images for benign and malignant cases.

7.4.3 Classification Strategy Using ROC Curves and SVM

We used ROC curve analysis for demonstrating the performance of our CAD system. The performances of the four strain features can be evaluated by examining the ROC area index, A_z. In addition, to improve validation of the classification results, discriminant analysis was applied in combination with jack-knifing. *Jack-knifing* means that each case was classified using an individual discriminant function trained with all cases except this one [46, 47].

In this SVM experiment, the ultrasonic images were divided into five groups by the jack-knifing procedure. At the beginning, the first group was used as the testing group, and the remaining four groups were used to train the SVM. After training, the SVM was then tested on the first group. Then the second group was used as another testing group, with the remaining four groups being the training set, and again the SVM was tested on the second group. This process was repeated until each group had been set as a testing group. A nonlinear SVM with Gaussian radial basis kernel was adopted as our classifier, where C and γ are 658 and 0.094, respectively.

Table 7.1 shows the mean value, standard deviation, and p-value using Student's t-test for 2-D and 3-D strain features in the benign and malignant lesions. Table 7.2 lists the experimental results of 2-D and 3-D SVM. Figure 7.15 shows the ROC analysis for all strain features using an SVM classifier on the 3-D strain

Table 7.1 The Mean Value, Standard Deviation and p-Value (Using Student's t-Test) of Four Strain Features: Volume Difference (V_d), Shape Difference (S_d), Contour Difference (C_d), and Shift Distance (M_d), for Benign and Malignant Lesions Using 3-D and 2-D Strain Methods

		Mean		p-Value (Using t-Test)	
Value	*Type*	*3-D*	*2-D*	*3-D*	*2-D*
V_d	Benign	19.43 ± 8.06%	19.08 ± 12.49%	<0.0001	<0.0001
	Malignant	6.62 ± 7.90%	10.07 ± 9.17%		
S_d	Benign	0.289 ± 0.138	0.245 ± 0.161	<0.0001	<0.0001
	Malignant	0.146 ± 0.12	0.145 ± 0.136		
C_d	Benign	32.53 ± 12.01%	25.53 ± 12.73%	<0.0001	<0.0001
	Malignant	13.15 ± 7.98%	15.27 ± 8.21%		
M_d	Benign	3.087 ± 0.425	2.979 ± 0.648	<0.0001	<0.0001
	Malignant	2.352 ± 0.453	2.458 ± 0.512		

Table 7.2 Classification of Breast Tumors by SVM with the Four Strain Features for 3-D and 2-D Strain Methods

	3-D Strain		2-D Strain	
Classification	*Benign*	*Malignant*	*Benign*	*Malignant*
Benign	TN 55	FN 4	TN 45	FN 10
Malignant	FP 5	TP 36	FP 15	TP 30
Total	60	40	60	40

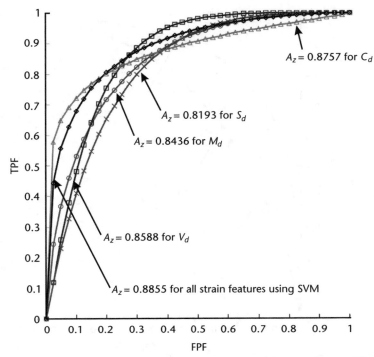

Figure 7.15 The ROC analysis of four strain features (contour difference, shape difference, shift distance, and volume difference) using an SVM classifier in 3-D strain imaging framework.

method. The total ROC analyses for the 3-D and 2-D strain methods are shown in Figure 7.16.

7.4.4 Figure of Merit for 2-D Versus 3-D: Performance Evaluation

To estimate the performances of the 2-D and 3-D strain images, five performance indices are compared by using Chi-square test. These indices are accuracy, sensitivity, specificity, PPV, and NPV and are mathematically given as:

$$\text{Accuracy} = (\text{TP} + \text{TN})/(\text{TP} + \text{TN} + \text{FP} + \text{FN})$$
$$\text{Sensitivity} = \text{TP}/(\text{TP} + \text{FN})$$
$$\text{Specificity} = \text{TN}/(\text{TN} + \text{FP})$$
$$\text{PPV} = \text{TP}/(\text{TP} + \text{FP})$$
$$\text{NPV} = \text{TN}/(\text{TN} + \text{FN})$$

(7.14)

where TP is the number of positives correctly classified as positive; TN is the number of negatives correctly classified as negative; FP is the number of negatives falsely classified as positive; and FN is the number of positives falsely classified as negative. In this experiment shown in Table 7.3, accuracy of the 2-D SVM is 75.00% (75 of 100), the sensitivity is 75.00% (30 of 40), the specificity is 75.00% (45 of 60), the PPV is 66.76% (30 of 45), and the NPV is 81.82% (45 of 55). The accuracy rate of

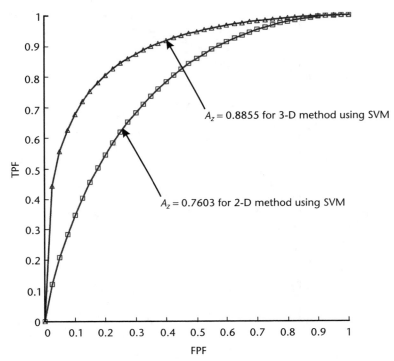

Figure 7.16 Comparison between 2-D and 3-D strain imaging method using ROC analysis. The ROC analyses of the four strain features are computed in SVM classifier frameworks. The *p*-value of the A_z values for 3-D and 2-D methods is 0.0119 by using *Z*-test.

Table 7.3 Results of Performance Indices and the *p*-Value of the Chi-Square Test for the SVM Based on the Proposed Features

Items	3-D Strain (%)	2-D Strain (%)	p-Value (Using Chi-Square Test)
Accuracy	91.00	75.00	0.0026
Sensitivity	90.00	75.00	0.0775
Specificity	91.67	75.00	0.0143
PPV	87.80	66.67	0.0205
NPV	93.22	81.82	0.0638

the 3-D SVM is 91.00% (91 of 100), the sensitivity is 90.00% (36 of 40), the specificity is 91.67% (55 of 60), the PPV is 87.80% (36 of 41), and the NPV is 93.22% (55 of 59). The *p*-values that result when using the Chi-square test are 0.0026, 0.0775, 0.0143, 0.0205, and 0.0638, respectively. We can conclude that the 3-D strain image analysis is better than the 2-D strain image analysis. In the next section, we present some of the reasons for better performance of the 3-D method compared to the 2-D method.

7.5   Discussions

7.5.1   Discussion About the Level Set Framework

In this chapter, segmentation is one of the critical components for extraction of the true irregularities in breast lesions that help with classification. Hence, we have tried several segmentation strategies ranging from deformable models to regional models. Though the level set does optimize in the sense that it has the ability to capture the irregularity of the lesions, in some cases even the level set is powerless. These are misdiagnosis cases; that is, those cases in which a benign case can be considered malignant and vice versa.

These types of misdiagnosis are a result of broken islands or patches of lesions that are not well connected to each other in ultrasound images. Such broken islands can result from the partial volume effect during the ultrasound image acquisition process (discussion of which is beyond the scope of this chapter). Another reason for such broken patches is that sometimes lesions are not concentrated at one location like a mass, but are instead scattered around in multiple masses with a fair number of gaps in between. Here our level set assumes that the lesion is more concentrated as a mass and hence shows a uniform intensity in ultrasound images. Thus, when the level set is made to run under the speed functions governed by the sigmoid enhancer, it does capture it accurately in a majority of the cases, resulting in high sensitivity. But when the level set is made to run in multiple concentrated masses, it fails because of its inability to find the same uniform propagating field while propagating. We anticipate working on this area in the near future, but the current system shows very promising results on central massed regions with large noise patterns.

7.5.2   Discussion About Strain Features

Four strain features were used for classifying the lesions into benign and malignant lesions. These strain features are the major protocols for tumor analysis and classification in this chapter. In general, benign tumors are usually softer than malignant tumors; a soft benign tumor becomes flatter than a stiffer malignant tumor after compression with a probe. This is because benign tumors generally have smooth regular borders and are loosely bound to the surrounding tissues, whereas malignant tumors are usually characterized by firm desmoplastic reactions with surrounding tissue. Therefore, benign tumors have larger values for all strain features than malignant ones, as shown in Table 7.1. Although there is a little overlap in the measured values for 3-D and 2-D between benign and malignant lesions, the p-values (<0.0001) can prove the validity of these statistical data.

In the ROC curve analysis in Figure 7.15, all A_z values for individual strain features are satisfactory. Besides, combining all strain features with the SVM classifier, the A_z value can be improved to approach 0.89 (0.8855). From five performance indices (accuracy, sensitivity, specificity, PPV, and NPV), we also can prove that all strain features with the SVM classifier can make the five performance values more efficacious for clinical diagnosis, as shown in Table 7.2. This is to say that the proposed strain features are useful for our 3-D ultrasound strain imaging framework.

7.5.3   Discussion About 3-D Versus 2-D Strain Methods

Two sets of protocols were used for our experiments with the datasets. One used the 3-D ultrasound acquisition method, and the other used the 2-D datasets acquired using a conventional 2-D method. The 2-D strain imaging method means that only one precompressed image slice and one postcompressed image slice were used for feature analysis. These two slices were selected from the middle slice of the precompressed and postcompressed images that contained tumor information. Because the current system utilizes only one slice in each condition (precompressed or postcompressed), it may lack statistical significance, thereby yielding errors. Therefore, according to the ROC curves analysis in Figure 7.16, the A_z values for all proposed strain features for the 3-D strain method are superior to 2-D method. This A_z value also shows the robustness of the 3-D strain method.

Because the 2-D strain image method is low in spatial content due to only one slice being used, it suffers from a lack of features. Therefore, all performance values in Table 7.3 indicate that the 3-D strain image analysis is better than the 2-D method. To test and verify the validity of this opinion, a statistical tool, the Chi-square test, was used. Hence, the robustness of the 3-D strain image method can be proven once more according to the lower p-values, especially in accuracy and specificity.

7.6   Conclusion

In this chapter, the level set segmentation approach was adopted to extract contour information about lesion. This information was then used for 3-D strain image analyses. According to the experiment results, the 3-D strain image analysis method can provide more robust and accurate results than the 2-D strain analysis method in which only two images are used. Four strain features—volume difference, shape difference, contour difference and shift distance—were proposed to analyze the strain characteristics of the lesions under the pressure of a probe. From the experimental results, the combination of the four 3-D strain features was better than the one of the 2-D strain features.

The SVM experimental results show that the accuracy of our proposed system is 91%, and the sensitivity of our diagnosis system is 90%. That is, if a tumor is malignant, the proposed CAD system can detect it with high probability. This CAD system will offer an efficient and accurate diagnosis for physicians to help reduce instance of misdiagnosis and the requirement for a biopsy. Hence, this system is expected to be a useful CAD tool for physicians trying to discriminate between benign and malignant lesions.

The main problem of dealing with 3-D strain images is how to register the precompressed and postcompressed lesions efficiently and accurately. Hence, other more advanced 3-D alignment methods are worth investigating and using in this system to further improve its performance. Furthermore, reducing the misdiagnosis rate and combining the system with other cancer diagnosis tools to create a more robust CAD system are new challenges that should be addressed in future work.

References

[1] American Cancer Society, *Cancer Facts and Figures 2004*, 2004.

[2] Kolb, T. M., J. Lichy, and J. H. Newhouse, "Comparison of the Performance of Screening Mammography, Physical Examination, and Breast US and Evaluation of Factors That Influence Them: An Analysis of 27,825 Patient Evaluations," *Radiology*, Vol. 225, No. 1, October 2002, pp. 165–175.

[3] Chang, R. F., et al., "Computer-Aided Diagnosis for 2-D/3-D Breast Ultrasound," in *Recent Advances in Breast Imaging, Mammography, and Computer-Aided Diagnosis of Breast Cancer*, J. S. Suri and R. M. Rangayyan, (eds.), Bellingham, WA: SPIE Press, 2006.

[4] Chen, D. R., R. F. Chang, and Y. L. Huang, "Breast Cancer Diagnosis Using Self-Organizing Map for Sonography," *Ultrasound Med. Biol.*, Vol. 26, No. 3, March 2000, pp. 405–411.

[5] Chen, D. R., et al., "Diagnosis of Breast Tumors with Sonographic Texture Analysis Using Wavelet Transform and Neural Networks," *Ultrasound Med. Biol.*, Vol. 28, No. 10, October 2002, pp. 1301–1310.

[6] Hamper, U. M., et al., "Three-Dimensional US: Preliminary Clinical Experience," *Radiology*, Vol. 191, No. 2, May 1994, pp. 397–401.

[7] Tong, S., et al., "Analysis of Linear, Area and Volume Distortion in 3-D Ultrasound Imaging," *Ultrasound Med. Biol.*, Vol. 24, No. 3, March 1998, pp. 355–373.

[8] Fenster, A., and D. B. Downey, "3-D Ultrasound Imaging: A Review," *IEEE Engineering in Medicine and Biology Magazine*, Vol. 15, No. 6, November 1996, pp. 41–51.

[9] Huang, S. F., et al., "Characterization of Spiculation on Ultrasound Lesions," *IEEE Trans. on Med. Imaging*, Vol. 23, No. 1, January 2004, pp. 111–121.

[10] Chen, D. R., et al., "Computer-Aided Diagnosis for 3-Dimensional Breast Ultrasonography," *Arch. Surg.*, Vol. 138, No. 3, March 2003, pp. 296–302.

[11] Chen, W. M., et al., "Breast Cancer Diagnosis Using Three-Dimensional Ultrasound and Pixel Relation Analysis," *Ultrasound Med. Biol.*, Vol. 29, No. 7, July 2003, pp. 1027–1035.

[12] Suri, J. S., and R. M. Rangayyan, *Recent Advances in Breast Imaging, Mammography and Computer Aided Diagnosis of Breast Cancer*, Bellingham, WA: SPIE Press, 2006.

[13] Chang, R. F., et al., "Improvement in Breast Tumor Discrimination by Support Vector Machines and Speckle-Emphasis Texture Analysis," *Ultrasound Med. Biol.*, Vol. 29, No. 5, May 2003, pp. 679–686.

[14] Kuo, W. J., et al., "Computer-Aided Diagnosis of Breast Tumors with Different US Systems," *Acad. Radiol.*, Vol. 9, No. 7, July 2002, pp. 793–799.

[15] Drukker, K., et al., "Computerized Detection and Classification of Cancer on Breast Ultrasound," *Acad. Radiol.*, Vol. 11, No. 5, May 2004, pp. 526–535.

[16] Horsch, K., et al., "Computerized Diagnosis of Breast Lesions on Ultrasound," *Med. Phys.*, Vol. 29, No. 2, February 2002, pp. 157–164.

[17] Drukker, K., et al., "Computerized Lesion Detection on Breast Ultrasound," *Med. Phys.*, Vol. 29, No. 7, July 2002, pp. 1438–1446.

[18] Horsch, K., et al., "Performance of Computer-Aided Diagnosis in the Interpretation of Lesions on Breast Sonography," *Acad. Radiol.*, Vol. 11, No. 3, March 2004, pp. 272–280.

[19] Chen, D. R., et al., "Texture Analysis of Breast Tumors on Sonograms," *Semin. Ultrasound CT MR*, Vol. 21, No. 4, August 2000, pp. 308–316.

[20] Chen, D. R., R. F. Chang, and Y. L. Huang, "Computer-Aided Diagnosis Applied to US of Solid Breast Nodules by Using Neural Networks," *Radiology*, Vol. 213, No. 2, November 1999, pp. 407–412.

[21] Drukker, K., M. L. Giger, and E. B. Mendelson, "Computerized Analysis of Shadowing on Breast Ultrasound for Improved Lesion Detection," *Med. Phys.*, Vol. 30, No. 7, July 2003, pp. 1833–1842.

[22]   Varghese, T., J. Ophir, and T. A. Krouskop, "Nonlinear Stress-Strain Relationships in Tissue and Their Effect on the Contrast-to-Noise Ratio in Elastograms," *Ultrasound Med. Biol.,* Vol. 26, No. 5, June 2000, pp. 839–851.

[23]   Doyley, M. M., et al., "A Freehand Elastographic Imaging Approach for Clinical Breast Imaging: System Development and Performance Evaluation," *Ultrasound Med. Biol.,* Vol. 27, No. 10, October 2001, pp. 1347–1357.

[24]   Garra, B. S., et al., "Elastography of Breast Lesions: Initial Clinical Results," *Radiology,* Vol. 202, No. 1, January 1997, pp. 79–86.

[25]   Ophir, J., et al., "Elastography: A Quantitative Method for Imaging the Elasticity of Biological Tissues," *Ultrason. Imaging,* Vol. 13, No. 2, April 1991, pp. 111–134.

[26]   Cespedes, I., et al., "Elastography: Elasticity Imaging Using Ultrasound with Application to Muscle and Breast *In Vivo,*" *Ultrason. Imaging,* Vol. 15, No. 2, April 1993, pp. 73–88.

[27]   Ophir, J., et al., "Elastography: Ultrasonic Estimation and Imaging of the Elastic Properties of Tissues," *Proc. Inst. Mech. Eng. [H] J. Eng. Med.,* Vol. 213, No. 3, 1999, pp. 203–233.

[28]   Moon, W. K., et al., "Solid Breast Masses: Classification with Computer-Aided Analysis of Continuous US Images Obtained with Probe Compression," *Radiology,* Vol. 236, No. 2, August 2005, pp. 458–464.

[29]   Suri, J. S., D. L. Wilson, and S. N. Laxminarayan, *Handbook of Medical Image Analysis: Segmentation Models,* Vols. I, II, and III, Berlin: Springer, 2005.

[30]   Rakotomamonjy, A., P. Deforge, and P. Marche, "Wavelet-Based Speckle Noise Reduction in Ultrasound B-Scan Images," *Ultrason. Imaging,* Vol. 22, No. 2, April 2000, pp. 73–94.

[31]   Marom, E., S. Kresic-Juric, and L. Bergstein, "Speckle Noise in Bar-Code Scanning Systems—Power Spectral Density and SNR," *Appl. Opt.,* Vol. 42, No. 2, January 2003, pp. 161–174.

[32]   Tang, H., T. Zhuang, and E. X. Wu, "Realizations of Fast 2-D/3-D Image Filtering and Enhancement," *IEEE Trans. on Med. Imaging,* Vol. 20, No. 2, February 2001, pp. 132–140.

[33]   Rapp, C. L., "Sonography of the Breast," *Proceedings of Society of Diagnostic Medical Sonography's 17th Annual Conference,* Dallas, TX, 2003, pp. 57–67.

[34]   Osher, S., and J. Sethian, "Fronts Propagating with Curvature-Dependent Speed: Algorithms Based on Hamilton-Jacobi Equations," *J. Comput. Phys.,* Vol. 79, No. 1, 1988, pp. 12–49.

[35]   Sethian, J. A., *The Level Set Methods and Fast Marching Methods: Evolving Interfaces in Computational Geometry, Fluid Mechanics, Computer Vision, and Materials Science,* 2nd ed., Cambridge, U.K.: Cambridge University Press, 1999.

[36]   Suri, J. S., and S. N. Laxminarayan, *PDE and Level Sets: Algorithmic Approaches to Static and Motion Imagery,* New York: Kluwer Academic/Plenum Publishers, 2002.

[37]   Suri, J. S., et al., "Modeling Segmentation via Geometric Deformable Regularizers, PDE and Level Sets in Still/Motion Imagery: A Revisit," *Int. J. Image and Graphics,* Vol. 1, No. 4, December 2001, pp. 681–734.

[38]   Suri, J. S., et al., "Shape Recovery Algorithms Using Level Sets in 2-D/3-D Medical Imagery: A State-of-the-Art Review," *IEEE Trans. on Inf. Technol. Biomed.,* Vol. 6, No. 1, March 2002, pp. 8–28.

[39]   Malladi, R., J. A. Sethian, and B. C. Vemuri, "Shape Modeling with Front Propagation: A Level Set Approach," *IEEE Trans. on Pattern Analysis and Machine Intelligence,* Vol. 17, No. 2, February 1995, pp. 158–175.

[40]   Suri, J. S., "Leaking Prevention in Fast Level Sets Using Fuzzy Models: An Application in MR Brain," *Proc. IEEE–EMBS Int. Conf. Information Technology Applications in Biomedicine,* 2000, pp. 220–225.

[41]   Richardson, I. E. G., *H. 264 and MPEG-4 Video Compression: Video Coding for Next-Generation Multimedia,* New York: John Wiley & Sons, 2003.

[42] Pontil, M., and A. Verri, "Support Vector Machines for 3-D Object Recognition," *IEEE Trans. on Patt. Anal. Mach. Intell.*, Vol. 20, No. 6, June 1998, pp. 637–646.

[43] Chapelle, O., P. Haffner, and V. N. Vapnik, "Support Vector Machines for Histogram-Based Image Classification," *IEEE Trans. on Neur. Networks*, Vol. 10, No. 5, September 1999, pp. 1055–1064.

[44] Vapnik, V. N., *The Nature of Statistical Learning Theory*, 2nd ed., New York: Springer-Verlag, 1999.

[45] Chang, R. F., et al., "Support Vector Machines for Diagnosis of Breast Tumors on US Images," *Acad. Radiol.*, Vol. 10, No. 2, February 2003, pp. 189–197.

[46] Besthorn, C., et al., "Discrimination of Alzheimer's Disease and Normal Aging by EEG Data," *Electroencephalogr. Clin. Neurophysiol.*, Vol. 103, No. 2, August 1997, pp. 241–248.

[47] Cichocki, A., et al., "EEG Filtering Based on Blind Source Separation (BSS) for Early Detection of Alzheimer's Disease," *Clin. Neurophysiol.*, Vol. 116, No. 3, March 2005, pp. 729–737.

CHAPTER 8

Automatic Classification of Breast Lesions in 3-D Ultrasound Images

Paulo S. Rodrigues, Gilson A. Giraldi, Marcia Provenzano, Marcelo D. Faria, Ruey-Feng Chang, and Jasjit S. Suri

8.1 Introduction

Per the American Cancer Society [1], breast cancer ranks second in the list of women's cancers. Even though the rate of breast cancer has risen since 1980, the mortality rates have declined since 1990. The reduction in mortality rate is due to early detection and improvements in treatment.

With the advances in ultrasound acquisition techniques [2] and the digital nature of 3-D volume acquisition, computer-aided diagnosis (CAD) has become even more powerful, thereby improving the workflow of breast imaging and reducing the physician's workload (see [3]). The process of breast lesion classification using efficient CAD-based methods with ultrasound is generally categorized into several techniques: Bootstrap [4], SVM [5], and neural networks [6, 7]. Along these lines, Sawaki et al. [8] proposed a CAD system that uses fuzzy inference for breast sonography, and they adopted six different criteria to classify lesions: lesion shape, border, edge shadows, internal echoes, posterior echoes, and halo. However, their system accuracy, sensitivity, and specificity were only 60.3%, 82.1%, and 42.9%, respectively.

However, the texture framework in ultrasound image processing is also an important tool for extracting interesting lesion features. Approaches underlining texture go back more than two decades. In this respect, Garra et al. [9] analyzed breast lesions in ultrasound images using a co-occurrence matrix to represent their textures. Chen et al. [10] proposed an autocorrelation coefficient to analyze the texture information. Because their implementations were presented on a small set of data, there is a fundamental weakness with texture-based strategies. The settings for ultrasound machine parameters have to be fixed when acquiring ultrasound images. If the ultrasound parameter settings changed, the CAD performance became very unstable [11]. Moreover, a CAD system trained by images from one ultrasound machine needs to be trained again for a different ultrasound machine due to different image resolution and image quality. Hence, Chen et al. [12] proposed nearly setting-independent features based on shape information. Their system was very robust and powerful because the statistical data using ROC curves were all greater than 0.95.

Note that all of these strategies were used on 2-D ultrasound images, which has limitations. For example, it did not allow 3-D spatial processing during the ultrasound data acquisition process. As a result, the shape and structure information of the breast lesions could not be reconstructed, and consequently it may be not possible to determine the growth of the cancer and its spatial relationships. Recently, 3-D ultrasound [13] has shown promising signs that overcome the limitations of traditional 2-D ultrasound, allowing the anatomy to be viewed interactively in three dimensions. Chen et al. [14] proposed a 3-D ultrasound texture features autocorrelation matrix and run difference matrix. However, as pointed out earlier, texture-based CAD is highly machine dependent and utilizes these settings all of the time.

The main reasons for investigating automatic analysis are as follows. The automatic diagnosis can assist new physicians in lesion investigation, and it can be used for training students in the area of medicine education because the results can be easily compared (manual with automatic ones). In addition, diagnostics can be carried out in a relatively short time and with better precision. Finally, an automatic analysis at a local computer or workstation generally is the first step before the development of parallel or grid computing systems so that, in the future, it will seriously diminish the computational overhead, allowing for more reliable and faster analyses.

A totally automatic process, however, is still not a reality; so much work remains, mainly in the early steps, which may involve segmentation, recognition, and extraction of lesions from the background. Specifically speaking, in the case of ultrasound images, the initial segmentation is a complex process because the images have a low SNR, low resolution, low contrast, and several small and spurious regions. Several other challenges in all steps remain. In particular, the initial segmentation is a fundamental task because the success of the remaining steps depends on it.

Despite the number and size of the regions inside the ultrasound images, in most cases the unique requirement is to extract the region of interest from its background. Then, algorithms to binarize images may be sufficient. However, the lesion's region is difficult to define in terms of its boundary location and texture pattern. Generally, the boundary represents a third (narrow) region that separates the inside from the outside of the lesion's region. This narrow boundary is generally composed of dead cells that later move to the inside of the lesion, where, due to pressure instability, they may be transformed to liquid material. These watery environments, more related to benign than malignant lesions, tend to generate anechoic regions due to liquid concentration. The contrary occurs with malignant regions. Then, the anechoicity is one of the main factors that influences the lesion analysis. Other important features are the presence of protuberances, circularity, homogeneity, and lesion area.

Algorithms for image segmentation based only on binarization are not completely sensitive to a lesion's pattern behavior and frequently fail to trace adequately the lesion boundary. On the other hand, traditional algorithms such as fuzzy c-means (FCM), k-means (KM), self-organized maps (SOMs), and Bootstrap can also be used and generally give good results because they can achieve several clusters adequately. Nonetheless, they have a high computational time because they use implementations with some kind of energy minimization, usually in an iterative loop. Other algorithms, such as those based on the watershed approach, may also generate several regions, but their disadvantage is the oversegmentation. Because algorithms based on thresholding do not have such problems, they may be a good alternative. However, the choice of the threshold is not always an easy task, because thresholds are generally based on some kind of contrast enhancement between the background and the foreground. In the narrow boundary around the lesion, the threshold is not easy to compute, so these algorithms usually do not provide a good boundary location.

Regarding the fact that patterns of ultrasound images are probability distributions of gray scales, the entropy associated with these distributions may be measured in order to maximize the differences in interclusters composed by the foreground and background pixels. In this respect, the algorithms of Pun [15], Kapur et al. [16], Abutaleb et al. [17], Li and Lee [18], Pal [19], and Sahoo et al. [20] can be used to achieve the desired threshold. The disadvantage of these methods, as mentioned, is the image binarization, which may be insufficient for US breast images in terms of expected segmentation, when a narrow boundary is important.

To overcome this problem, Albuquerque et al. [21] presented an algorithm for ultrasound image segmentation based on a threshold computed using entropy information. Their proposal considers the interpixel relationships as having log-time and log-space interactions. These interpixel features yield a new kind of entropy, called nonextensive entropy because it is a generalization of the traditional Boltzman-Gibbs (BG) entropy for extensive systems. Their work suggests good results for US breast images: It achieved better thresholds than the traditional BG entropy. However, the image binarization also is a limitation due to the issues mentioned earlier.

Our idea, as explained in this chapter, is as follows. (1) We propose a natural extension to the algorithm proposed in [21] which recursively achieves more than two regions in ultrasound images and also has a low computational time. (2) The proposed algorithm, called the Nonextensive Segmentation Recursive Algorithm (NESRA), is the first step of a five-step methodology for a CAD system for breast lesion classification. (3) This methodology uses an SVM framework to classify the breast regions as malignant or benign. This classification is carried out by taking five lesion features into account: lesion area, circularity, protuberance, heterogeneity, and anechoicity. These features are combined to achieve the best performance in terms of the system's accuracy, sensitivity, specificity, PPV, NPV, and ROC curves. The results shows better performance than the current literature documents.

In the next section we treat related works. Section 8.3 points out some theoretical background for further discussion. Section 8.4 presents our original approach, which is discussed in Section 8.5.

8.2 Related Works

Several US techniques are used to classify breast lesions, including Bootstrap [4], SVM [5], and neural networks [6, 7, 10]. These classifiers, however, are dependent on the features that are defined by radiologists. Such characteristics are those dependent on the tumor profile, say, the geometric features or texture and color patterns, as well as the features surrounding the lesion. These features, together with the classifiers, determine the efficiency of the CAD software.

Sawaki et al. [8] proposed a CAD system that uses fuzzy inference for breast sonography and adopted six different criteria to classify lesions: lesion shape, border, edge shadows, internal echoes, posterior echoes, and halo. However, their system accuracy, sensitivity, and specificity were only 60.3%, 82.1%, and 42.9%, respectively.

The use of a texture framework in US image processing goes back more than two decades. Garra et al. [9] analyzed the breast lesions in ultrasound images using a co-occurrence matrix of ultrasound images to represent their texture information. Chen et al. [10] proposed an autocorrelation coefficient to analyze the texture information. Because their implementations were presented on a smaller set of data, there is a fundamental weakness with texture-based strategies: The settings for the US machine parameters have to be fixed to acquire ultrasound images. If the ultrasound parameter settings were changed, CAD performance became very unstable [6]. Moreover, a CAD system trained with images from one ultrasound machine needs to be trained again for a different ultrasound machine because each machine has different image resolutions and image quality. Hence, Chen et al. [6] proposed nearly setting-independent features based on shape information. Their system was very robust and powerful because the statistical data using ROC curves were all greater than 0.95. Chang et al. [22] also used six shape features in their analysis: factor, roundness, aspect ratio, convexity, solidity, and extent. Note that all of the preceding strategies were used on 2-D US images. As a result, the shape and structure information of the breast lesions could not be reconstructed; consequently, it may not be possible to determine the growth of the cancer and its spatial relationships. Recently, 3-D US [13] has shown promising signs of overcoming the limitations of traditional 2-D US to allow physicians to view the anatomy in three dimensions interactively, instead of assembling the sectional images in their minds.

In the following, we discuss research that uses an autocorrelation matrix to extract texture feature and neural networks as classifiers.

8.2.1 Breast Cancer Diagnosis with Autocorrelation Matrix

Because the surface features and internal structure of a tumor are not easily demonstrated simultaneously using traditional 2-D US, Chen et al. [12, 14, 23] have investigated the use of texture features with 3-D US devices. They proposed use of an autocorrelation matrix that is running difference matrixes. They have studied pixel relation analysis techniques for use with 3-D US breast images and compares its performance to 2-D versions of images. The rectangular subimages of the volume of interest (VOI) were manually selected, and the selected VOIs were outlined to include the entire extent of the tumor margin. They found that using features from just a few slices were enough to provide good diagnostic results if the adopted features were modified from the 2-D features.

Because the speckle noise artifacts and low-intensity regions, generally from blood vessels, are present, it is not an easy task to extract the features from the US images by means of a common image processing technique [24]. One general feature of breast tumors is that the pixels in the tumor have a lower level intensity than the surrounding normal tissue [25]. Chen et al. [10] presented work that suggest thats the 2-D normalized autocorrelation coefficients are suitable for reflecting the interpixel relationships within an image and make it possible to differentiate between benign and malignant breast masses. Different tissues have significantly varying pixel relations in US images.

Classically, benign tumors are described as regular masses with homogeneous internal echoes, but carcinomas are described as having fuzzy borders with heterogeneous internal echoes. The correlation between neighboring pixels within the 2-D images is a patent feature of the tumor. This study is based on pixel relation analysis techniques developed by Chen et al. [10, 26, 27]. They have used normalized autocorrelation coefficients to reflect the interpixel correlation within a region. To generate similar autocorrelation characteristics for tumor regions under different brightnesses but with a similar pixel relation, the autocorrelation coefficients are further modified into mean-removed autocovariance coefficients. In their work, these autocovariance coefficients for each breast tumor image are found and taken as interpixel relation features in order to distinguish the differences between benign and malignant tumors.

Generally, the pixel per centimeter rates of each 3-D US image are not all the same and are decided by the magnifier control. In practice, the range of pixel rates in their database was between 20.35 and 110.56 pixels/cm. For an US image device, the autocorrelation computation will give a 5×5 autocorrelation matrix. Because the large variations in the resolution will affect and reduce the diagnostic result, the distances in the autocorrelation matrix should not be fixed for all cases and should be changed according to their pixel rates.

With 3-D US, the scanned slices are reconstructed into a single 3-D volume dataset. The original 2-D autocorrelation coefficients do not accord with the 3-D volume dataset. To overcome this problem, Chen et al. [14] proposed three new feature extraction methods to reduce the dimension of 3-D volume data from three to two to fit the 2-D extraction function and to attempt to compare the diagnostic results of these methods.

The first method is called the accumulative autocorrelation scheme (ACC-US). In this scheme, each 2-D image is first extracted from a 3-D volume dataset, and then the pixel relation coefficients are computed for each 2-D image. The autocorrelation matrix for the 3-D dataset will be accumulated into a new autocorrelation matrix. The accumulative matrixes for all of the 3-D training datasets are then fed into a neural network to compete the training process; then, the trained neural network could be used for breast tumor diagnosis.

There are two main stages. In the first segmentation stage, the extraction of the VOI is carried out and the autocorrelation matrix is computed. The procedure of VOI extraction is carried out in two steps. First, the physician must choose three key planes in the 3-D volume dataset, called the first, middle, and last planes. The first and last planes are the first and last slices that contain the tumor. The middle plane is the slice where the area of the tumor is the biggest. Second, the physician defines three ROIs in these three planes. This system then uses a linear interpolation method to select the ROIs of the other planes in the 3-D volume by relying on the three key ROIs and, then, all of the ROIs are combined to form a VOI. After the

VOI is extracted, the autocorrelation matrix may be computed and used as a characteristic vector. In the second stage, the characteristic vectors of the training tumors will be fed into the input layer of the neural network and the learning process starts. Also, a relatively simple feed-forward, error back-propagation artificial neural network (ANN) with one hidden layer was adopted. The output of each ANN will be a number between 0 and 1.

The coefficient matrix of each 2-D image within a 3-D US dataset will be calculated and added to an autocorrelation coefficient matrix. The pixel relation characteristic of tissues may be lost by such an accumulative operation and result in a false diagnosis. To improve the diagnostic accuracy, another pixel relation feature extraction scheme was proposed by Chen et al. [14]. This scheme of multiple autocorrelation matrixes (MULTI-US) explores more than one autocorrelation matrix for keeping the coefficients of each 2-D image in the 3-D volume. Because the number of frames in a 3-D volume dataset is large, it is impossible to use all of the coefficient matrixes. Hence, only five matrixes were used to reduce the computation time. Also, the diagnostic model needs to be modified to fit multiple coefficient matrixes.

In the first stage, the VOI extraction method is just like the previous scheme; however, the autocorrelation matrix is replaced by using five autocorrelation matrixes. Next, these five autocorrelation matrixes are fed into the neural network. In the second stage, the diagnosis method is modified. Each matrix is fed into the neural network and five neural network outputs will be obtained. These diagnosis outputs need to be combined into a final diagnostic output. In their study, if more than two diagnostic outputs were malignant, then the final diagnosis result was malignant.

The third scheme is a combination of the preceding two schemes. That is, the five matrixes of the second scheme are added into an accumulative matrix. The method can not only reduce the computation time of the first scheme, but can also avoid the problem of too many ROIs influencing the feature extraction result, because some ROIs may contain incomplete pixel relation information on the tumor. The hybrid scheme is called an autocorrelation matrix with five frames (ACC5-US).

Including the predefined threshold, the multilayer feed-forward neural network in their work [14] included 25 input nodes, 10 hidden nodes, and a single output node. The degree of freedom (i.e., the number of connections) was 260. The training process was stopped when the improvement of error distortion was smaller than 0.001 or when the number of training iterations was more than 20,000. The error is computed by the absolute difference between the desired output and the actual output of the neural network. Due to the small number of test cases, the cross-validation method was adopted. Because each set was used as the test set once, five sets of parameters for neural network were used.

The accuracy of this proposed ACC5-US scheme was 90.68% (146 of 161), the sensitivity 90.74% (49 of 54), the specificity 90.65% (97 of 107), the PPV 83.05% (49 of 59), and the NPV 95.1% (97 of 102).

The spiculation feature is used primarily to diagnose the breast tumors using 3-D US. In a study by Rotten et al. [28], the coronal section in a 3-D US image allows precise demonstration of the tissue surrounding the central lesion. A converging pat-

tern of the peripheral tissue is highly suspicious of malignancy. The converging pattern is a spiculation in which alternating hypoechoic and hyperechoic lines radiate
in multiple directions from the mass into the surrounding tissue. This feature is easily seen on mammograms, and malignancy cannot be excluded based on this finding. In their experiments, 91% (53 of 58) of malignant tumors and 6% (8 of 128) of
benign tumors have spiculations. Although several successful spiculation detection
techniques are available for mammograms [29–32], automatic spiculation detection on US images is not easily implemented due to the speckle noise in US images.
Moreover, a mature segmentation algorithm is needed to segment the tumor. Therefore, Chen et al. [14] provide another way to use 3-D US. Because the multilayer
feed-forward neural network can extract higher-order statistics by adding one or
more hidden layers in their model, it has become extremely popular in terms of classification and prediction. Hence, the multilayer feed-forward neural network was
adopted for classification. The computation time is rather critical for clinical applications. This proposed algorithm uses only the middle five consecutive frames in a
3-D US dataset to quickly obtain the pixel relation information. That is, their
proposed method can be used for real-time clinical applications.

Although all of the pixel relation information of a tumor was used in the first
scheme using an autocorrelation matrix, the first scheme's performance is worse
than that of the third scheme, in which only the pixel relation information of five
consecutive frames was adopted. This does not mean, however, that the whole pixel
relation information of a tumor is useless.

8.2.2   Solid Breast Nodules with Neural Networks

Chen et al. [27] have developed a CAD algorithm with setting-independent features
and ANNs to differentiate benign from malignant breast lesions. In their work, two
sets of breast sonograms were evaluated. The first set, containing 160 lesions, was
stored directly on the magnetic optic disks from the US system. Four different
boundaries were delineated by four persons for each lesion in the first set. The second set comprised 111 lesions that were extracted from the hardcopy images. Seven
morphologic features were used, five of which were newly developed. A multilayer
feed-forward neural network was used as the classifier, and the reliability,
extendibility, and robustness of the proposed CAD algorithm were evaluated. The
results of the proposed algorithm were compared with those from two previous
CAD algorithms. All performance comparisons were based on paired-samples
t-tests. The proposed CAD algorithm could effectively and reliably differentiate
between benign and malignant lesions. The proposed morphologic features were
nearly setting independent and could tolerate reasonable variation in boundary
delineation.

Chen et al.'s proposed CAD algorithm was composed of three essential components: feature extraction, feature selection, and classification. To relax the constraints on the system settings, the morphologic features rather than the regional
features were adopted. The potential dependence of the morphologic features on
the contour extraction process was minimized by capturing important topologic
properties of the lesions, which may not vary drastically with the delineated contour. Feature selection was necessary to alleviate dimensionality [33]. A set of essen-

tial morphologic features that yields the curve of the best performance was selected on the basis of stepwise logistic regression [34]. Classification was accomplished with a multilayer feed-forward neural network (MFNN) [35] on the basis of the essential morphologic features. The advantage of the MFNN is that arbitrarily complex convex separation surfaces can be approximated.

Seven morphologic features were extracted from each lesion to account for such sonographic features as shape, contour, and size. Five of these morphologic features were developed by including the number of substantial protuberances and depressions (NSPD), lobulation index (LI), elliptic-normalized circumference (ENC), elliptic-normalized skeleton (ENS), and long axis–to–short axis (L:S) ratio. The other two features were clinically useful indicators [36]: depth-to-width (D:W) ratio and size of the lesion.

- *NSPD:* The spiculation [37] and irregular shape and contour [38] of a lesion are two important sonographic features that characterize a malignant breast lesion. The NSPD is an effective descriptor in a lesion to quantify these two sonographic features. Consider this geographic analogy: A protuberance and a depression are like a peninsula and a bay, respectively. Because protuberances and depressions may easily result from a wobbly delineation process, only the substantial protuberances and depressions defined by the representative convex and concave points, respectively, were used to characterize a breast lesion.

- *LI:* This feature was devised to characterize the size distribution of the lobes in a lesion. A lobe is defined as the gray region enclosed by the lesion contour and the dashed line connected by two adjacent representative concave points. The size of the lobe is the area of the gray region. The LI can correctly characterize a benign lesion with multiple large lobes of similar sizes. This type of benign lesion may be misclassified as a malignant lesion with the NSPD.

- *ENC:* Anfractuosity is a common morphologic characteristic of malignant lesion boundaries that provides at least two visually appreciable geometric features. One feature is the multiple protuberances and depressions that may be well described with the NSPD. The other feature is the lengthened circumference due to the circuitous boundaries that define the protuberances and depressions. Because the boundary of a smaller lesion would appear to be more winding than that of a larger lesion with the same circumference, the circumference itself is not a good descriptor with which to characterize the anfractuosity of the lesion boundary. Alternatively, a more reasonable approach is quantification of the anfractuosity with the percentage of circumference increment relative to a lesion-dependent baseline. An ideal baseline would be a smooth curve such that the lesion boundary would look like oscillating around the curve. To quantify the anfractuosity of a lesion contour, the circumference ratio of the lesion and its equivalent ellipse is proposed, which is termed the ENC. The equivalent ellipse of a lesion [39, 40] is an ellipse with the same area and center of mass as those of the lesion when the interiors of the lesion and its equivalent ellipse are both set to the same constant gray level. Perceptually, one can see that the equivalent ellipse roughly captures the shape of the lesion, and the lesion boundary.

- *ENS:* A skeleton is an effective representation of a region [40] that is used frequently in such areas as computer vision and pattern recognition. The skeleton is sensitive to the anfractuous property of the lesion boundary. The more protuberances and depressions contained in the lesion boundary, the more complex the skeleton is. Therefore, it seems to be reasonable to quantify the shape complexity by the number of points in the skeleton. Nevertheless, the number of skeleton points is also a function of the lesion size. Just as for the ENC, to eliminate the size effect, it is suggested that the number of skeleton points be normalized by the circumference of the equivalent ellipse of the lesion, which gives the ENS.

In addition to the NSPD, LI, ENC, and ENS descriptors, which capture the contour and shape characteristics, three more mathematic features are considered to incorporate two clinically useful indicators. The first feature is the D:W ratio of the lesion. The depth and the width of a lesion are the horizontal and vertical edge lengths, respectively, of the minimal circumscribed rectangle of the lesion. The larger the D:W ratio, the more likely the lesion is malignant.

Because the D:W ratio may vary with the scanning angle and the compressing pressure, it is suggested that another quantity be used to describe the shape of the lesion, namely, the L:S ratio. The L:S ratio is the length ratio of the major (long) axis to the minor (short) axis of the equivalent ellipse of the lesion. Clearly, the L:S ratio is independent of the scanning angle but may be affected by the compressing pressure. The last feature is the size of the lesion (i.e., the area within the lesion boundary). Clinically, the larger the breast lesion, the more likely the lesion is malignant.

Feature selection was performed in two stages for each training dataset. In the first stage, the best NSPD value was selected from five candidate NSPD values. Then, in the second stage, the selected NSPD value along with the other six features were used to select the essential features that yield the best classification accuracy for the underlying training dataset. The classifier used in the present study is an MFNN. Once the essential features were selected by means of the logistic discrimination function for a set of training data, the training data were used to train the MFNN to divide the training data into benign and malignant categories. The MFNN used in this study was a two-layer feed-forward neural network with one hidden layer. The number of inputs for the MFNN was set to be the same as the number of essential features, and the number of neurons in the output layer was set to 1 for the underlying two-class classification. The number of neurons in the hidden layer was determined through exhaustive experiments to be 2 to 10 neurons. As a result, the number of neurons in the hidden layer was set to 2 because this number gave the best performance for almost all cases evaluated.

The training algorithm used to train the MFNN was the widely used error back-propagation training algorithm [35]. The training data with the essential features were fed into the MFNN in a cyclic manner. For each datum, the estimated output was computed on the basis of the synaptic weights determined in the previous iterations and the sigmoidal activation function. The discrepancy between the desired output and the estimated output was back propagated to modify the synaptic weights until the discrepancy was within the acceptable range. Because the final output was a number between 0 and 1, a threshold neural network (TNN) was required to assign the datum to the benign or the malignant category. In that work, the TNN was determined on the basis of the value that resulted in a dichotomization with the best classification accuracy for the training data, or TNN varied from 0 to 1 to generate the ROC curve.

The area under ROC curve (A_z) was 0.952 ± 0.014 for the first set, 0.982 ± 0.004 for the first set as the training set and the second set as the prediction set, 0.954 ± 0.016 for the second set as the training set and the first set as the prediction set, and 0.950 ± 0.005 for all 271 lesions. At the 5% significance level, the performance of the proposed CAD algorithm was shown to be extendible from one set of US images to the other set and robust for both small and large sample sizes. Moreover, the proposed CAD algorithm was shown to outperform the two previous CAD algorithms in terms of the A_z value.

However, as pointed out before, the texture-based CAD is highly machine dependent and utilizes these machine settings all the time.

As outlined earlier, this chapter proposes a five-step novel and automatic methodology for breast lesion classification in 3-D US images.

8.3 Theoretical Background

8.3.1 Tsallis Entropy

Entropy is an idea born in the heart of classic thermodynamics, not as something fundamentally intuitive, but as some fundamentally quantitative. It was defined by an equation and has been known as Boltzman-Gibbs (BG) entropy. Later, Shannon redefined the concept of BG entropy (now called BGS entropy) as an uncertainty measure associated with the content of system information. This traditional form of entropy is well known by:

$$S = -\sum_i p_i \ln(p_i) \tag{8.1}$$

Generically speaking, systems having statistics of the BGS type are called extensive systems and have an additive property, defined as follows. Let P and Q be two random variables, with probability densities functions $P = (p_1, ..., p_n)$ and $Q = (q_1,..., q_m)$, respectively, and S be the entropy associated with P or Q. If P and Q are independent, under the context of the probability theory, the entropy of the composed distribution[1] verifies the so-called additivity rule:

$$S(P*Q) = S(P) + S(Q) \tag{8.2}$$

This traditional form of entropy is well known and for years has achieved relative success in explaining several phenomena if the effective microscopic interactions are short ranged (i.e., close spatial connections) and the effective spatial microscopic memory is short-ranged (i.e., close time connections) and the boundary conditions are non(multi)fractal. Roughly speaking, the standard formalisms are applicable whenever (and probably only whenever) the relevant space–time is

---

1. We define the composed distribution, also called direct product of $P = (p_1, ..., p_n)$ and $Q = (q_1,..., q_m)$, as $P * Q = \{p_i q_j\}_{i,j}$, with $1 \le i \le n$ and $1 \le j \le m$.

non(multi)fractal. If this is not the case, some kind of extension appears to become necessary. We can make a complete analogy with Newtonian mechanics, when it becomes only an approximation (an increasingly bad one) when the involved velocities approach that of light or the masses are as small like, for instance, the electron mass; the standard statistical mechanics do not apply when the above requirements [short-range microscopic interactions, short-ranged microscopic memory and (multi)fractal boundary conditions] are not the case.

Recent developments, based on the concept of nonextensive entropy, also called Tsallis entropy, have generated new interest in the study of Shannon entropy for information theory [41, 42]. Tsallis entropy (or q-entropy) is a new proposal for the generalization of BG traditional entropy applied to nonextensive physical systems.

The nonextensive characteristic of Tsallis entropy has been applied through the inclusion of a parameter q, which generates several mathematical properties. The general equation is as follows:

$$S_q(p_1,\ldots,p_k) = \frac{1 - \sum_{i=1}^{k} p_i^q}{q-1} \qquad (8.3)$$

where k is the total number of possibilities of the whole system, and the real number q is the entropic index that characterizes the degree of nonextensiveness. In the limit $q \to 1$, (8.3) meets the traditional BGS entropy defined by (8.1). These characteristics give to q-entropy flexibility in explaining several physical systems. In addition, this new kind of entropy does not fail to explain traditional physical systems since it is a generalization.

Furthermore, a generalization of some theory may suppose the violation of one of its postulates. In the case of the generalized entropy proposed by Tsallis, the additive property described by (8.2) is violated in the form of (8.4), which applies if the system has a nonextensive characteristic. In this case, the Tsallis statistic is useful and the q-additivity better describes the composed system. In our case, the experimental results (Section 8.5) show that it is better to consider our systems as having nonextensive behavior:

$$S_q(P*Q) = S_q(P) + S_q(Q) + (1-q)S_q(P)S_q(Q) \qquad (8.4)$$

In this equation, the term $(1-q)$ stands for the degree of nonextensiveness.

Considering $S_q \geq 0$ in the pseudoadditive formalism of (8.4), the following classification for entropic systems is defined:

- Subextensive entropy ($q > 1$):

$$S_q(P*Q) > S_q(P) + S_q(Q)$$

- Extensive entropy ($q = 1$):

$$S_q(P*Q) = S_q(P) + S_q(Q)$$

- Superextensive entropy ($q < 1$):

$$S_q(P*Q) < S_q(P) + S_q(Q)$$

Taking into account the similarities between the formalisms of Shannon and BG entropy, it is interesting to investigate the possibility of the generalization of Shannon entropy to the case of information theory, as has been shown by Yamano [43]. This generalization may be extended to image segmentation tasks by applying Tsallis entropy, which has nonadditive information content.

Albuquerque et al. [21] proposed an algorithm that uses the concept of q-entropy to segment US images. Because this concept may be naturally applied over any statistical distribution, in this chapter we propose a natural extension of the algorithm proposed by Albuquerque et al. [21]. Our proposal is a recursive procedure of [21] in that, for each distribution P and Q, we applied the concept of q-entropy. We named our extended algorithm NESRA and also proposed to apply it in an initial segmentation of our proposed five-step methodology for CAD systems.

The motivations to use the q-entropy are: (1) managing only a simple parameter q yields a more controllable system; (2) as suggested in [21], mammographic images and possibly several other natural images have a nonextensive profile; and (3) it is simple and makes the implementation easy because it has a low computational overload.

In the following section, we fully describe the NESRA proposal.

8.3.2 The Nonextensive Segmentation Approaches

Applying the concept of entropy in order to segment a digital image is a common practice since Pun [15] showed that by maximizing the foreground and background Shannon entropy of a gray-level image good results were generated concerning the prominent regions. Then, other works following the same line were proposed. For example, Kapur et al. [16] maximized an upper bound of the total a posteriori entropy in order to obtain the threshold level. Abutaleb [17] extended the method using two-dimensional entropies. Li and Lee [18] and Pal [19] used the directed divergence of Kullback for the selection of the threshold, and Sahoo et al. [20] used the Reiny entropy model for image thresholding.

In 2004, Albuquerque et al. [21] presented the concept of nonextensive entropy applied to mmamographic gray-scale images. They assume a probability distribution, one for background and the other for foreground, and take the threshold that maximizes the nonadditivity characteristic given by (8.4). However, as with several methods designed to produce binary images, this approach does not work for multiregion segmentation. So we proposed an extension of the method presented in [21] by applying recursively the maximization of (8.4) over the background and the foreground in order to achieve multiple regions of homogenous gray-level or color distribution.

It can be argued that any parametric or nonparametric method with the objective of finding an ideal threshold to produce a binary image can be recursively used in the foreground and in the background in order to achieve multiple regions. However, our approach outperforms several well-known methods under the same conditions. In this section, we formalize the NESRA algorithm. First, we review the nonextensive procedure for image segmentation proposed in [21].

8.3.2.1   Nonextensive Segmentation Algorithm for Image Binarization

Suppose we have an image with k gray levels. Let the probability distribution of these levels be $P = \{p_i = p_1; p_2; \ldots ; p_k\}$. Then, we consider two probability distribution from P, one for the foreground (P_A) and another for the background (P_B). We can make a partition at luminance level t between the pixels from P into A and B. To maintain the constraints $0 \leq P_A \leq 1$ and $0 \leq P_B \leq 1$ we must to renormalize both distribution as:

$$P_A : \frac{p_1}{p_A}, \frac{p_2}{p_A}, \ldots, \frac{p_t}{p_A} \tag{8.5}$$

$$P_B : \frac{p_{t+1}}{p_B}, \frac{p_{t+2}}{p_B}, \ldots, \frac{p_k}{p_B} \tag{8.6}$$

where $p_A = \sum_{i=1}^{t} p_i$ and $p_B = \sum_{i=t+1}^{k} p_i$.

Now, following (8.3), we calculate the a priori Tsallis entropy for each distribution:

$$S_A = \frac{1 - \sum_{i=1}^{t} \left(\frac{p_i}{p_A}\right)^q}{q - 1} \tag{8.7}$$

$$S_B = \frac{1 - \sum_{i=t+1}^{k} \left(\frac{p_i}{p_B}\right)^q}{q - 1} \tag{8.8}$$

Observe that the Tsallis entropy represented by (8.3), (8.7), and (8.8) depends directly on the parameter t for the foreground and background, and it is formulated as the sum of each entropy, allowing the pseudoadditive property, given by (8.4), for statistically independent systems, defined by the following equation:

$$S_{A+B}(t) = \frac{1 - \sum_{i=1}^{t} \left(\frac{p_i}{p_A}\right)^q}{q - 1} + \frac{1 - \sum_{i=t+1}^{k} \left(\frac{p_i}{p_B}\right)^q}{q - 1}$$

$$+ (1 - q) \frac{1 - \sum_{i=1}^{t} \left(\frac{p_i}{p_A}\right)^q}{q - 1} \frac{1 - \sum_{i=t+1}^{k} \left(\frac{p_i}{p_B}\right)^q}{q - 1} \tag{8.9}$$

To accomplish the segmentation task, in [21] the information measure between the two classes (foreground and background) is maximized. In this case, the luminance level t is considered to be the optimum threshold value (t_{opt}), which can be achieved with a cheap computational effort of

$$t_{opt} = \arg \max \left[S_A(t) + S_B(t) + (1 - q) S_A(t) S_B(t) \right] \tag{8.10}$$

Note that the value of t that maximizes (8.10) depends primarily on the parameter q. This is an advantage due to its simplicity. Furthermore, the value of $q = 0.0001$, which generates t_{opt}, is not explicitly calculated and, in this work, it was empirically defined as that which generated the best CAD system performance. This value, $q < 1$, then suggests a system nonextensivity. An example of application of $q \neq 1$ (nonextensive system) against $q = 1.0$ (traditional extensive system) is shown in Figure 8.1.

Among the results presented in Figure 8.1, only parts (c) and (d) seem to be useful in a postprocessing module. With part (c) we can extract the central region and used its boundary curve to initialize some deformable model framework (level set, snake, dual snake, and so forth) in order to better determine the tumor's boundary location. The same can be thinking about part (d), but in this case we can initialize the deformable model from inside to outside, such as in [44, 45]. However, both initializations do not guarantee that the narrow boundary information (as explained in Section 8.1) will be included in the tumor region. Then, one more level of segmentation is needed, yielding to a multisegmentation process instead of the simple binarization. For this reason we create an extension of the extensive algorithm that is able to obtain the same result as in Figure 8.1 but with more levels of segmentation. This is a multisegmentation algorithm. Its advantages are that: (1) it does not oversegment the image; (2) it is as fast as the image binarization algorithm; (3) the number of regions is well controlled through the number of recursions; and (4) it preserves the original result of the binarization algorithm, adding a narrowband region, which means a transition between the foreground and background that can later be analyzed and included (or refuted) in the tumor region. This algorithm is formalized in the next section.

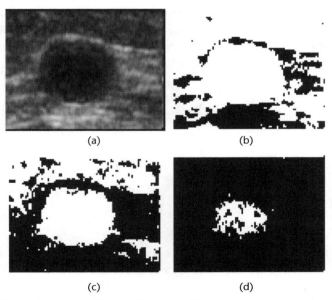

(a)   (b)

(c)   (d)

Figure 8.1   (a) Original ultrasound benign image and the nonextensive entropy segmentation results for different q values: (b) 1.0, (c) 6.0, and (d) 10.0, respectively.

8.3.2.2 The Nonextensive Segmentation Recursive Algorithm

Following the definitions and formulations from Section 8.3.2.1, we can take each distribution P_A and P_B and subdivide them into two news distribution, P_{A1}, P_{A2}, P_{B1}, and P_{B2}, as following:

$$P_{A1} : \frac{p_1}{p_{A1}}, \frac{p_2}{p_{A1}}, \ldots, \frac{p_t}{p_{A1}} \tag{8.11}$$

$$P_{A2} : \frac{p_{t+1}}{p_{A2}}, \frac{p_{t+2}}{p_{A2}}, \ldots, \frac{p_\partial}{p_{A2}} \tag{8.12}$$

$$P_{B1} : \frac{p_{\partial+1}}{p_{B1}}, \frac{p_{\partial+2}}{p_{B1}}, \ldots, \frac{p_v}{p_{B1}} \tag{8.13}$$

$$P_{B2} : \frac{p_{v+1}}{p_{B2}}, \frac{p_{v+2}}{p_{B2}}, \ldots, \frac{p_k}{p_{B2}} \tag{8.14}$$

having the constraints $p_{A1} = \sum_{i=1}^{t} p_i, p_{A2} = \sum_{t+1}^{\partial} p_i, p_{B1} = \sum_{\partial+1}^{v} p_i$, and p_{B2}.

For each one of the four preceding distributions, we can compute its respective nonextensive entropy as follows:

$$S_{A1} = \frac{1 - \sum_{i=1}^{t} \left(\frac{p_i}{p_{A1}} \right)^q}{q-1} \tag{8.15}$$

$$S_{A2} = \frac{1 - \sum_{i=t+1}^{\partial} \left(\frac{p_i}{p_{A2}} \right)^q}{q-1} \tag{8.16}$$

$$S_{B1} = \frac{1 - \sum_{i=\partial+1}^{v} \left(\frac{p_i}{p_{B1}} \right)^q}{q-1} \tag{8.17}$$

$$S_{B2} = \frac{1 - \sum_{i=v+1}^{k} \left(\frac{p_i}{p_{B2}} \right)^q}{q-1} \tag{8.18}$$

Using (8.4), (8.15) and (8.16) to compute $S(A) = S(A1 + A2)$ and, similarly (8.17) and (8.18) to compute $S(B) = S(B1 + B2)$, we have:

$$S_q(A+B) = \left\{ \frac{1 - \sum_{i=1}^{t}\left(\dfrac{p_i}{p_{A1}}\right)^q}{q-1} + \frac{1 - \sum_{i=t+1}^{\partial}\left(\dfrac{p_i}{p_{A2}}\right)^q}{q-1} \right.$$

$$+ (1-q) \cdot \frac{1 - \sum_{i=1}^{t}\left(\dfrac{p_i}{p_{A1}}\right)^q}{q-1} \cdot \frac{1 - \sum_{i=v+1}^{\partial}\left(\dfrac{q_i}{p_{A2}}\right)^q}{q-1} \left. \right\}$$

$$+ \left\{ \frac{1 - \sum_{i=\partial+1}^{v}\left(\dfrac{q_i}{q_{B1}}\right)^q}{q-1} + \frac{1 - \sum_{i=v+1}^{k}\left(\dfrac{q_i}{q_{B2}}\right)^q}{q-1} \right.$$

$$+ (1-q) \cdot \frac{1 - \sum_{i=\partial+1}^{v}\left(\dfrac{p_i}{q_{B1}}\right)^q}{q-1} \cdot \frac{1 - \sum_{i=v+1}^{\partial}\left(\dfrac{q_i}{q_{B2}}\right)^q}{q-1} \left. \right\} \tag{8.19}$$

$$+ (1-q) \cdot \left\{ \frac{1 - \sum_{i=1}^{t}\left(\dfrac{p_i}{p_{A1}}\right)^q}{q-1} + \frac{1 - \sum_{i=t+1}^{\partial}\left(\dfrac{p_i}{p_{A2}}\right)^q}{q-1} \right.$$

$$+ (1-q) \cdot \frac{1 - \sum_{i=1}^{t}\left(\dfrac{p_i}{p_{A1}}\right)^q}{q-1} \cdot \frac{1 - \sum_{i=t+1}^{\partial}\left(\dfrac{p_i}{p_{A2}}\right)^q}{q-1} \left. \right\}$$

$$\left\{ \frac{1 - \sum_{i=\partial+1}^{v}\left(\dfrac{q_i}{q_{B1}}\right)^q}{q-1} + \frac{1 - \sum_{i=v+1}^{k}\left(\dfrac{q_i}{q_{B2}}\right)^q}{q-1} \right.$$

$$\cdot$$

$$+ (1-q) \cdot \frac{1 - \sum_{i=\partial+1}^{v}\left(\dfrac{q_i}{q_{B1}}\right)^q}{q-1} \cdot \frac{1 - \sum_{i=v+1}^{k}\left(\dfrac{q_i}{q_{B2}}\right)^q}{q-1} \left. \right\}$$

In this case, to find the optimal luminance level, such as the equivalent in (8.10), we can compute the value for (8.19) by taking the argument that maximizes the next expression:

$$t_{opt} = \arg\max\Big[\big(S_{A1} + S_{A2} + (1-q) \cdot S_{A1} \cdot S_{A2}\big)$$
$$\times\big(S_{B1} + S_{B2} + (1-q) \cdot S_{B1} \cdot S_{B2}\big) \cdot (1-q) \qquad\qquad (8.20)$$
$$\times\big(S_{A1} + S_{A2} + (1-q) \cdot S_{A1} \cdot S_{A2}\big) \cdot \big(S_{B1} + S_{B2} + (1-q) \cdot S_{B1} \cdot S_{B2}\big)\Big]$$

In this case, instead of (8.10), $t_{opt} = \{t_{opt1}, t_{opt2}, t_{opt3}\}$ is a set of threshold values for two recursions, where t_{opt1} is the first threshold, and t_{opt2} and t_{opt3} are new thresholds for P_A and P_B distributions, respectively.

Equation (8.19) is simple, even though it has several terms. Developing a third recursion would yield a number of terms up to 16, making interpretation extremely difficult. However, two observations can be done. First, experimental results show that more than two or three recursions are not necessary in order to obtain results that are equal to or better than the traditional methods. Second, the growing number of recursions does not increase the algorithm complexity or computation, since this growing is accompanied by a reduction in the number of states to be computed at each recursion, as in the algorithms for binary partitions, which have an $O(n \lg_2 n)$ complexity. At each iteration NESRA generates 2^{r+1} regions, where r is the number of recursions. Note that for $r = 0$, NESRA is the simple image binarization algorithm.

The recursive algorithm for the previous formulation as follows:

```
Algorithmus 1 Nonextensive Segmentation Recursive Algorithm,
NESRA Procedure

Input: H = image histogram, i = first histogram bin, k = last
histogram bin

Call NESRA(H, i, k) procedure

if Histogram H is homogeneous then
 Goto FIM
end if

for all t = i until k do

   compute normalization for background
   compute normalization for foreground
   compute q-entropy for background according to (8.7)
   compute q-entropy for foreground according to (8.8)
   compute composed q-entropy according to (8.9) and (8.10)

end for
topt = argmax of the composed q-entropy
Call NESRA(H, i, topt) procedure
Call NESRA(H, topt + 1, k) procedure
FIM: there is nothing to do, return to calling procedure
```

After this segmentation, tumor's boundary must be extracted from the background. To accomplish this task, we apply a mathematical morphology approach. Because this approach may generate a coarse contour of the tumor, we apply a level set framework to smooth the contour. The level set theory is briefly presented in the next section.

8.3.2.3   The Level Set Formulation

A level set method, also called an implicit snake, is a numerical technique developed first by Osher and Sethian [46] to track the evolution of interfaces. These interfaces can develop sharp corners, break apart, and merge together. The technique has a wide range of applications, including problems in fluid mechanics, combustion, manufacturing of computer chips, computer animation, structure of snowflakes, the shape of soap bubbles, and image processing, including medical images.

Informally speaking, consider a boundary image interface separating one region of interest (the inside) from another (the outside), and a speed function F that gives the speed of each point of the boundary. This speed can depend on several physical effects such as temperature if the environment is ice (ROI) inside water. In this case, the boundary can shrink as the ice melts, or grow as the ice freezes. Then, the speed function F depends on the adopted model. However, the original level set formulation proposed by Osher and Sethian [46] takes a curve and builds it into a surface. The curve is evaluated in the surface, which intersects the xy plane exactly where the curve sits. The surface is called the level set function because it accepts as input any point in the plane and hands back its height as output. The curve that intersects the surface is called the zero level set, because it is the collection of all points that are at height zero. A complete level set study can be found in [47].

Generically speaking, a level set formulation is useful for image segmentation because, among other advantages, it handles naturally complex topologies, permits flexible models representing the boundary of ROI, can be robust to noise, and can be computationally efficient.

In our work, we use a level set formulation to smooth the ROI before the matching phase. To enhance the computational overhead, we compute the level set only inside a narrow band and the initial curve is given by the previous segmentation with Tsallis entropy.

More formally, the main idea of the 2-D level set method is to represent the deformable surface (or curve) as a level set $\{x \in \Re^2 | G(x) = 0\}$ of an embedding function:

$$G: \Re^2 \times \Re^+ \to \Re \tag{8.21}$$

such that the deformable surface (also called *front* in this formulation), at $t = 0$, is given by a surface S:

$$S(t = 0) = \left\{ x \in \Re^2 \middle| G(x, t = 0) = 0 \right\} \tag{8.22}$$

The next step is to find a Eulerian formulation for the front evolution.

Following Malladi et al. [48], let us suppose that the front evolves in the normal direction with velocity $\vec{F}$, where $\vec{F}$ may be a function of the curvature, normal direction, and so forth.

We need an equation for the evolution of $G(x, t)$, considering that the surface S is the level set given by:

$$S(t) = \left\{ x \in \Re^2 \middle| G(x, t) = 0 \right\} \tag{8.23}$$

Let us take a point $x(t)$, $t \in \Re^+$ of the propagating front S. From its implicit definition, given above, we have:

$$G(x(t),t) = 0 \tag{8.24}$$

Now, we can use the chain rule to compute the time derivative of this expression:

$$G_t + F|\nabla G| = 0 \tag{8.25}$$

where $F = \|\vec{F}\|$ is called the *speed function*. An initial condition $G(x, t = 0)$ is required.

A straightforward (and expensive) technique to define this function is to compute a signed-distance function as follows:

$$G(x,t = 0) = \pm d \tag{8.26}$$

where d is the distance from x to the surface $S(x, t = 0)$ and the signal indicates whether the point is interior (−) or exterior (+) to the initial front.

Finite difference schemes, based on a uniform grid, can be used to solve (8.25). The same entropy condition of T-Surfaces (*once a grid node is burnt, it stays burnt*) is incorporated to drive the model to the desired solution (in fact, T-Surfaces was inspired by the level sets model [49]).

In this higher-dimensional formulation, topological changes can be efficiently implemented. Numerical schemes are stable and the model is general in the sense that the same formulation holds for two and three dimensions, as well as for merges and splits. In addition, the surface geometry is easily computed. For example, the front normal and curvature are given by:

$$\vec{n} = \nabla G(x,t), \quad K = \nabla \cdot \left(\frac{\nabla G(x,t)}{\|\nabla G(x,t)\|} \right) \tag{8.27}$$

respectively, where the gradient and the divergent ($\nabla \cdot$) are computed with respect to x.

The initialization of the model through (8.26) is computationally expensive and not efficient if we have more than one front to initialize [50].

The narrowband technique is much more appropriate for this case. The key idea of this technique comes from the observation that the front can be moved by updating the level set function at a small set of points in the neighborhood of the zero set instead of updating it at all the points on the domain (see [48, 51] for details).

To implement this scheme, we need to preset a distance Δd to define the narrow band. The front can move inside the narrow band until it collides with the narrowband frontiers. Then, the function G should be reinitialized by treating the current zero set configuration as the initial one.

Also, this method can be made cheaper by observing that the grid points that do not belong to the narrow band can be treated as sign holders [48]; in other words, points belonging inside the narrow band are positive, and negative otherwise.

To set ideas, consider Figure 8.2(a), which shows the level set bounding the search space, and Figure 8.2(b), which shows a bidimensional surface in which the zero level set is the contours just presented.

The surface evolves such that the contour gets closer to the edge. To accomplish this goal, we must define a suitable speed function and an efficient numerical approach. For simplicity, we consider the 1-D version of the problem pictured in Figure 8.2. In this case, the evolution equation can be written as:

$$G_t + \frac{\partial G}{\partial x} F = 0 \tag{8.28}$$

The main point is to design the speed function F such that the desired result can be obtained if $G_t > 0$. For instance, if we set the sign of F opposite to the one of Gx we get $G_t > 0$, once:

$$G_t = -\frac{\partial G}{\partial x} F \tag{8.29}$$

Hence, the desired behavior can be obtained by the distribution of the sign of F, as shown in Figure 8.3.

Once our application focus is shape recovery in an image I, we must choose a suitable speed function F as well as a convenient stopping term S to be added to the right-hand side of (8.28). Among the possibilities [52], the following were suitable for our case:

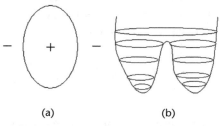

(a)   (b)

Figure 8.2   (a) Level set bounding the search space. (b) Initial function with the zero level set as the contour presented.

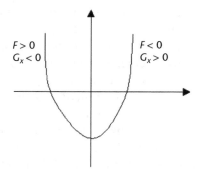

$F > 0$
$G_x < 0$

$F < 0$
$G_x > 0$

Figure 8.3   Sign of speed function.

$$F = \frac{1+\alpha k}{1-|\nabla I|^2} \qquad (8.30)$$

$$S = \beta \nabla I \cdot \nabla G \qquad (8.31)$$

where β is a scale parameter. Therefore, we are going to deal with the following level set model:

$$G_t = \left(\frac{1+\alpha k}{1+|\nabla I|^2}\right)|\nabla G| + \beta \nabla P \cdot \nabla G \qquad (8.32)$$

where $P = -|\nabla I|^2$. After initialization, the front evolves following (8.32).

Given the smooth contour of the level set output, we extract the region features mentioned in Section 8.1. These features are then input to an SVM classifier, which classifies the region as a malignant or benign tumor. The SVM theory is briefly discussed in the next section.

8.3.2.4   The Support Vector Machine Theory

SVM is a very good classification tool. It has been proven to be as a very effective method for many applications such as pattern recognition, machine learning, and data mining. The general idea is as follows. Let $\{x_i, y_i\}$, $i = \{1, ..., N\}$ be a training example set S; each example $x_i \in R^n$ belongs to a class labeled $y_i \in \{-1, 1\}$. The goal is to define a hyperplane that divides S, such that all of the points with the same label are on the same side of the hyperplane, while maximizing the distance between the two classes and the hyperplane. This means we need to find a pair (w, b) such that

$$y_i(w \cdot x_i + b) > 1, \, i = 1, ..., N \qquad (8.33)$$

where $w \in R^n$ and $b \in R$. After finding the pair (w, b), we can achieve a separating hyperplane (see Figure 8.4) and its equation is defined as follows:

$$w \cdot x + b = 0 \qquad (8.34)$$

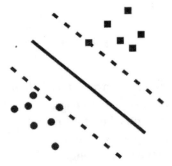

Figure 8.4   Separating hyperplane. The dashed lines identify the margin.

Maybe there are many separating hyperplanes; the optimal separating hyperplane (OSH) is a hyperplane for which the distance to the closest point is maximal. Hence, to find the OSH, we must minimize $\|w\|^2$ under constraint (8.33).

If we denote $\alpha = (\alpha_1, \alpha_2, ..., \alpha_N)$, where N is the nonnegative Lagrange multipliers associated with the constraints of (8.33), the problem of finding the OSH is equivalent to maximization of the function:

$$W(\alpha) = \sum_{i=1}^{N} \alpha_i - \frac{1}{2} \sum_{i,j=1}^{N} \alpha_i \alpha_j y_i y_j x_i \cdot x_j \qquad (8.35)$$

where $\alpha_i \geq 0$ and, under constraint,

$$\sum_{i}^{N} y_i \alpha_i = 0 \qquad (8.36)$$

Once we achieve the vector $\overline{\alpha} = \overline{\alpha}_1, \overline{\alpha}_2, ..., \overline{\alpha}_N$, the solution of (8.34), the OSH $(\overline{w}, b)$, has the following expansion:

$$\overline{w} = \sum_{i=1}^{N} \overline{\alpha}_i y_i x_i \qquad (8.37)$$

while $\overline{b}$ can be determined from $\overline{\alpha}$ and from the Kuhn-Tucker condition [53]:

$$\overline{\alpha}_i \left(y_i \left(\overline{w} \cdot x_i + \overline{b} \right) - 1 \right) = 0, i = 1, 2, ..., N \qquad (8.38)$$

The decision function of classifying a new data point can be written as:

$$f(x) = \text{sgn}\left(\sum_{i=1}^{N} \overline{\alpha}_i y_i x_i \cdot x + \overline{b} \right) \qquad (8.39)$$

If the S is not linearly separable, we must introduce N nonnegative slack variables $\xi = (\xi_1, \xi_2, ..., \xi_N)$, such that

$$y_i (w \cdot x_i + b) \geq 1 - \xi_i, i = 1, 2, ..., N \qquad (8.40)$$

Hence, we can regard the generalized OSH as the solution of the following minimization problem:

$$\frac{1}{2} w \cdot w + C \sum_{i=1}^{N} \xi_i \qquad (8.41)$$

where C is a regularization parameter. When the parameter C is small, the OSH tends to maximize the distance $1/\|w\|$; in contrast, a larger C will lead the OSH to minimize the number of misclassified points.

The training example set that we want to classify is usually linearly nonseparable. To achieve better generalization performance, the input data can first be mapped into a high-dimensional feature space. Then the OSH is constructed in the feature space. If $\phi(x)$ denotes a mapping function that maps x into a high-dimensional feature space, (8.34) can then be rewritten as follows:

$$W(\alpha) = \sum_{i=1}^{N} \alpha_i - \frac{1}{2} \sum_{i,j=1}^{N} \alpha_i \alpha_j y_i y_j \phi(x_i) \cdot \phi(x_j) \qquad (8.42)$$

Now, let $K(x_i, x_j) = \phi(x_i) \cdot \phi(x_j)$, we can rewrite (8.40) as

$$W(\alpha) = \sum_{i=1}^{N} \alpha_i - \frac{1}{2} \sum_{i,j=1}^{N} \alpha_i \alpha_j y_i y_j K(x_i, x_j) \qquad (8.43)$$

where K is a kernel function and must satisfy Mercers theorem [54]. Finally, the decision function becomes:

$$f(x) = \text{sgn}\left(\sum_{i=1}^{N} \overline{\alpha}_i y_i K(x_i, y_i) + \overline{b} \right) \qquad (8.44)$$

In this study, we use nonlinear SVM with a B-spline basis kernel as our classifier. The five tumor features discussed earlier are used as the inputs to an n-dimensional vector to find the OSH for distinguishing benign tumors from malignant ones.

In the next section we put to use all of the theories presented in this section in order to summarize our five-step methodology for breast image classification.

8.3.3 Morphological Chains: Mathematical Morphology Definitions

The use of mathematical morphology for initializing deformable models is a common practice with promising results and general application. Nevertheless, it is an open area of research and there is room for improvement. For the particular case of medical images, the general idea is to isolate objects of interest—such as lungs, arteries, heart, or bones—in the scene and to work with them individually while avoiding neighboring interference from other objects, noises, spurious artifacts, and background noise.

Mathematical morphology is a set of mathematical tools used in the digital image processing area to perform linear transformations on the shapes of image regions to obtain results such as merging, splitting, noiseless, or smoothing of an ROI. These results are accomplished throughout several sets of operations between the original image and specific sets, called structuring elements. However, two basic operations are always used: erosion and dilation. Suppose the image X and the structuring element B are represented as sets in 2-D Euclidian space. Let B_x denote the translation of B so that its origin is located at x. Then the erosion of X by B is defined as the set of all points x such that B_x is included in X, that is,

$$\text{Erosion}: X \ominus B = \{x : B_x \subset X\} \qquad (8.45)$$

Similarly, the dilation of X by B is defined as the set of all points x such that B_x hits X, that is, they have a nonempty intersection:

$$\text{Dilation:} X \oplus B = \left\{ x : B_x \cap X \neq \phi \right\} \tag{8.46}$$

These two operations are the base for all other, more complex operations. As an example, a well-known operation, directly derived from (8.45) and (8.46), is the opening, which consists of an erosion on a given image followed by a dilation of the result. As an effect of an opening, small connections between regions can be accomplished. This is particularly useful when we wish to disconnect two different regions to treat them separately. The dual of the opening is the close operation, which consists of an erosion over the dilation's result. The effect of closing an image is, correctly, the opposite of an opening: It connects weak separated regions.

Other operations based on opening and closing are defined in the specific literature. For example, we can find boundary, convex hull, skeleton, and thinning operations. For a complete approach to the issue, see [55].

Specific problems, such as those encountered in medical images, can require specific solutions, with different morphological operations and several structuring elements. A sequence of morphological operations is called a morphological chain. The chain to use depends on specific factors, such as the quality of the images (noise, illumination, size, resolution, and so forth), objects of interest (bones, lungs, heart, and so forth), computational time consumed (real-time applications or off-line), and our goals.

In our proposed five-step methodology, we use mathematical morphology after the initial segmentation of the image in order to extract the ROI from the background. This approach is described in detail next.

8.4 The Proposed Five-Step Methodology

As outlined, we put all of the theoretical background discussed in the preceding sections together to generate a five-step protocol that can be used to classify 3-D breast images. After application of NESRA, we extract the lesion area from the background. Figure 8.5 shows an original benign image example and the NESRA result.

This result was obtained with two NESRA recursions, which means that the gray-scale distribution was partitioned into two new distributions and each one was partitioned into two others, generating four distinct regions. Although we wish to split the image into two regions only—and in this case no recursion will be needed—the use of one recursion in this case helps to delimit a tumor's nucleus and also a boundary as a narrow region around the nucleus. Note in Figure 8.5(b) the white region around the image center (tumor) and the intermediate gray-level region around the tumor's region (the transition region between the tumor's region and background). This transition region can be isolated and help to delimit accurately the lesion boundary. As mentioned in Section 8.1, this transition region is made of a mix of dead and alive cells. These dead cells are generally swallowed to inside region and the tumor grows. For the result of Figure 8.5(b) we used $q = 0.5$. In this case, the ROI is the tumor's nucleus and its transition region.

In the second step we use a morphological chain approach to extract the ROI from the background. This is accomplished through the following rule. Considering the binary image generated by NESRA [e.g., Figure 8.5(b)], let α and β be the total

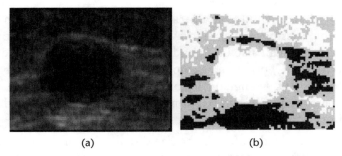

(a)   (b)

Figure 8.5   (a) Original ultrasound benign image and (b) the NESRA results with two recursions and $q = 0.5$.

ROI's area and the total image area, respectively. If $\alpha \geq \xi\beta$ an erosion is carried out; and if $\alpha \leq \delta\beta$, a dilation is carried out. After, assuming that the ROI has a geometric center near the image center, we apply a region growing algorithm that defines the final ROI's boundary. In our experiments, we fixed $\xi = 0.75$ and $\delta = 0.25$ to correctly extract most of the ROIs. The result of this morphological rule applied in the image of Figure 8.5(b) is shown in Figure 8.6(a).

The region generated by the morphological chain rule is a coarse representation of the lesion region. Then, as a third step of our method, we apply a level set framework using as the initialization this region's boundary [52]. However, because this initial ROI is near the real boundary, there is no need for several iterations, as normally occurs in several level set applications. Then, in our experiments, we have applied 10 iterations only. It was sufficient to get good results, which generate low computational overhead. Figure 8.6(b) shows the final lesion boundary after the level set computation.

After the preceding boundary tracing process, the next step is feature extraction of the ROI. Three radiologists have defined five features that have a high probability of working well as discriminators between malignant and benign lesions. In our work, we have used these five features and tested them in order to achieve a combination that yields good empirical results.

- *AR:* The first feature is the lesion area. Because malignant lesions generally have large areas in relation to benign ones, this characteristic is a power discriminant. We have normalized it by the total image area.

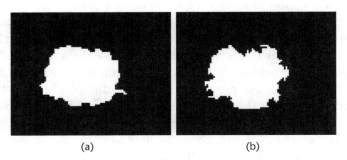

(a)   (b)

Figure 8.6   (a) ROI after morphological chain application. (b) ROI after level set application.

- *CT*: The second characteristic is related to the region circularity. Because benign lesions generally have more circular areas compared with malignant ones, this can also be a good discriminant. Then, we take the ROI's geometric center (GC) point and compute the distance from each boundary point (x_i, y_i) to the GC. Malignant lesions tend to have high standard deviations of the average distances in relation to the benign ones. Also, this feature is normalized by total image area.

- *PT*: The third feature is the size distribution of the lobes in a lesion. A boundary's lobe is a protuberant region on the boundary. We compute the convex hull of the ROI and the lobe as a protuberance between two valleys. The lobe areas are computed and only those that are greater than 10% of the lesion area are considered. This feature is taken as the average area of the lobes. Malignant lesions have high average area in relation to benign ones.

- *HT*: The next feature is related to the homogeneity of the lesion. Malignant lesions tend to be less homogeneous than benign ones. Then, we take the BGS entropy—taken over the gray-scale histogram—relative to the maximum entropy as the fourth discriminant feature. The higher the relative entropy, the more homogeneous the lesion region, and, consequently, the higher the chance that the lesion is benign.

- *AS*: The last feature is related to an interesting characteristic of the lesions: the acoustic shadow. When benign lesions have many water particles, the formation of an acoustic reinforcement below it is probable. However, when the lesion is more solid (a malignant characteristic), there is a tendency to form an acoustic shadow. Then, by comparing the lesion region and the region below the lesion, we can get an idea of whether the lesion is a benign or malignant one. If the region has an acoustic shadow, it tends to be more white than if it has an acoustic reinforcement. Then, we compute the gray-scale histograms of both regions and compare them. The darker the region below the lesion, the greater the acoustic reinforcement and, consequently, the higher the probability that the lesion is benign. We have computed the relative darkness between both areas (tumor's area and area below the tumor) and use it as the fifth lesion feature.

These five features are composed into a five-dimensional feature vector. The space of all feature vectors is then finally input for a nonlinear SVM in order to classify the lesion area as being a malignant or benign region. SVM is very good discriminant tool, primarily when we do not have a linear separation between the data. Because this feature space does not have a linear separation, we have used a B-spline curve as a kernel. Because some characteristics are better discriminants than others, we have also combined them into the SVM framework. To justify the use of a B-spline as a kernel for the SVM in this nonlinear space, we compare in Figure 8.7 the performance of our proposal under three different kernels: B-spline, polynomial, and exponential. In this figure, the area under the ROC curve (92% of the total) is clearly superior compared to the polynomial (area 84%) and exponential (area 65%) kernels. Also, Figure 8.8 shows how a polynomial kernel separates our feature space for two features only: AR and HT.

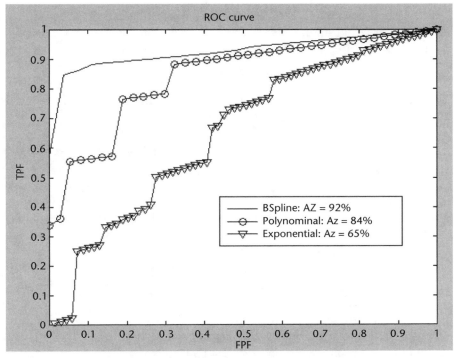

Figure 8.7   ROC curves for the proposed CAD system for three different kernel functions used in the SVM classifier.

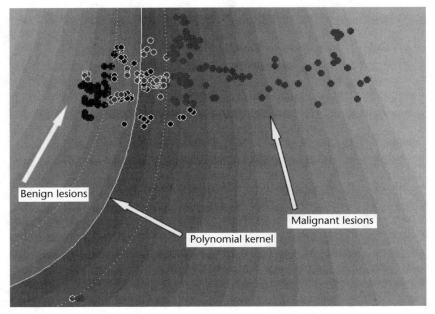

Figure 8.8   SVM output with a polynomial kernel. The white line is the polynome separating malignant tumors (gray circles) from benign ones (black circles). The circles with the white boundaries are those inside the confident region.

In contrast, when we use a B-spline kernel instead of polynomial, we have the aspect shown in Figure 8.9. In the case of B-spline kernel, note that in Figure 8.9 a better separation of the feature space is seen compared with the polynomial kernel shown in Figure 8.8.

Figure 8.10 presents three different examples of the proposed method for benign lesions. Figure 8.10(a) is the original image; Figure 8.10(d) is the result after NESRA application; Figure 8.10(g) is the result after background extraction with a morphological chain; and Figure 8.10(j) is the final result after level set application.

Figure 8.11 presents three different examples of the proposed method for malignant lesions. Figure 8.11(a) is the original image; Figure 8.11(d) is the results after NESRA application; Figure 8.11(g) is the result after background extraction with a morphological chain; and Figure 8.11(j) is the final result after level set application. Note in Figure 8.11(g) that the morphological step did not adequately extract the tumor from the background; however, the final step with level set application did converge to the desired boundary.

In Figure 8.12 we show five ROC curves for five different combinations of the defined tumor's features. According to these curves, the best performance is reached when the area (AR) + heterogeneity (HT) and acoustic shadow (AS) features are combined. This combination generated an A_z of 92%. All other combinations have resulted in similar performance behavior and A_z inferior to AR = HT + AS combination. A better discussion of these results is given in the next section.

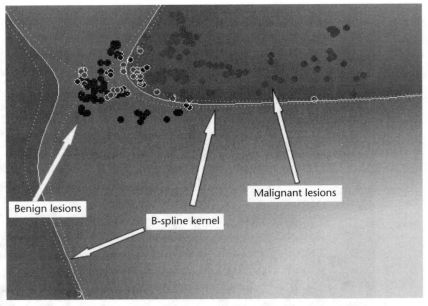

Figure 8.9 SVM output with a B-spline kernel. The white line is the polynome separating malignant tumors (gray circles) from benign ones (black circles). The circles with the white boundaries are those inside the confident region.

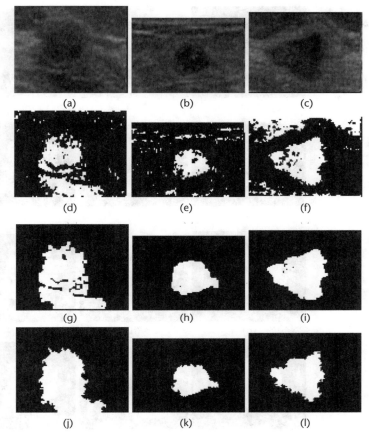

Figure 8.10   (a–l) The first three steps of the proposed methodology. Each column represents a different example of a benign lesion.

8.5   Performance Evaluation

To test our proposed method we used 50 pathology-proven cases—20 benign and 30 malignant. Each case is a sequence of five images of the same lesion. Then, we tested 100 images of benign lesions and 150 of malignant ones. Because the detection of a malignant lesion between five images of the same case indicates a malignant case, it is reasonable to consider 250 different cases.

Because our database is small, we have improved the results through a cross-validation method. Then, these ultrasonic images are randomly divided into five groups. We first set the first group as a testing group and use the remaining four groups to train the SVM. After training, the SVM is then tested on the first group. Then, we set the second group as a testing group and the remaining four groups are trained and then the SVM is tested on the second. This process is repeated until all five groups have been set in turn as the testing group.

To estimate the performance of the experimental result, five objective indices are used: accuracy, sensitivity, specificity, PPV, and NPV. In our experiment, the accuracy of the SVM with B-spline kernel for classification malignancies is 95.2%

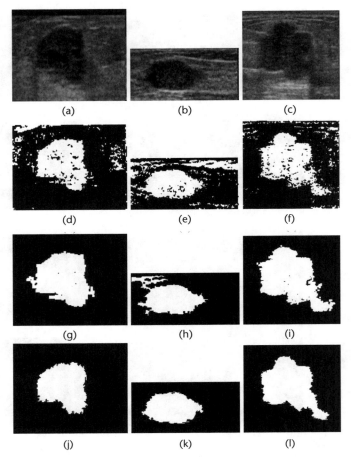

Figure 8.11   (a–l) The first three steps of the proposed methodology. Each column represents a different example of a malignant lesion.

(238/250), the sensitivity is 97% (97/100), the specificity is 94% (141/150), the PPV is 91.51% (97/106), and the NPV is 97.92% (141/144).[2]

Figure 8.12 shows the SVM classification output for our methodology when we consider two dimensions only.

8.6   Discussion and Conclusions

Geometrical and textured information from the lesion area in ultrasound images provide important discriminants for CAD systems. Because ultrasound images generally have complex characteristics between pixels, it is interesting to study them from the point of view of nonextensive entropy. Because the traditional Shannon entropy provides image segmentation between foreground and background only, it does not guarantee that transition areas (which normally occur on the boundary of

2.   TP = true positive; TN = true negative; FP = false positive; FN = false negative; Accuracy = (TP+TN)/(TP+TN+FP+FN); Sensitivity = TP/(TP+FN); specificity = TN/(TN+FP); positive predictive value = TP/(TP+FP); and negative predictive value = TN/(TN+FN).

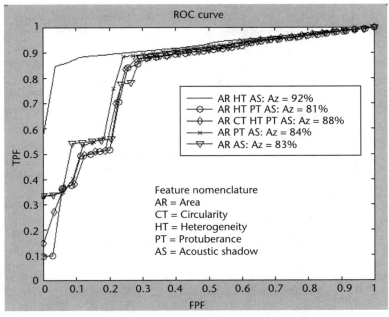

Figure 8.12   ROC curves for the proposed CAD system for several combinations of tumor features.

such lesions) will be adequately analyzed. So our proposed algorithm is a good option to initial segmentation. This algorithm, NESRA, includes the same results as the traditional entropy calculations, but with the improvement that it is able to isolate intermediate regions, such as the lesion's boundary transitions. This is one of the advantages of NESRA when applied on this kind of image. A CAD system can decide if the transition area will be included or not in the ROI based on a careful analysis.

On the other hand, the use a level set framework helps to find the correct boundary position without increasing the computational time, but with an improvement in system precision.

Regarding the chosen features, this chapter does not discuss why some feature combinations, such as AR + HT + AS, generate better CAD system performance than others. We chose to test empirically which combination is better, which can be seen in the graphic of Figure 8.12.

Our work proposes an SVM diagnostic system that uses as input five-dimensional feature space characteristics. These characteristics are based on geometrical and textured information and should be combined in order to tune the system. In our experiments, the best combination was achieved when only area, heterogeneity, and shadow were used. All of these features are affected by the segmentation process. For example, if the region of a benign lesion will not be adequately extracted from the background, it can be present as an area (AR), which is greater than the reality, yielding the inference that it is a malignant case. The same may occur with the circularity (CT) or protuberance (PT) features. In all cases, the use of a level set framework helps to revert a bad segmentation situation. For example, suppose that a lesion was erroneously segmented to include parts of background in the lesion region. In this case, because the segmented region is near the boundary, the level set approach has great chances to stop over the real boundary.

Another aspect (not addressed in this chapter) is the use of a dual approach for the level set framework, such as in Giraldi et al. [56, 57]. In this case, two embedded curves evolve toward the lesion boundary, one from inside and the other outside the lesion region.

Finally, the use of SVM with a B-spline kernel is one of the main contributions of our paper and one of the main features responsible for the system's performance (up to 92%), because this kind of curve allows more flexibility over our feature space, as seen in Figure 8.9. The use of a B-spline clearly improves the classification, as seen in Figure 8.7.

Actually, the greater challenge of the CAD systems is the delimitation of an initial region, which is normally rectangular. This delimitation is manually accomplished by radiologists. A possible future direction is to investigate tracking approaches. For example, an ultrasound analysis will be done over a frame sequence. In the first frame, several candidate regions can be automatically extracted, and then, a tracking system is accomplished over the remainder of the frames, where each candidate region is separately analyzed.

Acknowledgments

The authors are grateful to Coordenaçao de Aperfeiçoamento de Pessoal de Nível Superior, Conselho Nacional de Desenvolvimento Científico e Tecnològico, and FAPERJ, Brazilian agencies for scientific financing, for the support of this work.

References

[1] American Cancer Society, *Cancer Facts and Figures*, 2004.

[2] Suri, J, S., and R. M. Rangayyan, *Emerging Technologies in Breast Imaging and Mammography*, Los Angeles, CA: ASP Press, 2006.

[3] Suri, J. S., and R. M. Rangayyan, *Recent Advances in Breast Imaging, Mammography and Computer Aided Diagnosis of Breast Cancer*, Bellingham, WA: SPIE Press, 2006.

[4] Chen, D. R., et al., "Use of the Bootstrap Technique with Small Training Sets for Computer-Aided Diagnosis in Breast Ultrasound," *Ultrasound Med.*, Vol. 28, No. 7, July 2002, pp. 897–902.

[5] Chang, R. F., et al., "Improvement in Breast Tumor Discrimination by Support Vector Machines and Speckle-Emphasis Texture Analysis," *Ultrasound Med. Biol.*, Vol. 29, No. 5, 2003, pp. 679–686.

[6] Chen, C. M., et al., "Breast Lesion on Sonograms: Computer-Aided Diagnosis with Neural Setting-Independent Features and Artificial Neural Networks," *Journal of Radiology*, Vol. 226, 2003, pp. 504–514.

[7] Kuo, W. J., et al., "Computer-Aided Diagnosis of Breast Tumors with Different US Systems," *Acad. Radiol.*, Vol. 9, No. 7, July 2002, pp. 793–799.

[8] Sawaki, A., et al., "Breast Ultrasonography: Diagnostic Efficacy of Computer-Aided Diagnostic System Using Fuzzy Inference," *Radiat. Med.*, Vol. 17, No. 1, January 1999, pp. 41–45.

[9]   Garra, B. S., et al, "Improving the Distinction Between Benign and Malignant Breast Lesions: The Value of Sonographic Texture Analysis," *Ultrasound Imaging*, Vol. 15, No. 4, 1993, pp. 267–285.

[10] Chen, D. R., R. F. Chang, and Y. L. Huang, "Computer-Aided Diagnosis Applied to US of Solid Breast Nodules by Using Neural Networks," *Journal of Radiology*, Vol. 213, 1999, pp. 407–412.

[11] Chen, C. M., et al., "Breast Lesion on Sonograms: Computer-Aided Diagnosis with Neural Setting-Independent Features and Artificial Neural Networks," *Journal of Radiology*, Vol. 226, 2003, pp. 504–514.

[12] Chen, D. R., et al., "Computer-Aided Diagnosis for 3-Dimensional Breast Ultrasonography," *Arch. Surg.*, Vol. 138, No. 3, March 2003, pp. 296–302.

[13] Chang, R. F., et al., "Computer-Aided Diagnosis for 2-D/3-D Breast Ultrasound," in *Recent Advances in Breast Imaging, Mammography, and Computer-Aided Diagnosis of Breast Cancer*, J. S. Suri and R. M. Rangayyan, (eds.), Bellingham, WA: SPIE Press, 2006.

[14] Chen, W. M., et al., "Breast Cancer Diagnosis Using Three-Dimensional Ultrasound and Pixel Relation Analysis," *Ultrasound Med. Biol.*, Vol. 29, No. 7, July 2003, pp. 1027–1035.

[15] Pun, T., "Entropic Thresholding: A New Approach," *Comput. Graphics Image Process*, Vol. 16, 1981, pp. 210–239.

[16] Kapur, J. N., P. K. Sahoo, and A. K. C. Wong, "A New Method for Gray-Level Picture Thresholding Using the Entropy of the Histogram," *Comput. Graphics Image Process*, Vol. 29, 1985, pp. 273–285.

[17] Abutaleb, A. S., "A New Method for Gray-Level Picture Thresholding Using the Entropy of the Histogram," *Comput. Graphics Image Process*, Vol. 47, 1989, pp. 22–32.

[18] Li, C. H., and C. K. Lee, "Minimum Cross Entropy Thresholding," *Pattern Recognition*, Vol. 26, 1993, pp. 617–625.

[19] Pal, N. R., "On Minimum Cross Entropy Thresholding," *Pattern Recognition*, Vol. 26, 1996, pp. 575–580.

[20] Sahoo, P. K., S. Soltani, and A. K. C. Wong, "A Survey of Thresholding Techniques," *Comput. Vis. Graphics Image Process*, Vol. 41, 1988, pp. 233–260.

[21] Albuquerque, M. P., et al., "Image Thresholding Using Tsallis Entropy," *Pattern Recognition Letters*, Vol. 25, 2004, pp. 1059–1065.

[22] Chang, R. F., et al., "Automatic Ultrasound Segmentation and Morphology Based Diagnosis of Solid Breast Tumors," *Breast Cancer Res. Treat.*, Vol. 89, No. 2, January 2005, pp. 179–185.

[23] Chen, W. M., et al., "3-D Ultrasound Texture Classification Using Run Difference Matrix," *Ultrasound Med. Biol.*, Vol. 31, No. 6, June 2005, pp. 763–770.

[24] Cheng, X. -Y., et al., "Breast Tumor Diagnosis System Using Three Dimensional Ultrasonic Echography," *Proc. of 19th Annual International Conference of the IEEE*, Chicago, IL, 1997.

[25] Cheng, X. -Y., et al., "Automated Detection of Breast Tumors in Ultrasonic Images Using Fuzzy Reasoning," *Proc of the 1997 Int. Congress on Image Processing (ICIP '97)*, Vol. 3, 1997, p. 412.

[26] Chen, D. R., R. F. Chang, and Y. L. Huang, "Breast Cancer Diagnosis Using Self Organizing Map for Sonography," *Ultrasound Med. Biol.*, Vol. 26, No. 3, 2000, pp. 405–411.

[27] Chen, D. R., et al., "Texture Analysis of Breast Tumors on Sonograms," *Semin. Ultrasound CT MR*, Vol. 21, No. 4, 2000, pp. 308–316.

[28] Rotten, D., J. M. Levaillant, and L. Zerat, "Analysis of Normal Breast Tissue and of Solid Breast Masses Using Three-Dimensional Ultrasound Mammography," *Ultrasound Obstet. Gynecol.*, Vol. 14, No. 2, 1999, pp. 114–124.

[29] Karssemeijer, N., and G. M. te Brake, "Detection of Stellate Distortions in Mammograms," *IEEE Trans. on Med. Imaging*, Vol. 15, No. 5, 1996, pp. 611–619.

[30] Kobatake, H., and Y. Yoshinaga, "Detection of Spicules on Mammogram Based on Skeleton Analysis," *IEEE Trans. on Med. Imaging*, Vol. 15, No. 3, 1996, pp. 235–245.

[31] Vyborny, C. J., et al., "Breast Cancer: Importance of Spiculation in Computer-Aided Detection," *Radiology*, Vol. 215, No. 3, 2000, pp. 703–707.

[32] Liu, S., C. F. Babbs, and E. J. Delp, "Multiresolution Detection of Spiculated Lesions in Digital Mammograms," *IEEE Trans. on Image Processing*, Vol. 10, No. 6, 2001, pp. 874–884.

[33] Cherkassky, V., and F. Mulier, *Learning from Data: Concepts, Theory, and Methods*, New York: Wiley, 1998.

[34] Dillon, W. R., and M. Goldstein, *Multivariate Analysis: Method and Applications*, New York: Wiley, 1984.

[35] Zurada, J. M., *Introduction to Artificial Neural Systems*, Boston, MA: PWS, 1992.

[36] Tohno, E., D. O. Cosgrove, and J. P. Sloane, *Ultrasound Diagnosis of Breast Diseases*, Edinburgh, Scotland: Churchill Livingstone, 1994.

[37] Stavros, A. T., et al., "Solid Breast Nodules: Use of Sonography to Distinguish Between Benign and Malignant Lesions," *Radiology*, Vol. 196, 1995, pp. 123–134.

[38] Skaane, P., and K. Engedal, "Analysis of Sonographic Features in the Differentiation of Fibroadenoma and Invasive Ductal Carcinoma," *AJR Am. J. Roentgenol.*, Vol. 170, 1998, pp. 109–114.

[39] Reeves, A. P., et al., "Three-Dimensional Shape Analysis Using Moments and Fourier Descriptors," *IEEE Trans. on PAMI*, Vol. 10, 1988, pp. 937–943.

[40] Haralick, R. M., and L. G. Shapiro, *Computer and Robot Vision*, Reading, MA: Addison-Wesley, 1993.

[41] Tsallis, C., "Nonextensive Statistical Mechanics and Its Applications," *Series Lecture Notes in Physics*, Berlin: Springer, 2001.

[42] Shannon, C., and W. Weaver, *The Mathematical Theory of Communication*, Urbana, IL: University of Illinois Press, 1948.

[43] Yamano, T., "Information Theory Based in Nonadditive Information Content," *Entropy*, Vol. 3, 2001, pp. 280–292.

[44] Giraldi, G. A., E. Strauss, and A. A. F. Oliveira, "Dual-T-Snakes Model for Medical Imaging Segmentation," *Pattern Recognition Letters*, Vol. 24, No. 7, 2003.

[45] Giraldi, G. A., and A. A. F. Oliveira, "Invariant Snakes and Initialization of Deformable Models," *International Journal of Image and Graphics*, Vol. 4, No. 3, 2004, pp. 363–384.

[46] Osher, S. J., and J. A. Sethian, "Fronts Propagation with Curvature Dependent Speed: Algorithm Based on Hamilton-Jacobi Formulations," *Journal of Computational Physics*, Vol. 79, 1988, pp. 12–49.

[47] Sethian, J. A., "Level Set Methods and Fast Marching Methods: Evolving Interfaces in Computational Geometry, Fluid Mechanics," in *Computer Vision and Materials Science*, 2nd ed., New York: Cambridge University Press, 1999.

[48] Malladi, R., J. A. Sethian, and B. C. Vemuri, "Shape Modeling with Front Propagation: A Level Set Approach," *IEEE Trans. on Pattern Anal. Mach. Intell.*, Vol. 17, No. 2, 1995, pp. 158–175.

[49] McInerney, T. J., "Topologically Adaptable Deformable Models for Medical Image Analysis," Department of Computer Science, University of Toronto, 1997.

[50] Niessen, W. J., B. M. ter Haar Romery and M. A. Viergever, "Geodesic Deformable Models for Medical Image Analysis," *IEEE Trans. on Medical Imaging*, Vol. 17, No. 4, August 1998, pp. 634–641.

[51] Sethian, J. A., "Level Set Methods: Evolving Interfaces in Geometry," in *Fluid Mechanics, Computer Vision, and Materials Sciences*, New York: Cambridge University Press, 1996.

[52] Suri, J. S., et al., "Shape Recovery Algorithms Using Level Sets in 2-D/3-D Medical Imagery: A State-of-the-Art Review," *IEEE Trans. on Information Technology in Biomedicine*, Vol. 6, No. 1, March 2002, pp. 8–28.

[53] Kuhn, H. K., and A. W. Tucker, "Nonlinear Programming," *Berkeley Symp. Math. Stat. Probab.*, 1961, pp. 481–492.

[54] Vapnik, V., and O. Chapelle, "Bounds on Error Expectation for Support Vector Machines," *Neural Comput.*, Vol. 12, No. 3, 2000, pp. 2013–2026.

[55] Jain, A. K., *Fundamentals of Digital Image Processing*, Upper Saddle River, NJ: Prentice-Hall, 1989.

[56] Giraldi, G. A., and A. A. F. Oliveira, "Dual-Snake Model in the Framework of Simplicial Domain Decomposition," *International Symposium on Computer Graphics, Image Processing and Vision (SIBGRAPI'99)*, Campinas, Sao Paulo, Brazil, 1999, pp. 103–106.

[57] Giraldi, G. A., L. M. G. Gonalvez, and A. A. F. Oliveira, "Dual Topologically Adaptable Snakes," *Proc. of the 5th Joint Conference on Information Sciences (JCIS 2000), 3rd Intl. Conf. on Computer Vision, Pattern Recognition and Image Processing*, Vol. 2, 2000, pp. 103–106.

PART IV

3-D Cardiac Ultrasound Imaging

CHAPTER 9

Cardiac Motion Analysis Based on Optical Flow of Real-Time 3-D Ultrasound Data

Qi Duan, Elsa D. Angelini, Olivier Gerard, Kevin D. Costa, Jeffrey W. Holmes, Shunichi Homma, and Andrew F. Laine

With relatively high frame rates and the ability to acquire volume datasets with a stationary transducer, 3-D ultrasound systems based on matrix phased-array transducers provide valuable 3-D information, from which quantitative measures of cardiac function can be extracted. Such analyses require segmentation and visual tracking of the myocardial borders. Due to the large size of the volumetric datasets, manual tracing of the endocardial border is tedious and impractical for clinical applications. In addition, manual tracing usually requires slicing the 3-D dataset into 2-D images, resulting in a loss of some of the spatial continuity and making manual boundary detection more error prone. Therefore, the development of automatic methods for tracking 3-D endocardial motion is essential.

In this study, we evaluate a 4-D optical flow motion tracking algorithm to determine its ability to follow the left ventricular borders in 3-D ultrasound data through time. The optical flow method was implemented using a 3-D correlation. We tested the algorithm on an experimental open-chest dog dataset and a clinical dataset, both of which were acquired with a Philips iE33 3-D ultrasound machine. Initialized with left ventricular endocardial data points obtained from manual tracing at end diastole, the algorithm automatically tracked these points frame by frame through the whole cardiac cycle. A finite element surface was fitted through the data points obtained by both optical flow tracking and manual tracing from an experienced observer for quantitative comparison of the results. Parameterization of the finite element surfaces was performed, and maps displaying relative differences between the manual and semiautomatic methods were compared.

The results showed good consistency between manual tracing and optical flow estimation on 73% of the entire surface with less than a 10% difference. In addition, the optical flow motion tracking algorithm greatly reduced the processing time (about 94% reduction compared to human involvement per cardiac cycle) for analyzing cardiac function in 3-D ultrasound datasets. A displacement field was computed from the optical flow output, and a framework for computation of dynamic cardiac information was introduced. The method was applied to a clinical dataset from a heart transplant patient, and dynamic measurements agreed with physiologic knowledge as well as experimental results.

9.1   Real-Time 3-D Echocardiography

Developments in 3-D echocardiography started in the late 1980s with the introduction of off-line 3-D medical ultrasound imaging systems. The evolution of 3-D ultrasound acquisition systems can be divided into three generations: free-hand scanning, mechanical scanning, and matrix phased arrays. Many review articles have been published during the past decade that assessed the progress and limitations of 3-D ultrasound technology for clinical screening [1–10].

Development of real-time 3-D (RT 3-D) echocardiography started in the late 1990s by Volumetrics [11] based on matrix phased-array transducers. Recently, a new generation of RT 3-D transducers was introduced by Philips Medical Systems (Best, the Netherlands) with the SONOS 7500 transducer followed by the iE33, which can acquire a fully sampled cardiac volume in four cardiac cycles. This technical design enabled a dramatic increase in spatial resolution and image quality, which makes such 3-D ultrasound techniques increasingly attractive for daily cardiac clinical diagnoses. Because RT 3-D ultrasound acquires volumetric ultrasound sequences with fairly high temporal resolution and a stationary transducer, it can capture the complex 3-D cardiac motion very well.

Advantages of using 3-D ultrasound in cardiology include the possibility to display a 3-D dynamic view of the beating heart, and the ability for the cardiologist to explore the 3-D anatomy at arbitrary angles to localize abnormal structures and assess wall motion. During the past decade, this technology has been shown to provide more accurate and reproducible screening for quantification of cardiac function for two main reasons: the elimination of assumptions about ventricular geometry and the improved selection of the visualization planes for performing the ventricular volume measurements. It was validated through several clinical studies for quantification of left ventricle (LV) function as reviewed in [12]. Nevertheless, full exploitation of 3-D ultrasound data for qualitative and quantitative evaluation of cardiac function remains suboptimal for two reasons: lack of appropriate display and lack of automatic boundary detection. Manual tracing of myocardial borders is a tedious task that requires the intervention of an expert cardiologist familiar with the ultrasound machine. Also, slicing the 3-D dataset into 2-D images results in a loss of some of the spatial continuity and makes manual boundary detection more error prone. For this reason, ventricular volumes are commonly estimated via visual inspection of 2-D B-scan images or semiautomated segmentation for difficult cases. Existing commercialized semiautomatic segmentation programs include TomTec (TomTec, Germany) and QLAB (Philips, the Netherlands).

9.2   Anisotropic Diffusion

The presence of speckle noise patterns makes the interpretation of ultrasound images, either by a human operator or with a computer-based system, very difficult. It is highly desirable for certain applications, such as automatic segmentation, to apply some denoising prior to scan conversion in order to remove speckle noise artifacts and improve signal homogeneity within distinct anatomic tissues.

A number of methods have been proposed to denoise and improve the ultrasound image quality including temporal averaging, median filtering, maximum amplitude writing (temporal dilation), adaptive speckle reduction (ASR) (statistical enhancement) [13–17], adaptive weighted median filter (AWMF) [18], homomorphic Wiener filtering, and wavelet shrinkage (WS) [19, 20]. Most of these methods suffer from insufficient denoising, image quality degradation, or high computational costs. Furthermore, some of them require raw RF data, available prior to logarithmic compression [21].

Our group has presented previous work on applying brushlet denoising in spherical coordinates to RT 3-D cardiac ultrasound [22]. Experiments on phantom and clinical cardiac datasets have shown excellent performance for the method. However, the main limitation of this type of denoising remains the computational costs, which currently prevent its implementation for real-time visualization applications in clinical practice.

In this context, in [23], we investigated the performance of a more computationally efficient denoising filter based on anisotropic diffusion for data represented in spherical coordinates. A similar framework can be found in the work of Abd-Elmoniem et al. [21, 24] who used 2-D anisotropic filtering in radial coordinates. Anisotropic diffusion methods are very efficient for speckle reduction in ultrasound and radar images. Yu and Acton [25, 26] applied their speckle reducing filter to synthetic aperture radar images and compared their performance to Lee and Kuan filters and to Frost filters. These filters are all derived from anisotropic diffusion. Finally, Montagnat et al. [27] applied a 3-D anisotropic diffusion filter for rotational cardiac 3-D ultrasound data.

Anisotropic diffusion methods apply the following heat diffusion type of dynamic equation to the gray levels of a given 3-D image data $I(x, y, z, t)$:

$$\frac{\partial I}{\partial t} = \text{div}\big(c(x, y, z, t)\nabla I\big) \tag{9.1}$$

where $c(x, y, z, t)$ is the diffusion parameter, div denotes the divergence operator, and ∇I denotes the gradient of the image intensity. In the original work of Perona and Malik [28, 29], the concept of anisotropic diffusion was introduced with the selection of a variable diffusion parameter, as a function of the gradient of the data:

$$c(x, y, z, t) = g\big(\left|\nabla I(x, y, z, t)\right|\big) \tag{9.2}$$

We used the diffusion function proposed by Weickert [30], defined as follows:

$$g(x, \lambda) = \begin{cases} 1 & x \leq 0 \\ 1 - e^{\frac{3.315}{(x/\lambda)^4}} & x > 0 \end{cases} \tag{9.3}$$

The parameter λ serves as a gradient threshold, defining edge points x_k as locations where $\left|\nabla I_{x_k}\right| > \lambda$. This bell-shaped diffusion function acts as an edge-enhancing filter, with high diffusion values in smooth areas and low values at edge points. The structure of the diffusion tensor with separate weights for each dimension enables it to control the direction of the diffusion process, with flows parallel to edge contours.

In the case of ultrasound, as the diffusion process evolves, image data properties change dramatically, and it is desirable to modify the gradient threshold parameter value. In their paper, Montagnat et al. [27] reported a decrease in the value of significant edges as the homogeneous regions in the ultrasound data are filtered. Therefore, they chose to decrease the threshold gradient in time and proposed values based on a fraction of the cumulative histograms of the data gradients recomputed at each iteration of the diffusion process. In our case, we used a linear model in [23] where:

$$\lambda(t) = \lambda_0 + at \qquad\qquad (9.4)$$

with λ_0 as the initial gradient value, a is a slope parameter, and t is the time iteration index. Parameters were set empirically for the datasets processed. Specifically, in [23], we chose an increasing threshold in order to smooth out sampling artifacts and remove speckle noise.

Filtering performance was assessed in terms of visual quality and for quantitative measurements on a phantom object in [23]. Our quantitative study showed that very high measurement accuracy could be achieved, but required suitable parameter settings for the scan conversion method, whereas the visual quality was similar for all interpolation kernels.

9.3 Tracking of LV Endocardial Surface on Real-Time 3-D Ultrasound with Optical Flow

Clinical evaluation of 3-D ultrasound data for assessment of cardiac function is performed via interactive inspection of animated data, along selected projection planes. Facing the difficulty of inspecting a 3-D dataset with 2-D visualization tools, it is highly desirable to assist the cardiologist with quantitative tools for analysis of 3-D ventricular function. Complex and abnormal ventricular wall motion, for example, can be detected, at a high frame rate, via quantitative 4-D analysis of the endocardial surface and computation of local fractional shortening [31]. Such preliminary studies showed that RT 3-D ultrasound provides unique and valuable quantitative information about cardiac motion, when derived from manually traced endocardial contours. Recent software tools provide interactive segmentation capabilities for the endocardium using a 3-D deformable model that alleviates the need for full manual tracing of the endocardial border. To assist the segmentation process over the entire cardiac cycle, we evaluated the use of optical flow (OF) tracking between segmented frames and tried to answer the following questions in [32]: Can OF track the endocardial surface between end diastole (ED) and end systole (ES) with reliable positioning accuracy? How does dynamic information derived from OF tracking on RT 3-D ultrasound compare to the manual tracing method, given the high intervariability and intravariability of segmentation by experts? Can OF be used as a dynamic interpolation tool for tracking the endocardial surface?

Cardiac motion analysis from images has been an active research area during the past decade. However, most research efforts were based on CT and MRI data.

Previous efforts using ultrasound data for motion analysis include intensity-based OF tracking, strain imaging, and elastography. Intensity-based OF tracking methods described in [33–38] combine a local intensity correlation with specific regularizing constraints (e.g., continuity). For strain imaging or elastography, strain calculation and motion estimation are typically derived from autocorrelations and cross-correlations of RF data. The commercialized strain imaging package 2-D Strain from General Electric [39] uses such a paradigm. Most published papers on strain imaging or elastography [39–43] are limited to 1-D or 2-D images. Early studies [44] used simple simulated phantoms, whereas recent research [45] used a 3-D ultrasound data sequence for LV volume estimation. The presence of speckle noise in ultrasound prevents the use of gradient-based methods, although relatively large region-matching methods are reasonably robust to the presence of noise. In this study, we propose a surface tracking technique based on a 4-D correlation-based OF method on 3-D volumetric ultrasound intensity data.

9.3.1 Correlation-Based Optical Flow

Optical flow tracking refers to the computation of the displacement field of objects in an image, based on the assumption that the intensity of the object remains constant. In this context, motion of the object is characterized by a flow of pixels with constant intensity. The assumption of intensity conservation is typically unrealistic for natural movies and medical imaging applications, motivating the argument that OF can only provide *qualitative* estimation of object motions. There are two global families of OF computation techniques: (1) differential techniques [46–48] that compute velocity from spatiotemporal derivatives of pixel intensities, and (2) region-based matching techniques [49, 50], which compute OF via identification of local displacements that provide optimal homogeneity measurements between two consecutive image frames. Compared to differential OF approaches, region-based methods using homogeneity measures are less sensitive to noisy conditions and fast motion [51] but assume that displacements in small neighborhoods are similar. For 3-D ultrasound, this latter approach appeared more appropriate and was selected for this study. Given two datasets from consecutive time frames, $(I(\mathbf{x}, t), I(\mathbf{x}, t + \Delta t))$, the displacement vector $\Delta \mathbf{x}$ for each pixel in a small neighborhood Ω around a pixel x is estimated via maximization of the cross-correlation coefficient, which is defined as:

$$r = \frac{\sum_{\mathbf{x} \in \Omega} \left(I(\mathbf{x}, t) I(\mathbf{x} + \Delta \mathbf{x}, t + \Delta t) \right)}{\sqrt{\sum_{\mathbf{x} \in \Omega} I^2(\mathbf{x}, t) \sum_{\mathbf{x} \in \Omega} I^2(\mathbf{x} + \Delta \mathbf{x}, t + \Delta t)}} \tag{9.5}$$

In [32], correlation-based OF was applied to estimate the displacement of selected voxels between two consecutive ultrasound volumes in the cardiac cycle. The search window Ω was centered about every $(5 \times 5 \times 5)$ pixel volume and was set to size $(7 \times 7 \times 7)$. To increase the robustness of the estimation, the final estimation of the displacement for each point is the average within a 6-connected neighborhood (i.e., considering, for each voxel, the elements connected in the 3-D space).

9.3.2   3-D Ultrasound Datasets

The tracking approach was tested on three datasets acquired with the SONOS 7500 3-D ultrasound machine mentioned earlier:

1. Two datasets on an anesthetized open-chest dog were acquired before (baseline) and 2 minutes after induction of ischemia via occlusion of the proximal left anterior descending coronary artery. These datasets were obtained by positioning the transducer directly on the apex of the heart, providing high image quality and a small field of view. Spatial resolution of the analyzed data was 0.56 mm^3, and 16 frames were acquired per cardiac cycle.

2. One transthoracic clinical dataset was acquired from a heart transplant patient. Spatial resolution of the analyzed data was 0.8 mm^3, and 16 frames were acquired for one cardiac cycle. Because of the smaller field of view used to acquire the open-chest dog data and the positioning of the transducer directly on the dog's heart, image quality was significantly higher in this dataset, with some fine anatomic structures visible. Cross-sectional views at ED from the open-chest baseline dataset and the patient dataset are shown in Figure 9.1.

9.3.3   Surface Tracing

The endocardial surface of the LV was extracted with two methods. (1) An expert performed manual tracing of all time frames in the datasets, on rotating B-scan views (long-axis views rotating around the central axis of the ventricle) and C-scan views (short-axis views at different depths). (2) QLAB software was used to segment the endocardial surface. Initialization was performed by a human expert, and a parametric deformable model was fit to the data at each time frame. Segmentation results were reviewed by the same expert and adjusted manually for final correc-

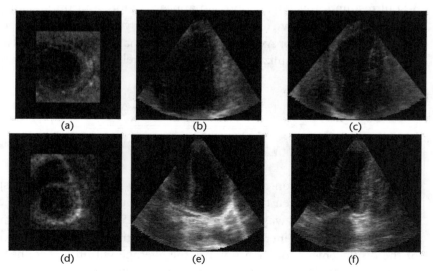

| (a) | (b) | (c) |
| (d) | (e) | (f) |

Figure 9.1   Cross-sectional views at ED for (a–c) open-chest dog data, prior to ischemia, and (d–f) patient with transplanted heart; (a, d) axial, (b, e) elevation, and (c, f) azimuth views.

tions. We emphasize here that QLAB is used as a semiautomated segmentation tool. The QLAB software was designed to process human clinical datasets. Because significant anatomic differences between canine and human hearts could lead to misbehavior of the segmentation software, we decided to apply the software tool only to clinical datasets.

9.3.4   Surface Tracking with Optical Flow

Tracking of the endocardial surface with OF was applied after initialization using the manually traced surfaces (for the dog data and clinical data) and the QLAB segmented surfaces (for the clinical data). Starting with a set of endocardial surface points (about 3,000 points, roughly 1 mm apart for manual tracing, and about 800 points, roughly 3 mm apart for QLAB) defined at ED, the OF algorithm was used to track the surface in time through the whole cardiac cycle. Because the correlation-based OF method is very sensitive to speckle noise, all datasets were presmoothed with edge-preserving anisotropic diffusion as developed in [23] and described in Section 9.2. We emphasize here that OF was not applied as a segmentation tool but as a surface tracking tool for a given segmentation method.

9.3.5   Evaluation

We evaluated OF tracking performance via visualization and quantification of dynamic ventricular geometry compared to segmented surfaces. Usually comparison of segmentation results is performed via global measurements such as volume difference or MSE. To provide local comparisons, we proposed a novel comparison method in [52] based on a parameterization of the endocardial surface in prolate spheroidal coordinates [53] and previously used for comparison of ventricular geometries from two 3-D ultrasound machines in [54].

The endocardial surfaces were registered using three manually selected anatomic landmarks: the center of the mitral orifice, the endocardial apex, and the equatorial midseptum. The data were fitted in prolate spheroidal coordinates (λ, μ, θ), projecting the radial coordinate λ to a 64-element surface mesh with bicubic Hermite interpolation, yielding a realistic 3-D endocardial surface. The fitting process (illustrated in Figure 9.2 for a single endocardial surface) was performed using custom routines written in MATLAB. In this figure, we can observe the initial positioning of the data points and the surface mesh, and the finite element surface after fitting with very high agreement between the data and the mesh. The inset provided for a small region shows the quality of agreement between the fitted surface and the points resulting from region-based global optimization of radial projections.

The fitted nodal values and spatial derivatives of the radial coordinate, λ, were then used to map relative differences between two surfaces, $\varepsilon = (\lambda_{\text{seg}} - \lambda_{\text{OF}})$, using custom software. Hammer mapping was used to flatten the endocardial surface via an area-preserving mapping [55]. For each time frame, the RMSE of the difference in λ summed over all nodes on the endocardial surface, were computed between OF and individual segmentation methods. Ventricular volumes were also computed from the segmented and the tracked endocardial surfaces. Finally relative λ difference maps were generated for ES, providing a direct quantitative comparison of ventricu-

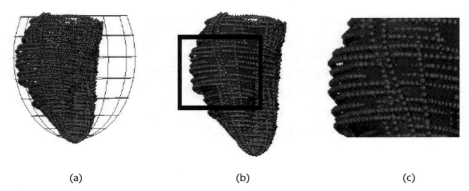

(a)   (b)   (c)

Figure 9.2 Fitting process of the endocardial surface at ES. (a) Initial FEM mesh and data points. (b) Fitted FEM surface and data points. (c) Inset details for a small region with the FEM fitted surface and the data points.

lar geometry. These maps are visualized in Figure 9.3 with isolevel lines, quantified in fractional values of radial difference.

9.3.6 Results

9.3.6.1 Dog Data

On the dog datasets, RMSE results reported a maximum radial absolute difference of 0.19 (average radial coordinate value was 0.7 ± 0.2 at ED and 0.6 ± 0.3 at ES) at frame 11 (start of diastole) on the baseline dataset and 0.08 (average radial coordinate value was 0.7 ± 0.3 at ED and 0.6 ± 0.2 at ES) at frame 12 (start of diastole) on the postischemia dataset. Maximum LV volume differences were less than 7 mL on baseline data and 5 mL on the postischemia dataset. RMSE values were smaller for OF tracking on larger volumes. On the radial difference maps in Figure 9.3, we observe similar difference patterns in the baseline and the postinfarct data except for a dark region near the apical lateral region, demonstrating repeatability of the OF tracking performance on a given ventricular geometry but with different contractility patterns. An area with large error in the baseline comparison localized on the anterior lateral wall disappeared in postischemia tracking. This error is caused by a small portion of tracked points that were confused by acquisition artifacts at the boundary between the first and second quadrants of acquisition. Errors were rather evenly distributed over the endocardial surface with overall shape agreement. Similar maps can be used to examine local fractional shortening using the technique developed by the Cardiac Biomechanics Group at Columbia University [55]; such maps revealed similar patterns of abnormal wall motion after ischemia using OF tracked surface or manual tracing, corroborating the accuracy of OF tracking to provide dynamic functional information.

9.3.6.2 Clinical Data

OF tracking was run with initialized surfaces provided by either manual tracing or the QLAB segmentation tool on the clinical dataset. Because of lower image quality

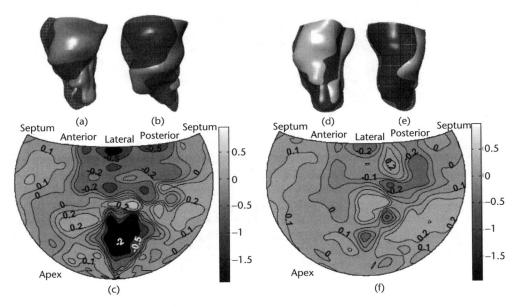

Figure 9.3   Endocardial surfaces from open-chest dog datasets at ES. (a–c) Results on baseline data. (d–f) Results on postischemia data. The 3-D renderings of endocardial surfaces were generated from manual tracings (dark gray) and OF tracking (light gray) for (a, d) lateral views and (b, e) anterior views. (c, f) Relative difference maps between OF and manual tracing surfaces.

on the clinical dataset, compared to the open-chest dog data, we performed two sets of additional experiments. First, we checked if the time frame selected for initialization had an influence on the tracking quality.

Based on manual tracing, we initialized OF tracking for the whole cardiac cycle with ED (forward tracking) or ES (backward tracking) and compared RMSE over the entire cycle. The results plotted in Figure 9.4(a) show very comparable performance, confirming that the OF seems to be repeatable and insensitive to initialization setup. Therefore, we selected the first volume in the sequences, which always corresponds to ED in our experiments. A second experiment evaluated the agreement between QLAB and OF tracking when increasing the number of reference surfaces used to reinitialize OF over the cardiac cycle. The results plotted in Figure 9.4(b) show that agreement of OF tracking and QLAB segmentation increases with reinitialization frequency and reaches RMSE levels similar to the experiment with manual tracing for reinitialization (i.e., reload QLAB segmentation for that frame instead of using the tracing result from the previous frame) every other frame. We point out that strong smoothing constraints, applied by the deformable model of the QLAB segmentation, lead to surface positioning that did not always correspond to the apparent high-contrast interface. Finally, we compared RMSE values from forward tracking and from averaging forward and backward tracked shapes. We observed a large increase in agreement with the QLAB smooth segmentation when averaging tracked surfaces.

As shown in Figure 9.5, experiments showed that OF tracking initialized with manual tracing provides ventricular endocardial surfaces similar to that obtained by manual tracing, with less than 0.1 maximum absolute differences in RMSE and

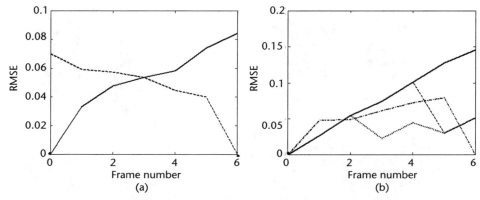

Figure 9.4 Results of clinical data: (a) RMSE between OF tracking and manual tracing: forward (solid line) and backward (dashed line). (b) RMSE between OF tracking and QLAB segmentation: forward tracking without reinitialization (solid line), forward tracking with reinitialization every fourth frame (dashed line), forward tracking with reinitialization every second frame (dotted line), and average result from forward and backward tracking without reinitialization (dashed-dotted line).

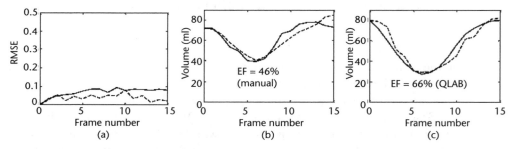

Figure 9.5 Results of clinical data: (a) RMSE of radial difference for OF initialized with manual tracing (solid line) and QLAB segmentation with two-frame reinitialization (dashed line). LV volumes over one cardiac cycle: (b) manual tracing (solid line) and OF initialized with manual tracing (dashed line); and (c) QLAB segmentation (solid line) and OF initialized with QLAB segmentation (dashed line).

maximum LV volume differences below 10 mL. When initialized with QLAB, OF tracking with reinitialization shows results with less than 0.08 maximum RMSE difference and less than 13 ml for LV volume differences. These differences are similar to interobserver and intraobserver variability for measurement of LV volume by echocardiography [56, 57].

Ventricular geometries are illustrated in Figure 9.6. We again observed high overall agreement between endocardial geometries provided by manual tracing and OF tracking. Radial differences were distributed over the entire surface, with higher values on the lateral-posterior wall. The QLAB segmentation provided very smooth surfaces, well tracked by the OF. Larger errors were again observed on the lateral posterior wall. Comparison of the two experiments shows that OF over one time

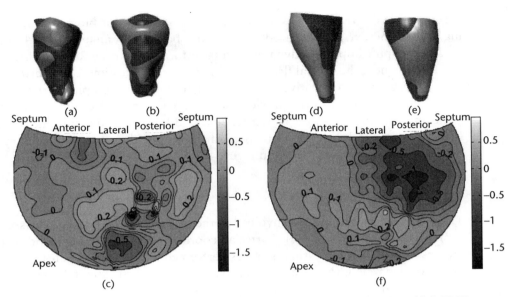

Figure 9.6   Endocardial surfaces from clinical data at ES. (a–c) Manual tracing; (d–f) QLAB segmentation. The 3-D rendering of endocardial surfaces from the segmentation method (dark gray) and OF tracking (light gray): (a, d) lateral view; and (b, e) anterior view. Relative radial difference maps between OF tracking and the segmentation method.

frame can preserve the smoothness of the surface, but will tend toward more convoluted surfaces during temporal propagation of the tracking process.

The time needed for computing optical flow is about 30 seconds per frame, compared with 5 to 10 minutes per frame with manual tracing. With optical flow, the processing time for model-based 3-D cardiac motion analysis can be cut from 30 to 60 minutes to 3 minutes, which makes the application of 3-D cardiac motion analysis much more practical in clinical applications.

Based on the high agreement with manual tracing, we can infer that OF might be a good candidate for guiding a deformable model with high smoothness constraints to better adapt to the ultrasound data and incorporate temporal information in the segmentation process. On the other hand, OF tracking could be adapted to these smoothness constraints, which are better ranked by cardiologists, to track larger spatial windows around the endocardial surface.

9.4   Dynamic Cardiac Information from Optical Flow

In [58], we extended our approach to the extraction of motion fields, generated from the optical flow algorithm that efficiently described complex 3-D myocardial deformations. Traditional approaches convert displacement information recovered from the image data in Cartesian coordinates into polar (2-D) or cylindrical (3-D) coordinates to adapt to the natural shape of the left ventricle. Most efforts to quantify cardiac motion from echocardiography focus on radial and circumferential displacements, but ignore gradients of displacements, like thickening and twist. These gradients are of great diagnostic interest and are critical for biomechanical modeling. In this context, we proposed a framework based on semiautomatic 4-D optical flow to compute important dynamic cardiac information using RT 3-D ultrasound.

In this study, the optical flow algorithm sequentially estimated the displacement field between two consecutive frames throughout the ejection phase from ED to ES on the clinical dataset in the previous section. Myocardial motion is estimated via optical flow tracking using a similar scheme in [32]. Cardiac dynamic measurements (displacements and their derivatives) are then computed. A flowchart of the computational framework is provided in Figure 9.7.

9.4.1 Coordinate Systems

Three coordinate systems are involved in the computational framework (see Figure 9.8): pixel coordinates (i, j, k), Cartesian coordinates (x, y, z), and cylindrical coordinates (r, θ, z). The OF estimation was performed in pixel coordinates. For computation of dynamic information, displacements in pixel coordinates were converted into Cartesian coordinates and centered inside the ventricular cavity so that the z-axis is aligned with the long axis of the left ventricle. This coordinate transform is performed via rigid transformation:

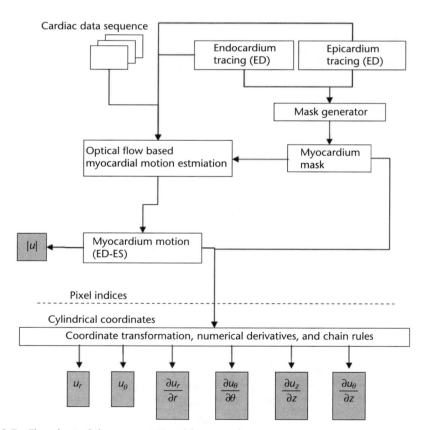

Figure 9.7 Flowchart of the computational framework.

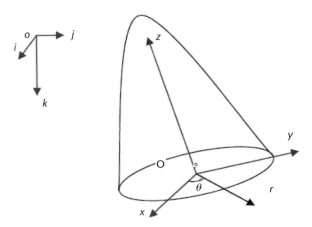

Figure 9.8 Coordinate systems for data acquisition and computation.

$$\begin{bmatrix} x \\ y \\ z \end{bmatrix} = \mathbf{R} \begin{bmatrix} i \\ j \\ k \end{bmatrix} + \mathbf{T} \begin{bmatrix} r_{11} & r_{12} & r_{13} \\ r_{21} & r_{22} & r_{23} \\ r_{31} & r_{32} & r_{33} \end{bmatrix} \begin{bmatrix} i \\ j \\ k \end{bmatrix} + \begin{bmatrix} -O_i \\ -O_j \\ -O_k \end{bmatrix} \tag{9.6}$$

where $\mathbf{R}$ is a rotation matrix, and $\mathbf{T}$ is a translation vector, equal to the negative pixel coordinates of the origin O of the Cartesian coordinate system. The ventricular axis was defined as the axis connecting the center of the mitral orifice and the endocardial apex. This axis has a very stable position during the whole cardiac cycle [55]. Based on the Cartesian coordinate system, a corresponding cylindrical coordinate system is established with the r-θ plane corresponding to the x-y plane and with the x-axis used as the reference for θ.

9.4.2 Dynamic Cardiac Information Measurements

Besides displacement (u_x, u_y, u_z) in Cartesian coordinates, we computed the following dynamic measurements:

- Flow magnitude $|u|$ (mm);
- Radial displacement u_r (mm);
- Circumferential displacement u_θ (mm);
- Thickening $\partial u_r/r$;
- Circumferential stretch $\partial u_\theta/\partial\theta$;
- Longitudinal stretch $\partial u_z/\partial z$;
- Twist $\partial u_\theta/\partial z$.

Gradient values were computed directly in pixel coordinates and converted into the cylindrical coordinate system via the chain rule. Derivatives in pixel coordinates were approximated by central difference operators to accommodate second-order continuity of the flow field after RBF interpolation.

9.4.3   Data

We used the heart transplant clinical dataset described in the previous section. Due to the heart transplant surgery procedure, the patient had reduced cardiac function and his septum had significant motion reduction. Due to field-of-view limitations, the apical epicardial surface was not visible in the ultrasound volume. Therefore, only the basal and middle parts of the left ventricle were used for dynamic analysis. In fact, the myocardial shape in this region can be well approximated by a cylinder, which can reduce geometric errors in radial displacement estimation.

9.4.4   Results and Discussion

We present results for computation of the myocardial flow field (Figure 9.9), radial displacement (Figure 9.10), thickening (Figure 9.11), and twist (Figure 9.12) during the systolic phase.

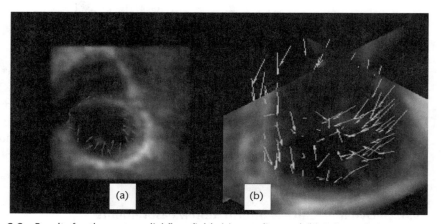

Figure 9.9   Results for the myocardial flow field: (a) one slice and (b) 3-D rendering.

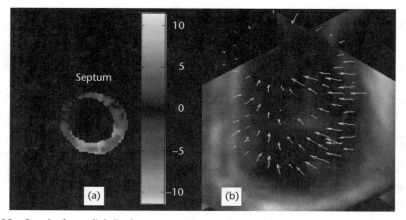

Figure 9.10   Results for radial displacement: (a) one slice and (b) 3-D rendering.

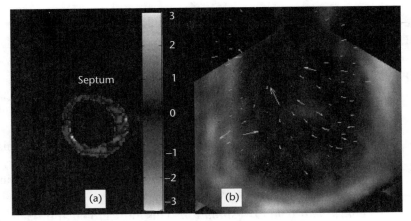

Figure 9.11   Results for thickening: (a) one slice and (b) 3-D rendering.

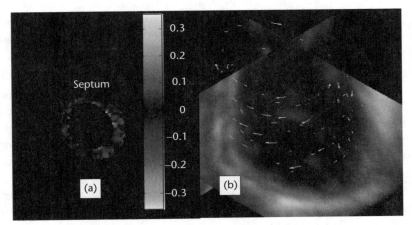

Figure 9.12   Results for twist: (a) one slice and (b) 3-D rendering.

Most of the radial displacement components showed inward motion (negative displacement) of the ventricular wall, except on the septal side, where reduced amplitude and outward motion were observed (Figure 9.10). These findings were in agreement with clinical observations on the dataset and typical findings after heart transplant surgery. The gradient of radial displacement, or thickening, yielded positive thickening at the endocardial surface except for the septal wall where zero or small thinning at the epicardium border was observed (Figure 9.11). Such a pattern agrees with experimental findings [60, 61]. Regarding twist, most parts of the wall exhibited clockwise twist patterns relative to the base, when looking from base to apex. This result also agrees with experimental findings [60, 61] of positive (clockwise) twist during the systolic phase. However, we observed negative twist values in the septal wall. For most parts of the wall, twist values increased radially from the epicardial to the endocardial surface, which concurs with theoretical and experimental results [61].

9.4.5   Discussion About Estimation of the Myocardial Field

Different schemes exist to estimate the myocardial motion field using optical flow in real-time 3-D echocardiography. In [62], four different optical flow–based schemes were investigated under a generalized framework:

Scheme 1: Boundary tracking with RBF interpolation [59]: This is the scheme we proposed in [58].

Scheme 2: Direct tracking within a myocardial mask: This scheme is a straightforward alternative to scheme 1. It tracks with OF every voxel within a myocardial mask defined by the myocardial surfaces, instead of interpolating the OF tracking result of these surfaces.

Scheme 3: Full-field OF estimation: This more global approach consists of using OF to estimate the motion field for all of the voxels in the input volumes. The myocardial motion field is then extracted by masking the motion field of the voxels belonging to the myocardium.

Scheme 4: Full-field OF estimation with smoothing: The RBF interpolation of scheme 1 provided a second-order continuity that is not used in direct OF estimation. We tested an alternative to scheme 3 by adding smoothing via cubic spline regularization on the full-field OF computation.

Experimental results showed the following: The radial displacement fields derived from the four different schemes were similar except for fine details within the myocardium. This was expected because all the methods depended on the OF tracking results. For the radial thickening, schemes 1, 3, and 4 provided similar results, whereas scheme 2 produced flawed results due to the derivative calculation across the boundary.

The thickening values of the normal part of the wall from schemes 1, 3, and 4 (around 0.1 to 0.25) were close to the normal values of 0.1 to 4 reported in [63] and 20% to 40% reported in [64].

The segmental averaged thickening result showed that anterior and lateral segments had normal motion, whereas the septal segment had outward motion and negative thickening (i.e., thinning) values; the posterior and anteroseptal segments had reduced motion; the inferior segments had very small deformation or thickening.

All schemes required the same amount of manual initialization, that is, endocardial and epicardial tracing at ED. In terms of accuracy, scheme 3 provided more accurate displacement estimations; Scheme 4, however, was more robust when estimating thickening, and it benefitted from its intrinsic smoothness constraints.

One important thing to be pointed out is that, although on some "normal" datasets the interpolated scheme (scheme 1) and the full-field schemes (schemes 3 and 4) may provide similar results, we still recommend using a full-field scheme (e.g., scheme 4 or other, more sophisticated field fitting techniques) for myocardial deformation estimation instead of using an interpolated version from the ventricular boundary, in order to capture the abnormal motion patterns within the myocardium. An interpolated scheme should not be used in clinical settings.

9.5 Summary

Real-time 3-D echocardiography provides valuable 3-D information, from which quantitative measures of cardiac function can be extracted. In this chapter, we proposed an optical flow–based method to extract the ventricular boundaries semiautomatically. This method was validated on experimental and clinical datasets. Information extracted by optical flow can be fed into model-based motion analysis tools. With a huge savings in processing time, the optical flow method makes cardiac motion analysis on RT 3-D echocardiography more practical in clinical applications. The myocardial motion field can also be estimated based on the optical flow estimation, from which clinical meaningful cardiac dynamic metrics can be derived.

Acknowledgments

This work was funded by National Science Foundation grant BES-02-01617, American Heart Association No. 0151250T, Philips Medical Systems, New York State NYSTAR/CAT Technology Program, and the Louis Morin Fellowship program. The authors also would like to thank Dr. Todd Pulerwitz (Department of Medicine, Columbia University), Susan L. Herz, and Christopher M. Ingrassia.

References

[1] Ofili, E. O., and N. C. Nanda, "Three-Dimensional and Four-Dimensional Echocardiography," *Ultrasound Medical Biology*, Vol. 20, 1994, pp. 663–675.

[2] Fenster, A., and D. B. Downey, "Three-Dimensional Ultrasound Imaging," in *Handbook of Medical Imaging, Volume 1: Physics and Psychophysics*, H. L. K. J. Beutel and R. L. Metter, (eds.), Bellingham, WA: SPIE Press, 2000, pp. 463–510.

[3] Rankin, R. N., et al., "Three-Dimensional Sonographic Reconstruction: Technique and Diagnostic Applications," *American Journal of Radiology*, Vol. 161, 1993, pp. 695–702.

[4] Belohlavek, M., et al., "Ultrasound Imaging: A New Era for Echocardiography," *Mayo Clinic Proceedings*, Vol. 68, 1993, pp. 221–240.

[5] Warmath, J. R., et al., "Ultrasound 3-D Volume Reconstruction from an Optically Tracked Endorectal Ultrasound (TERUS) Probe," *Proc. of SPIE*, Vol. 5367, San Diego, CA, 2004, pp. 228–236.

[6] Managuli, R., et al., "Advanced Volume Rendering Algorithm for Real-Time 3-D Ultrasound: Integrating Pre-Integration into Shear-Image-Order Algorithm," *Proc. of SPIE*, Vol. 6147, San Diego, CA, 2006, pp. 10–17.

[7] Zhang, H., et al., "Freehand 3-D Ultrasound Calibration Using an Electromagnetically Tracked Needle," *Proc. of SPIE*, Vol. 6147, San Diego, CA, 2006, pp. 775–783.

[8] Yu, H., M. S. Pattichis, and M. Beth Goens, "Multi-View 3-D Reconstruction with Volumetric Registration in a Freehand Ultrasound Imaging System," *Proc. of SPIE*, Vol. 6147, San Diego, CA, 2006, pp. 45–56.

[9] Sanches, J., J. M. Bioucas-Dias, and J. S. Marques, "Minimum Total Variation in 3-D Ultrasound Reconstruction," *ICIP 2005, IEEE Intl. Conf. on Image Processing*, Vol. 3, Genova, Italy, 2005, pp. 597–600.

[10] Xu, J., et al., "Texture-Based 3-D Ultrasound Real-Time Volume Rendering," *Jisuanji Gongcheng/Computer Engineering*, Vol. 32, 2006, pp. 231–232.

[11] Ramm, O. T. V., and S. W. Smith, "Real Time Volumetric Ultrasound Imaging System," *Journal of Digital Imaging*, Vol. 3, 1990, pp. 261–266.

[12] Krenning, B. J., M. M. Voormolen, and J. R. T. C. Roelandt, "Assessment of Left Ventricular Function by Three-Dimensional Echocardiography," *Cardiovasc. Ultrasound*, Vol. 1, No. 1, 2003.

[13] Bamber, J. C., and C. Daft, "Adaptive Filtering for Reduction of Speckle in Ultrasound Pulse-Echo Images," *Ultrasonics*, 1986, pp. 41–44.

[14] Bamber, J. C., and G. Cook-Martin, "Texture Analysis and Speckle Reduction in Medical Echography," *SPIE Intl. Symp. Pattern Recog. Acous. Imaging*, Vol. 768, 1987, pp. 287–288.

[15] Bamber, J. C., and J. V. Philips, "Real-Time Implementation of Coherent Speckle Suppression in B-Scan Images," *Ultrasonics*, Vol. 29, 1991, pp. 218–224.

[16] Crawford, D. C., D. S. Bell, and J. C. Bamber, "Implementation of Ultrasound Speckle Filters for Clinical Trial," *Proc. IEEE Ultrasonic Symp.*, 1990, pp. 1589–1592.

[17] Crawford, D. C., D. S. Bell, and J. C. Bamber, "Compensation for the Signal Processing Characteristics of Ultrasound B-Mode Scanners in Adaptive Speckle Reduction," *Ultrasound Med. Biol.*, Vol. 19, 1993, pp. 469–485.

[18] Loupas, T., W. N. Mcdicken, and P. L. Allan, "An Adaptive Weighted Median Filter for Speckle Suppression in Medical Ultrasonic Images," *IEEE Trans. on Circuits and Systems*, Vol. 36, 1989, pp. 129–135.

[19] Hao, X., S. Gao, and X. Gao, "A Novel Multiscale Nonlinear Thresholding Method for Ultrasonic Speckle Suppressing," *IEEE Trans. on Medical Imaging*, Vol. 18, 1999, pp. 787–794.

[20] Zong, X., A. F. Laine, and E. A. Geiser, "Speckle Reduction and Contrast Enhancement of Echocardiograms via Multiscale Nonlinear Processing," *IEEE Trans. on Medical Imaging*, Vol. 17, 1998, pp. 532–540.

[21] Abd-Elmoniem, K. Z., A.-B. M. Youssef, and Y. M. Kadah, "Real-Time Speckle Reduction and Coherence Enhancement in Ultrasound Imaging Via Nonlinear Anisotropic Diffusion," *IEEE Trans. on Biomedical Engineering*, Vol. 49, 2002, pp. 997–1014.

[22] Angelini, E. D., et al., "LV Volume Quantification Via Spatiotemporal Analysis of Real-Time 3-D Echocardiography," *IEEE Trans. on Medical Imaging*, Vol. 20, 2001, pp. 457–469.

[23] Duan, Q., E. D. Angelini, and A. Laine, "Assessment of Visual Quality and Spatial Accuracy of Fast Anisotropic Diffusion and Scan Conversion Algorithms for Real-Time Three-Dimensional Spherical Ultrasound," *SPIE International Symposium on Medical Imaging*, San Diego, CA, 2004.

[24] Abd-Elmoniem, K. Z., Y. M. Kadah, and A.-B. M. Youssef, "Real Time Adaptive Ultrasound Speckle Reduction and Coherence," *Proc. 2000 Int. Conf. Image Processing*, Vancouver, BC, Canada, 2000.

[25] Yu, Y., and S. T. Acton, "Segmentation of Ultrasound Imagery Using Anisotropic Diffusion," *Asilomar Conference on Signals, Systems and Computers*, Pacific Grove, CA, 2001, pp. 4–7.

[26] Yu, Y., and S. T. Acton, "Speckle Reducing Anisotropic Diffusion," *IEEE Trans. on Image Processing*, Vol. 11, 2002, pp. 1260–1270.

[27] Montagnat, J., et al., "Anisotropic Filtering for Model-Based Segmentation of 4D Cylindrical Echocardiographic Images," *Pattern Recognition Letters*, Vol. 24, 2003, pp. 815–828.

[28] Perona, P., and J. Malik, "Scale Space and Edge Detection Using Anisotropic Diffusion," *IEEE Workshop on Computer Vision*, 1987.

[29] Perona, P., and J. Malik, "Scale-Space and Edge Detection Using Anisotropic Diffusion," *IEEE Trans. on Pattern Anal. Machine Intell.*, Vol. 12, 1990, pp. 629–639.

[30] Weickert, J., B. M. T. H. Romeny, and M. A. Viergever, "Efficient and Reliable Schemes for Nonlinear Diffusion Filtering," *IEEE Trans. on Image Processing*, Vol. 7, 1998, pp. 398–410.

[31] Herz, S., et al., "Parameterization of Left Ventricular Wall Motion for Detection of Regional Ischemia," *Annals of Biomedical Engineering*, Vol. 33, 2005, pp. 912–919.

[32] Duan, Q., et al., "Tracking of LV Endocardial Surface on Real-Time Three-Dimensional Ultrasound with Optical Flow," *3rd International Conference on Functional Imaging and Modeling of the Heart 2005*, Barcelona, Spain, 2005.

[33] Tsuruoka, S., et al., "Regional Wall Motion Tracking System for High-Frame Rate Ultrasound Echocardiography," *Proc. 4th Int. Workshop on Advanced Motion Control, Part 1*, Tsu, Japan, 1996.

[34] Mikic, I., S. Krucinski, and J. D. Thomas, "Segmentation and Tracking in Echocardiographic Sequences: Active Contours Guided by Optical Flow Estimates," *IEEE Trans. on Medical Imaging*, Vol. 17, 1998, pp. 274–284.

[35] Boukerroui, D., J. A. Noble, and M. Brady, "Velocity Estimation in Ultrasound Images: A Block Matching Approach," *Lecture Notes in Computer Science*, Vol. 2732, 2003, pp. 586–598.

[36] Yu, W., et al., "Motion Analysis of 3-D Ultrasound Texture Patterns," *Lecture Notes in Computer Science*, Vol. 2674, 2003, pp. 252–261.

[37] Paragios, N., "A Level Set Approach for Shape-Driven Segmentation and Tracking of the Left Ventricle," *IEEE Trans. on Medical Imaging*, Vol. 22, 2003, pp. 773–776.

[38] Bardinet, E., L. D. Cohen, and N. Ayache, "Tracking and Motion Analysis of the Left Ventricle with Deformable Superquadrics," *Medical Image Analysis*, Vol. 1, 1996, pp. 129–149.

[39] Behar, V., et al., "The Combined Effect of Nonlinear Filtration and Window Size on the Accuracy of Tissue Displacement Estimation Using Detected Echo Signals," *Ultrasonics*, Vol. 41, 2004, pp. 743–753.

[40] Bang, J., et al., "A New Method for Analysis of Motion of Carotid Plaques from RF Ultrasound Images," *Ultrasound in Medicine and Biology*, Vol. 29, 2003, pp. 967–976.

[41] Rabben, S. I., et al., "Ultrasound-Based Vessel Wall Tracking: An Auto-Correlation Technique with RF Center Frequency Estimation," *Ultrasound in Medicine and Biology*, Vol. 28, 2002, pp. 507–517.

[42] D'Hooge, J., et al., "Deformation Imaging by Ultrasound for the Assessment of Regional Myocardial Function," *2003 IEEE Ultrasonics Symposium*, Honolulu, HI, 2003.

[43] Konofagou, E. E., et al., "Myocardial Elastography—Comparison to Results Using MR Cardiac Tagging," *IEEE Ultrasonics Symposium*, Honolulu, HI, 2003.

[44] Gutierrez, M. A., et al., "Computing Optical Flow in Cardiac Images for 3-D Motion Analysis," *Proc. Conf. Computers in Cardiology*, London, U.K., 1993.

[45] Shin, I.-S., et al., "Left Ventricular Volume Estimation from Three-Dimensional Echocardiography," *Proc. SPIE: Ultrasonic Imaging and Signal Processing*, San Diego, CA, 2004.

[46] Lucas, B. D., and T. Kanade, "An Iterative Image Registration Technique with an Application to Stereo Vision," *International Joint Conference on Artificial Intelligence (IJCAI)*, Vancouver, Canada, April 1981, pp. 674–679.

[47] Horn, B. K. P., and B. G. Schunck, "Determining Optical Flow," *Artificial Intelligence*, Vol. 17, 1981, pp. 185–203.

[48] Nagel, H., "Displacement Vectors Derived from Second-Order Intensity Variations in Image Sequences," *Computer Vision Graphics Image Processing*, Vol. 21, 1983, pp. 85–117.

[49] Anandan, P., "A Computational Framework and an Algorithm for the Measurement of Visual Motion," *International journal of Computer Vision*, Vol. 2, 1989, pp. 283–310.

[50] Singh, A., "An Estimation-Theoretic Framework for Image-Flow Computation," *International Conference on Computer Vision*, Osaka, Japan, 1990, pp. 168–177.

[51] Barron, J. L., D. Fleet, and S. Beauchemin, "Performance of Optical Flow Techniques," *Int. Journal of Computer Vision*, Vol. 12, 1994, pp. 43–77.

[52] Duan, Q., et al., "Evaluation of Optical Flow Algorithms for Tracking Endocardial Surfaces on Three-Dimensional Ultrasound Data," *SPIE International Symposium on Medical Imaging 2005*, San Diego, CA, 2005.

[53] Ingrassia, C. M., et al., "Impact of Ischemic Region Size on Regional Wall Motion," *Proc. 2003 Annual Fall Meeting of the Biomedical Engineering Society,* Nashville, TN, October 1–4, 2003.

[54] Angelini, E. D., et al., "Comparison of Segmentation Methods for Analysis of Endocardial Wall Motion with Real-Time Three-Dimensional Ultrasound," *Computers in Cardiology*, Memphis, TN, 2002.

[55] Herz, S., et al., "Novel Technique for Quantitative Wall Motion Analysis Using Real-Time Three-Dimensional Echocardiography," *Proc. 15th Annual Scientific Sessions of the American Society of Echocardiography,* San Diego, CA, June 26–30, 2004.

[56] Schiller, N. B., et al., "Left Ventricular Volume from Paired Biplane Two-Dimensional Echocardiography," *Circulation*, Vol. 60, 1979, pp. 547–555.

[57] Folland, E. D., et al., "Assessment of Left Ventricular Ejection Fraction and Volumes by Real-Time, Two-Dimensional Echocardiography. A Comparison of Cineangiographic and Radionuclide Techniques," *Circulation*, Vol. 60, 1979, pp. 760–766.

[58] Duan, Q., et al., "Dynamic Cardiac Information from Optical Flow Using Four Dimensional Ultrasound," *27th Annual International Conference IEEE Engineering in Medicine and Biology Society (EMBS)*, Shanghai, China, 2005.

[59] Baxter, B. J. C., "The Interpolation Theory of Radial Basis Functions," Ph. D. thesis, Trinity College, 1992.

[60] Humphrey, J. D., *Cardiovascular Solid Mechanics: Cells, Tissues, and Organs*, New York: Springer, 2002.

[61] Buchalter, M. B., et al., "Noninvasive Quantification of Left Ventricular Rotational Deformation in Normal Humans Using Magnetic Resonance Imaging Myocardial Tagging," *Circulation*, Vol. 81, 1990, pp. 1236–1244.

[62] Duan, Q., et al., "Comparing Optical-Flow Based Methods for Quantification of Myocardial Deformations on RT3-D Ultrasound," *IEEE International Symposium on Biomedical Imaging (ISBI)*, Arlington, VA, 2006, pp. 173–176.

[63] Waldman, L. K., Y. C. Fung, and J. W. Covell, "Transmural Myocardial Deformation in the Canine Left Ventricle. Normal In Vivo Three-Dimensional Finite Strains," *Circ. Res.*, Vol. 57, 1985, pp. 152–163.

[64] Nieminen, M., et al., "Serial Evaluation of Myocardial Thickening and Thinning in Acute Experimental Infarction: Identification and Quantification Using Two-Dimensional Echocardiography," *Circulation*, Vol. 66, 1982, pp. 174–180.

CHAPTER 10

Echocardiography: A Tool for LVSD Identification

Manivannan Jayapalan

10.1   Introduction

The human heart is a hollow, muscular, four-chamber organ that pumps blood through the blood vessels by repeated rhythmic contractions. Ventricular dysfunction is the result of a diversity of structural and/or physiologic abnormalities of myocardial relaxation or contraction that alters the ventricular flow. The cardiac cycle contains two phases, diastole and systole, and both may be affected by diseases of the heart. Thus, ventricular dysfunction can be diastolic or systolic, or both.

It is essential to consider the ventricular dysfunction in the left-hand side because the left ventricle plays an important role in circulating the blood to the whole body. Systolic dysfunction results from a decrease in global or regional myocardial contractility. The fraction of blood that is expelled in relation to the blood volume in the heart in diastole is an expression of systolic function and is called the ejection fraction (EF). During systole, the heart empties and an impairment of contractility causes less blood to be expelled. thus decreasing the stroke volume but increasing the end-systolic volume as a whole EF that ends in progressive pump failure. The impaired capacity of the left ventricle to fill and to maintain its stroke volume without a compensatory increase of atrial filling pressures represents diastolic dysfunction. Ejection of a lesser volume of blood because of a decrease in contractility and reduced ventricular filling because of the impairment of myocardial muscles are shown in the Figure 10.1.

Reduced ability of the left ventricle to eject blood, called left ventricle systolic dysfunction (LVSD), is a common predecessor of heart failure (HF). The prognosis for heart failure is poor and the morbidity is high with frequent hospitalizations, especially among the elderly. The prevalence of HF is increasing and threatens to become a major logistic and economic burden to global health care systems. HF currently contributes to 30.9% of global mortality and 10.3% of global morbidity [1] that can be reduced by optimal treatment, requiring objective evaluation of LV function and cardiac anatomy. Detection and treatment of LVSD is essential to reduce the incidence of cardiac failure and its consequences. The major cause of

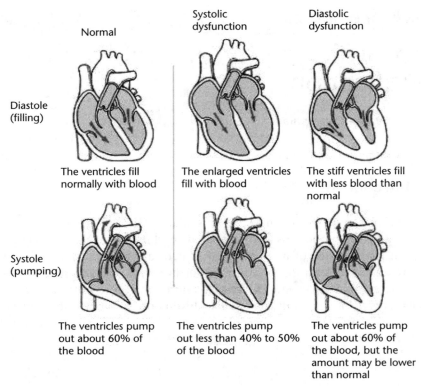

Figure 10.1 Left ventricular dysfunction.

LVSD is blockage in the coronary arteries that supply blood to the heart surface, thus reducing the oxygen supply, which leads to necrosis followed by the impaired contractility of cardiac muscles [2].

10.1.1 Left Ventricular Systolic Dysfunction

The inability of the LV to eject sufficient blood, referred as systolic dysfunction, is a complex process involving improper contractions of subendocardial, midwall, subepicardial, and papillary muscle fibers. In the presence of LVSD, the heart contracts less forcefully and cannot pump out as much of the blood that is returned to it as it normally does. As a result, more blood remains in the lower chambers of the heart (ventricles). Blood then accumulates in the veins. The muscle fibers of the midwall LV are arranged mainly circumferentially, whereas subepicardial, subendocardial, and papillary fibers are arranged mainly in a longitudinal and/or oblique fashion. The shortening of these muscle fibers results in a thickening of the ventricular walls, as well as a movement of the mitral annulus toward the LV apex. In addition, the oblique fibers produce a twisting movement of the LV. These events result in a volume reduction of the LV lumen, a reduction of the stroke volume, and, hence, a reduction in the ejection fraction. The interaction of the LV, the other cardiac chambers, and the entire vascular system results in decreased cardiac output, which is the product of stroke volume and heart rate.

10.1.2   Determinants of LV Systolic Performance

The forces acting on the myocardium during filling and contraction are important determinants of LV systolic performance. These forces are affected by chamber size and shape and can be described according to the concept of wall stress. The law of Laplace states that the stress in the walls of a cylinder is directly proportional to the transmural pressure and the radius of the cylinder and inversely proportional to the thickness of the cylinder walls. So an increased radius of the LV, that is, dilation, causes increased stress on the chamber walls, if the transmural pressure and wall thickness are constant. By increasing the wall thickness (i.e., LV hypertrophy), the wall stress can be reduced.

The wall stress acting to stretch the noncontracting muscle in diastole can be viewed as preload. An increase in preload produces an increase in stroke volume by lengthening the fundamental contractile units—the sarcomeres—in the myocyte (the length–tension relationship). Frank first described this phenomenon in the frog heart in 1895, and Starling described it in the mammalian heart in 1914. It is therefore often referred to as the Frank-Starling mechanism. The myocardial wall stress that builds up during the systolic contraction can be viewed as afterload. In addition to chamber size and wall thickness, this wall stress depends on the peripheral vascular resistance, arterial compliance, and intraventricular pressure. An increase in afterload reduces the myofiber shortening velocity, resulting in a decrease in stroke volume.

Last but definitely not least, contractility is the fundamental quality of cardiac muscle that determines the systolic performance independent of loading conditions. Increased contractility produces a greater rate of contraction to reach a greater peak force. Figure 10.2 shows the left ventricle *P-V* relationship in a normal heart along with systolic and diastolic dysfunctions.

10.2   Assessment of Left Ventricular Function—An Overview

Systolic function of the left ventricle represents a complex interaction between preload, afterload, and contractility. Preload is the force that stretches the

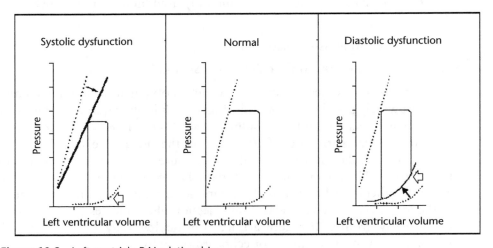

Figure 10.2   Left ventricle P-V relationship.

myofibrils in diastole. The balance of two opposing forces determines the end-diastolic sarcomere length. These forces are end-diastolic wall stress, which is an estimate of the force stretching the myocardial fibers at end diastole, and the intrinsic myocardial stiffness or elasticity. These two opposing forces will eventually determine the left ventricular end-diastolic volume and pressure. Afterload is the force that resists myofibril shortening in systole and varies throughout systole as blood is ejected from the ventricles. The left ventricular systolic wall stress is the measure of the force resisting myofibril shortening in systole in humans and, therefore, represents afterload. It is directly proportional to the radius of and the pressure inside the chamber and is inversely proportional to the wall thickness [3].

Contractility is an intrinsic property of the myocardium and generally means the strength of the shortening of the myofibrils when they contract. Preload, afterload, and contractility are intimately related in vivo and alterations in any one of these three parameters will affect the other two. For example, an increase in preload generally increases contractility according to the Frank-Starling mechanism. However, an increase in preload may also result in an increase in afterload because the left ventricular end-diastolic volume will increase.

Therefore, the overall increase in contractility as a result of an increase in preload is somewhat blunted because the increased afterload (that accompanies an increase in preload) will decrease contractility. Assessment of ventricular function, therefore, requires careful consideration of this complex interaction between preload, afterload, and contractility, and ventricular function is not synonymous with contractility.

In common usage, left ventricular function usually refers to left ventricular systolic function. Assessment of left ventricular function remains a vital part of management of almost all cardiovascular diseases because left ventricular function carries important diagnostic and prognostic information. The ideal parameter of ventricular function should be accurate; independent of loading conditions, heart rate, cardiac chamber shape and size; easily measured; reproducible; and sensitive only to intrinsic changes of contractile function. Numerous indices of cardiac function have been proposed and tested. However, none of the currently available indices fulfills the criteria of the ideal index. In the following sections, the rationale, benefits, and shortcomings of the commonly used indices are discussed.

10.2.1 Ejection Phase Indices

10.2.1.1 Ejection Fraction and Fractional Shortening

The left ventricular ejection fraction, which is obtained from volume data only, is the most widely used index of systolic function. It is often quoted as a measure of left ventricular function in heart failure trials [4–6]. Fractional shortening is the one-dimensional surrogate of EF and has the advantage of relative simplicity because its calculation depends only on measurement of a single dimension of the left ventricle. The velocity of circumferential fiber shortening [7] is the rate of shortening of a theoretical myocardial fiber in the circumferential plane at the midpoint of the long axis of the ventricle. These ejection phase indices can be measured relatively easily from either contrast left ventriculography or noninvasively using radionuclide ventriculography or echocardiography.

Except for the velocity of circumferential fiber shortening, both EF and fractional shortening are extremely popular. Their popularity stems from the fact that these parameters are easily obtainable, quantitative, easy to understand, and reproducible. However, ejection phase indices are very dependent on both preload and afterload and cannot be considered a reliable index of ventricular contractility, especially in conditions of altered loading conditions. For example, in patients with mitral regurgitation, the ejection fraction is increased as a result of the increase in preload and the decrease in afterload. The increase in EF is not indicative of increased ventricular function and it can, in effect, mask the development of ventricular dysfunction.

10.2.1.2 Myocardial Performance Index

Tei et al. [8, 9] advocated the use of a Doppler-derived myocardial performance index, which is defined as the sum of the isovolumetric relaxation and isovolumetric contraction times divided by the ejection time. They studied 100 patients with different degrees of left ventricular dysfunction and 70 normal subjects with Doppler echocardiography. The researchers found that the mean value of the index was significantly different between normal and patients with moderate or severe left ventricular dysfunction (0.39 ± 0.05, 0.59 ± 0.10, and 1.06 ± 0.24, respectively; $p < 0.001$ for all) with a small degree of overlap [8]. Furthermore, the value of the index did not appear to be related to heart rate, mean arterial pressure, or the degree of mitral regurgitation. In a subsequent study, the authors performed simultaneous cardiac catheterization and Doppler echocardiography in 34 patients [9] and found good correlations between the Doppler index and simultaneously recorded systolic peak $+dp/dt$ ($r = 0.821$; $p < 0.0001$), diastolic peak $-dp/dt$ ($r = 0.833$; $p < 0.001$), and tau (τ, the relaxation time constant; $r = 0.680$, $p < 0.0001$).

The Doppler index can be measured easily and is reproducible and appears to be promising in the assessment of global ventricular function. However, it does not differentiate between systolic and diastolic function and its use in patients with significant mitral regurgitation has not been validated. Moreover, in patients with significant mitral regurgitation, the isovolumetric contraction time cannot be measured accurately. Therefore, the index may have limited applicability in these patients.

10.2.2 Isovolumetric Indices

The maximum rate of rise of left ventricular systolic pressure ($+dp/dt$) is the maximum of the first derivative of ventricular pressure and has been used as an indicator of contractility. It was first measured by means of a catheter-tip micromanometer. More recently, dp/dt has been successfully measured noninvasively with the use of Doppler echocardiography [10]. Using continuous wave Doppler, Bargiggia et al. found good correlations between catheter-derived and Doppler-derived dp/dt. Interventions known to increase myocardial contractility, such as exercise, infusion of adrenaline, and tachycardia also increased left ventricular dp/dt. However, left ventricular dp/dt is not completely load independent. Studies have shown that left ventricular dp/dt increased with a moderate increase in preload [11] and showed little change with changes in afterload [12]. Whereas changes in *dp/dt* reflect changes in contractility in a given individual, its use in comparing one individual with another is limited, especially when there has been chronic pressure or volume overload.

10.2.3 Pressure-Volume Analysis

The relationship between left ventricular pressure and volume can be expressed in a plot of left ventricular pressure versus volume. One cardiac cycle can be represented by a complete pressure-volume loop (Figure 10.3). In Figure 10.3, segment b is isovolumetric contraction, segment c is left ventricular ejection, and d is isovolumetric relaxation.

10.2.3.1   End Systolic Pressure-Volume Relationships (ESPVR)

The fundamental principle underlying ESPVR is that such a relationship is unique for a given level of contractility and is independent of loading conditions [13]. The ratio of simultaneous pressure and volume defines the stiffness or the elastance of the left ventricle. During the cardiac cycle, this ratio increases to a maximum during systole and decreases to a minimum during diastole. By varying the loading conditions, a series of left ventricular pressure-volume loops can be generated. The line joining the upper left-hand corners of each individual loop is the end-systolic pressure-volume line, characterized by a slope, the end-systolic elastance (E_{es}), an x-axis intercept, and an extrapolated volume at zero pressure (V_0) (Figure 10.4).

Therefore, ESPVR is represented by:

$$P = E_{es}(V - V_0) \tag{10.1}$$

It is generally agreed that an increase in the slope of the ESPVR line (i.e., an increase in end-systolic elastance) is indicative of an increase in contractility and vice versa [13, 14]. Measurement of ESPVR involves simultaneous measurements of left ventricular pressure and volume. Left ventricular pressures can be measured with a catheter-tip micromanometer, and measurement of left ventricular volume

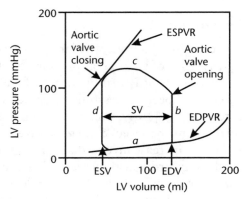

Figure 10.3   A typical left ventricular pressure-volume loop.

LV pressure (mmHg)

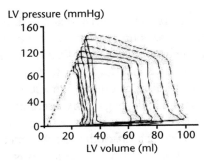

Figure 10.4   Determination of ESPVR.

can be made by contrast ventriculography, radionuclide techniques [15], echocardiography [16], or with the use of an impedance catheter [17].

An inflatable balloon inserted into the inferior vena cava near the right atrial junction usually achieves manipulation of loading conditions. With the use of either contrast or radionuclide ventriculography to measure volume, alteration of loading conditions is more commonly achieved by pharmacological means (phenylephrine to increase afterload and sodium nitroprusside to decrease afterload and preload). With the use of pharmacological agents, autonomic blockade [18] or temporary pacing at a slightly higher heart rate [15] may be necessary to eliminate the reflex changes in heart rate during such manipulations.

Whether the ESPVR is linear has been extensively investigated. A linear ESPVR has been suggested in isolated hearts [19] and in conscious dogs [20]. It was also supported by studies involving humans [21]. This linearity is important because, theoretically, the ESPVR line can be constructed from only two to three pressure-volume loops [22, 23]. This is especially important for investigators using contrast ventriculography to measure ventricular volume, because the number of pressure-volume loops that can be generated from such measurements is usually limited.

Experimental and clinical studies often determined end-systolic elastance over a limited range of loading conditions and inotropic states and, therefore, may have failed to demonstrate the nonlinearity of such relationships. However, subsequent investigators have suggested that the ESPVR is curvilinear at higher afterloads [24] and that the curvilinearity of ESPVR is also contractility dependent [25]. With increased contractility (i.e., high-end systolic elastance, E_{es}), the ESPVR tended to be concave to the volume axis. At low contractility, ESPVR tended to be convex to the volume axis. However, over the range of end-systolic elastance values that are encountered in normal states of contractility, a curvilinear fit was not statistically better than a linear fit for the ventricular volumes that could be examined in the animal model [25]. This curvilinearity has been used to explain why the extrapolated V_0 is sometimes a negative number [26]. It is important to realize that the linearity of the ESPVR is a convenience that allows determination of such relationships to be carried out in simple terms. Although the ESPVR is in fact curvilinear, it does not alter the fact that such relationships define the systolic performance of the ventricle at a given state.

End-systolic elastance has also been suggested to be influenced by heart size [27]. In an animal model of mitral regurgitation, Berko et al. [27] found that the end-systolic elastance decreased significantly after creation of mitral regurgitation, but the end-systolic elastance normalized by end-diastolic volume remained unchanged. However, it was unclear whether there was a true decline in contractility or whether the decline in end-systolic elastance was purely a function of the increased preload. In another study by Starling et al. [15], only the volume axis intercepts showed a linear relationship with left ventricular end-diastolic volume, and end-systolic elastance was negatively correlated with left ventricular mass.

Assessment of ESPVR Using Echocardiographic Automated Border Detection
Echocardiographic automated border detection is a new ultrasound technique that provides real-time online endocardial border detection and is commercially available as the Acoustic Quantification software package (AQ, Hewlett-Packard/ Agilent Technologies, Andover, Massachusetts). The technique works by differentiating the acoustic backscatter characteristics of blood from myocardial tissue within an operator-determined region of interest. It then computes an automated border between blood and myocardial tissue.

The area bound by the automated border is then computed and displayed in the form of a real-time waveform and values. The area is then extrapolated to estimate the 3-D volume of the chamber according to the modified Simpson's method [28]. This technique of computing ventricular volumes and function has been extensively validated against other standard techniques including conventional echocardiography [29], radionuclide [30] and contrast ventriculography [31], computed tomography [32], MRI [33], and electromagnetic flow meters [34].

Transesophageal echocardiography provides better endocardial border definition than transthoracic echocardiography because of the higher frequency transducers and the closer proximity of the transducers to the heart. Therefore, signals obtained with automated border detection from the transesophageal window are usually of a superior quality. Logically, automated border detection has been used with transesophageal echocardiography to quantitate ventricular dimensions [29]. A major drawback with the transesophageal window in extrapolating ventricular volumes is that the four-chamber view obtained is often foreshortened, resulting in erroneous calculations.

Transesophageal echocardiographic imaging of the left ventricle at the midventricular short-axis plane is often used to monitor ventricular volume and function in the intraoperative settings. However, there is no validated formula to extrapolate the cross-sectional area into ventricular volumes. Gorcsan et al. [34] have shown a close linear relationship between changes in the cross-sectional area and changes in stroke volume measured with an aortic electromagnetic flow probe. Therefore, the cross-sectional area change of the left ventricle as imaged from the transgastric view can be used as a surrogate for changes in ventricular volumes.

Volume or area data from automated border detection has been combined with simultaneous left ventricular pressures to enable construction of real-time left ventricular pressure and volume loops, thus enabling assessment of ESPVR. In a canine model, Gorcsan et al. [34] combined area data obtained from echocardiographic automated border detection from the transgastric left ventricular short-axis view and pressure data from a high-fidelity left ventricular pressure catheter to construct pressure-area loops with preload modified by inferior vena caval occlusion cathe-ters. These pressure-area loops compared favorably with simultaneous pressure-volume loops with volume data obtained with an electromagnetic flow probe placed in the ascending aorta.

Changes in stroke area with automated border detection correlated highly with changes in stroke volume over a wide range of loading conditions. The end-systolic pressure-area relationships changed appropriately with dobutamine and propranolol infusions. The authors used the term *stroke force* to represent the area bound by the pressure-area curve to distinguish it from stroke work, which is obtained from volume data. Similarly, the symbol E' was used for elastance instead of the usual symbol E.

In another canine model, Gorcsan et al. [35] compared pressure-volume rela-tionships obtained from automated border detection with those from conductance catheter techniques. They found that the relative changes in echocardiographic vol-ume were linearly related to conductance volume at steady state with changes in end-diastolic volume, end-systolic volume, and stroke work with caval occlusion also significantly correlated. Although there was an overall bias for absolute echocardiographic volume to be less and end-systolic and maximal elastance values to be higher, the direction and relative magnitude of changes in elastance with inotropic modulation were similar.

10.2.3.2   Stroke Work

The area bound by the left ventricular pressure-volume loop represents the left ven-tricular external work (Figure 10.3). It is a measure of the total chamber function and can be considered a measure of contractility as shown in Figure 10.5. However, it is only applicable as a measure of contractility when the left ventricle is relatively homogeneous and is of limited use in coronary artery disease with regional wall motion abnormalities. Left ventricular stroke work is increased in states of chronic volume overload such as mitral or aortic regurgitation. Attempts have been made to normalize stroke work to left ventricular volume: This relationship between left ventricular stroke work and end-diastolic volume is termed preload recruitable stroke work (PRSW).

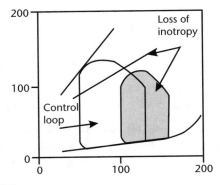

Figure 10.5   Loss of inotropy.

10.2.3.3   Preload Recruitable Stroke Work

The PRSW to left ventricular end-diastolic volume relationship is highly linear and is considered a reliable index of contractility [36]. The slope of such a relationship (M_{SW}) is dependent on ventricular contractility but independent of heart size and shape. Furthermore, this relationship is sensitive to changes in contractility but not to changes in loading conditions [36]. The preload recruitable stroke work relationship can be represented by:

$$SW = M_{SW}(E_{DV} - V_W) \tag{10.2}$$

where SW is left ventricular stroke work, E_{DV} is end-diastolic volume, and V_W, the x-axis intercept, is the extrapolated end-diastolic volume at which the external stroke work is zero.

Another advantage of PRSW is that it diminishes at a faster rate with reduction in preload compared with end-systolic pressure and dp/dt, therefore giving rise to a greater range of values [37]. As a result, the preload recruitable stroke work relationship or M_{SW} is much more reproducible and robust than ESPVR and maximum dp/dt to end-diastolic volume relationship. Furthermore, Takeuchi et al. [38] suggested that PRSWR was a more linear and reliable indicator of left ventricular contractility in man than ESPVR, because the latter tended to change with variations in afterload, whereas PRSWR remained unchanged. A potential drawback of PRSWR in assessing systolic function is that it does not separate systolic performance from diastolic performance as the whole pressure-volume loop is taken into consideration in deriving the PRSW.

The derivation of both ESVPR and PRSW involves manipulation of loading conditions, either by means of transient occlusion of the inferior vena cava or by pharmacological means. This may prove to be cumbersome; therefore, various investigators have attempted to derive ESPVR [39] and PRSWR [40] with data obtained from a single heartbeat. The basic principle of single-beat determination of PRSW is that V_W in (10.2) has been found to be essentially constant in an individual regardless of any short-term changes in the loading or inotropic conditions [37]. It is then apparent from the equation that M_{SW} can be calculated once the value of V_W is known for that particular individual without the need for loading condition manipulation. In a canine model, Karunanithi et al. [40] validated single-beat measurement of PRSW against multiple-beat measurements and found excellent correlations. Furthermore, the authors found that single-beat determination of PRSW was less load dependent and more reliable and more reproducible than other single-beat measurements such as single-beat elastance. However, more widespread use of these methods, especially in the clinical settings, is still pending.

10.2.4   Left Ventricular Stress Strain Relationship

Another approach to assess myocardial contractility is to measure the amount of myocardial shortening in relation to the force resisting such shortening (i.e., the afterload). The left ventricular end-systolic wall stress is usually considered a reliable estimate of afterload, and various measures, such as EF and fractional shortening, have been used as markers of myocardial shortening. If the myocardial fiber shortening on the y-axis is plotted against end-systolic wall stress on the x-axis, a more or less linear inverse relationship is seen [41]. An upward shift of this relationship is seen with increased contractility, for example, with dobutamine infusion [41]. A decrease in ejection fraction as a result of increased afterload, such as that seen in aortic stenosis, is not considered a result of impaired contractility, because there is no shift in the end-systolic wall stress to EF relationship. Instead, it is seen as a result of afterload mismatch.

An obvious advantage of the left ventricular stress strain relationship over ESPVR is that the former takes into account ventricular geometry and muscle mass. This will be advantageous in situations of chronic alterations in loading conditions such as those seen in valvular heart disease. Another advantage of assessing end-systolic wall stress is that it can be estimated noninvasively with echocardiography [3]. End-systolic wall stress is related directly to systolic blood pressure and left ventricular end-systolic dimension and is inversely related to wall thickness in end systole.

10.3   Echocardiographic Examinations

Comprehensive echocardiographic examinations were performed on study patients. Echocardiographic examinations were performed using commercially available equipment (Hewlett Packard Sonos 2500, 5500, Chennai, India). Second harmonic imaging was used in all later studies on the Sonos 5500. All echocardiographic studies were recorded on VHS videotapes. These recordings were converted into individual frames of bitmap images using the video capture card Pinnacle Mp30. All echocardiographic examinations were performed from standard views with the patients in a left lateral decubitus position. The standard views included parasternal long- and short-axis views, the apical four- and two-chamber views, and subcostal views. From the parasternal long-axis view, M-mode measurements, including left ventricular end-systolic, end-diastolic dimensions, and left ventricular posterior wall thickness in end systole, were measured (Figure 10.6). For patients where properly aligned M-mode dimensions could not be measured, 2-D guided measurements were used. Left ventricular outflow tract diameters were also measured in this view, using the zoom function to magnify the views when appropriate.

From the apical four-chamber view, left ventricular end-diastolic and end-systolic volumes were measured using the modified Simpson's rule [42]. Continuous and pulsed wave Doppler examination of the mitral inflow was performed at the

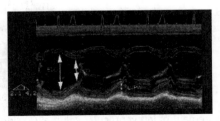

Figure 10.6   Parasternal long-axis view for measurements. EDD, end-diastolic dimension; ESD, end-systolic dimension; LVPW, left ventricular posterior wall thickness in end systole.

apical four-chamber view with the sampling volume placed at the tips of the mitral valve leaflets. The mitral leaflets were carefully imaged and examined in all views to ascertain the mechanisms and the degree of mitral regurgitation. The severity of mitral regurgitation was graded semiquantitatively according to the criteria of Helmcke et al. [43].

10.3.1 Echocardiographic Measurements and Calculations

10.3.1.1 Echocardiographic Measurement of Left Ventricular Volume

Left ventricular diastolic and systolic volumes were measured from the apical four-chamber view according to the modified Simpson's rule (Figure 10.7). The endocardial border was traced excluding the papillary muscles. The single-plane method was used in all studies except for patients with functional mitral regurgitation; for those patients, a biplane method was used. The use of the biplane method in these patients with coronary artery disease resulted in more accurate estimation of left ventricular volumes as the enlargement of the left ventricle may be asymmetrical.

Only representative cycles were measured and the average of at least three measurements was taken. The frame captured at the R wave of the electrocardiogram was considered the end-diastolic frame, and the frame with the smallest left ventricular cavity was considered the end-systolic frame. The end-diastolic and end-systolic volumes were normalized to the body surface area to give the respective index. The EF was calculated from the left ventricular end-systolic and end-diastolic volume.

According to Simpson's rule, the volume of the large figure is the sum of the volumes of a series of similar, smaller figures. The Simpson method is sometimes referred to as the *disc summation method* because the left ventricle is assumed to be made up of a series of discs stacked on top of each other. The Simpson method is attractive because it does not require the ventricle to conform to any particular shape or geometry. A simplified model Simpson's rule, which used only images from the apical four-chamber view, has been used [44]. In a study by Erbel et al. [44], estimation of left ventricular volume by the single-plane disc method yielded accurate results compared with direct volume measurements only in left ventricles without asymmetry. For asymmetric hearts, such as those seen with coronary artery disease and previous myocardial infarction, the biplane disc method was more accurate than the single-plane method. The volumes measured for a normal subject and one with systolic dysfunction subject are shown in Figure 10.8.

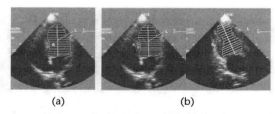

(a)   (b)

Figure 10.7 Measurement of left ventricular volume: (a) single-plane method and (b) biplane method.

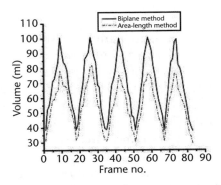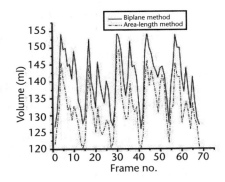

Figure 10.8   Volume variation in (a) normal and (b) systolic dysfunction.

Interobserver and Intraobserver Variability

To determine the interobserver variability in measuring left ventricular volumes, we randomly selected 10 studies each from the resting phase, immediate postexercise phase, and postoperative phase. Two experienced observers measured the left ventricular end-systolic and end-diastolic volumes. The interobserver variability was measured by the mean difference of the two measurements ± the standard deviation. Pearson's correlation coefficients between the two measurements were also calculated. To determine the intraobserver variability, one single observer measured the ventricular volumes on 10 randomly selected sets of images on two occasions. The intraobserver variability was measured by the mean difference of the two measurements ± the standard deviation. The correlation coefficients between the two measurements were also calculated. The interobserver variability in the measurement of resting left ventricular end-diastolic volume (EDV) and end-systolic volume (ESV) was 3 ± 8.5 mL ($r = 0.98$) and 0.7 ± 8.2 mL ($r = 0.94$), respectively, corresponding to a variability in resting EF of $0 \pm 5\%$. Immediately after exercise, the variability was 1.2 ± 14.5 mL for EDV ($r = 0.94$), -1.8 ± 4.7 mL for ESV ($r = 0.96$), and $1 \pm 3\%$ for EF. After surgery, it was -1.5 ± 5.7 mL for EDV ($r = 0.99$), 2 ± 9.2 mL for ESV ($r = 0.97$), and $0 \pm 5\%$ for EF.

The intraobserver variability in the measurement of resting left ventricular EDV and ESV was 4.9 ± 10.4 mL ($r = 0.97$) and -0.9 ± 3.5 mL ($r = 0.99$), respectively, corresponding to a variability in ejection fraction of $-2 \pm 4\%$. Immediately after exercise, variability was -0.8 ± 9.5 mL for EDV ($r = 0.98$), -0.9 ± 3.2 mL for ESV ($r = 0.99$), and $-1 \pm 2\%$ for EF. Postoperatively, it was -2.4 ± 6.3 mL for EDV ($r = 0.99$), -1.2 ± 4.1 mL for ESV ($r = 0.99$), and $1 \pm 3\%$ for EF.

The interobserver and intraobserver variability in the measurement of left ventricular volumes should be evaluated. Measurement of ventricular volumes and ejection fraction has been reported to show significant variability [45]. In a study by Himelman et al. [45], the variability in the estimation EF was 7%, whereas the variability in the estimation of ventricular volumes was up to 10%. The variability in these measurements could be acknowledged by taking the average of at least three measurements for each of the parameters assessed.

10.3.1.2   Left Ventricular *dp/dt*

The maximum rate of rise of left ventricular pressure during the isovolumetric contraction phase (left ventricular *dp/dt*) has been used as an indicator of left ventricular contractility. It can be measured echocardiographically in the presence of mitral regurgitation by continuous-wave Doppler examination of the mitral regurgitant jet [10]. From the apical four-chamber view, continuous-wave Doppler examination of the mitral regurgitant jet was performed. The time taken for the velocity of the mitral regurgitant jet to increase from 1 to 3 m/s was measured from the spectral display (Figure 10.9). Assuming an unchanged left atrial pressure, the rise in intraventricular pressure associated with an increase of mitral regurgitant jet velocity from 1 to 3 m/s would be 32 mmHg. The left ventricular *dp/dt* in mmHg/s was given by dividing 32 mmHg by the measured time interval. Continuous-wave Doppler examination of the mitral regurgitant jet with a clear envelope on the spectral display was not possible in all patients because it required proper alignment of the regurgitant jet and the Doppler beam. This is difficult and sometimes impossible in the presence of a very eccentric regurgitant jet such as that seen in patients with mitral valve prolapses or flail mitral leaflets.

Continuous-wave Doppler examination of the mitral regurgitant jet from the apical four-chamber view. The time it takes for the mitral regurgitant jet velocity to increase from 1 to 3 m/s can be measured from the spectral display. In this example, the time interval is 20 ms; therefore, the left ventricular *dp/dt* is 32 mmHg/0.02 second, that is, 1600 mmHg/s.

10.3.1.3   Left Ventricular End-Systolic Wall Stress

Left ventricular end systolic wall stress (ESWS) at rest was calculated from noninvasively obtained parameters from

$$ESWS = \frac{0.334\,BPsys/LVESD}{PWsys\left[1 + \dfrac{PWsys}{LVESD}\right]}$$

which has been validated by [3]. BPsys is the resting systolic blood pressure measured by a sphygmomanometer in mmHg, LVESD is the left ventricular end systolic dimension and PWsys is the posterior left ventricular wall thickness measured echocardiographically at end systole from the parasternal long-axis view. The average of three measurements was taken.

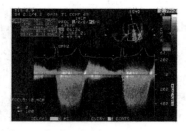

Figure 10.9   Measurement of left ventricular *dp/dt*.

10.3.1.4 Left Ventricular Forward Stroke Volume and Cardiac Output

Pulsed wave examination of the left ventricular outflow tract was performed from the apical four-chamber view with the sampling volume placed just below the aortic valve. The time velocity integral (in centimeters) was obtained from the spectral display by tracing the modal velocity (Figure 10.10). The diameter of the left ventricular outflow tract was measured from the parasternal long-axis view. The cross section of the left ventricular outflow tract was assumed to be circular, and the area calculated from the radius accordingly. The left ventricular forward stroke volume was obtained by determining the product of the cross-sectional area and the time velocity integral. The cardiac output was given by the product of the forward stroke volume and the heart rate. Resting cardiac output was calculated with these parameters measured at rest, and maximum exercise cardiac output was calculated with the same parameters measured at peak exercise or immediately afterward.

10.3.1.5 Mitral Regurgitant Volume

Mitral regurgitant volume can be estimated echocardiographically. Mitral regurgitant volume can be calculated as the difference between left ventricular forward stroke volume in systole and the total mitral inflow volume in diastole. More recently, the proximal convergence method in calculating mitral regurgitant volume has been used with increasing popularity [10, 46]. It has been shown that the proximal flow convergence method is associated with a significant overestimation of mitral regurgitant volume when the proximal flow acceleration field is constrained and that a geometric correction is necessary for accurate estimation [47].

A proximal flow acceleration field constraint was commonly seen in patients with mitral valve prolapse or flail mitral leaflets. Furthermore, proper visualization of the proximal flow acceleration field immediately after exercise was almost impossible. As a result, estimation of mitral regurgitant volume immediately after exercise was not possible with the proximal flow convergence method. In the method used by Blumlein et al. [48], mitral regurgitant volume was obtained by the difference between the total stroke volume and the left ventricular forward stroke volume. The total stroke volume was given by the difference between the left ventricular end-diastolic and end-systolic volumes obtained from the apical four-chamber view. Because the total stroke volumes and the left ventricular forward stroke volumes were measured both at rest and immediately after exercise, the mitral regurgitant volumes could be estimated in the resting and exercise states.

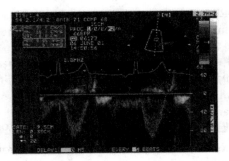

Figure 10.10 Measurement of left ventricular outflow tract time velocity integral.

10.3.1.6 Wall Motion Index

Wall motion scoring is based on a systematic visual judgment of the LV wall motion. In each echocardiographic imaging plane, the LV wall is 16 divided into several segments. Segmental models involving between 5 and 20 segments have been described [42, 49–52]. Each segment is assigned a numerical value that indicates the degree of wall contraction. In previous recommendations the endocardial motion of a segment was the only property to be rated [44]. However, recent guidelines from the American Society of Echocardiography (ASE) recommend that the systolic thickening of a segment, in addition to its endocardial motion, be evaluated [53]. The ASE standard assigns a score of 1 for normal contraction, 2 for hypokinesia, 3 for akinesia, 4 for dyskinesia, and 5 for aneurysmal bulging. Berning et al. proposed a reverted scoring system in which a lower score indicated worse function and, in addition, they included a score for hyperkinetic motion and excluded the aneurysmal score (+3 = hyperkinesia, +2 = normokinesia, +1 = hypokinesia, 0 = akinesia, −1 = dyskinesia) [54]. A global wall motion score, referred to as the wall motion index (WMI), can be calculated by averaging the readings in all segments. The WMI has shown good correlation with the LVEF as measured by radionuclide angiography (RNA) or contrast ventriculography, and the agreement with the reference method was comparable to that for the Simpson-LVEF or visual estimations of the LVEF [55]. Several studies have reported a strong prognostic performance by the WMI (obtained according to ASE or Berning) in terms of mortality or cardiac events in patients with heart failure or acute myocardial infarction [56, 57].

10.3.1.7 Mitral Annulus Motion

The mitral annulus is part of the fibrous skeleton of the heart and consists of a ring of collagenous tissue that surrounds and supports the left atrioventricular orifice [58]. It forms the attachment of the mitral valve leaflets and myocardial fibers of the LV and left atrium. The contraction of longitudinal/oblique LV myocardial fibers pulls the mitral annulus in a caudal direction toward the apex. Moreover, the myocardial tissue volume is regarded as noncompressible and the total volume of the heart is fairly constant during the entire heart cycle [59].

The displacement of myocardial tissue toward the apex caused by long-axis shortening will therefore inevitably cause a volume reduction of the inner LV diameter, in addition to the volume reduction caused by the annular excursion. The long-axis shortening of the LV can be evaluated by M-mode echocardiography as the maximum amplitude of mitral annulus motion (MAM). Measuring MAM is easy to learn and rapid to perform, even in patients with poor image quality, and it has advantageous reproducibility [60]. The amplitude of MAM is load dependent, which has to be considered when it is used for evaluations of LV systolic function. Early studies of the maximum amplitude of MAM revealed a high correlation with the LVEF, and MAM was therefore proposed as an LVEF surrogate [61].

10.3.2 Real-Time Left Ventricular Volume by Automated Boundary Detection

Real-time online measurement of left ventricular volume during the cardiac cycle can be performed by using the automated boundary detection algorithm (Acoustic

Quantification, Agilent, Andover, Massachusetts). The study was performed from the apical four-chamber view in the second-harmonic mode. The depth of the imaging was minimized to ensure that as large an image of the left ventricle as possible was obtained. An artifact-free electrocardiogram was obtained and the 2-D images of the left ventricle were optimized by adjusting the overall gain and transmitting function, time gain compensation, and the lateral gain controls so as to obtain a uniformly gray picture. The acoustic quantification function was then activated.

Further fine adjustments of the gain and transmitting function were performed to ensure proper tracking of the endocardial borders throughout the cardiac cycle and minimization of noise artifacts within the blood pool of the left ventricle. A region of interest that included the entire left ventricle was drawn and the long axis of the left ventricle was determined. Real-time display of ventricular volumes and EF was then obtained by activating the waveform function. Finally, scaling of the display was adjusted for optimal display of the waveforms (Figure 10.11).

10.4   Left Ventricle Systolic Dysfunction Identification

In 2003, an automated edge detection algorithm using fuzzy logic for the delineation of left ventricles was developed. From such delineated left ventricles, the systolic dysfunction was quantified and the region of dysfunction was identified [62].

10.4.1   Delineation of Left Ventricle

Each image was preprocessed before doing measurements. Histogram equalization was performed on each frame to increase the contrast. The speckle noise was removed by means of a median filter. The mask size used in filtering was 5×5. The greater the mask size, the greater the blur in the image. A Mamdani-type fuzzy logic–based edge detection algorithm was framed to detect the left ventricle contours. It has eight inputs and one output, with three member functions each at the input side and two member functions at the output side. Gaussian filters were adopted for all membership functions (Figure 10.12).

The function adopted to implement the AND and OR operations were the minimum and maximum functions, respectively. The Mamdani method was chosen as

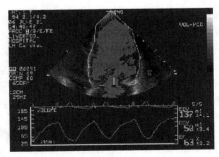

Figure 10.11   Real-time left ventricular volume by automated boundary detection. The region of interest is drawn around the left ventricle in the apical four-chamber view. The real-time ventricular volume and EF are displayed on the bottom right-hand corner of the display and the waveform is displayed together with the electrocardiogram.

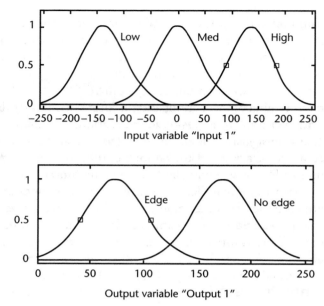

Figure 10.12 Fuzzy membership functions.

the defuzzification procedure, which means that the fuzzy sets obtained by applying each inference rule to the input data were joined through the add function; the output of the system was then computed as the centroid of the resulting membership function. The fuzzy inference rules were defined in such a way that the system output ("Edges") was high only for those pixels belonging to edges in the input image. In all, 45 rules were written to guarantee edge detection even in low-contrast areas of a noisy echocardiogram image.

From each image, the input circular gradient vector (CGV) was computed as the derivatives between the current pixel $p(x_i, y_i)$ and its eight neighborhood pixels in a 3×3 neighborhood. This input vector is weighted in the fuzzy system according to the membership functions and the rules. The current pixel is then assigned the category of "edge" or "no edge" based on the weight it received. After the edge detection process, groupings were done using a distance transform from which the left ventricle alone is delineated (Figure 10.13).

The central idea behind the use of a fuzzy system for edge detection is to avoid interobserver and intraobserver variability in identifying the endocardial borders and to automate the entire process. To illustrate the performance of the implemented fuzzy system, this method was applied to a test dataset that contained apical long-axis view sequences of varying quality and frame rates. It was found that the approach copes well with the varying image quality and frame rates. The results were also compared with existing edge detection algorithms, and we found that the Sobel and Perwitt operator fails to detect edges in the low-contrast region, whereas in case of the Canny operator, the pixels not belonging to the edges are also included. But the proposed method detects the edges and hence the endocardial borders clearly even in low-contrast regions, as shown in Figure 10.14.

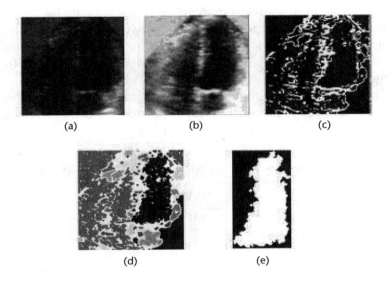

Figure 10.13   (a) Cropped image, (b) preprocessed image, (c) fuzzy edge detected image, (d) after grouping, and (e) left ventricle.

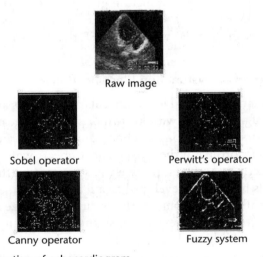

Figure 10.14   Edge detection of echocardiogram.

10.4.2   Quantification of Systolic Dysfunction

The left ventricle as whole is a nonrigid object whose shape changes periodically over each cardiac cycle due to the global and local deformations during the pumping of the heart. Previous studies of shape abnormalities have demonstrated that changes in ventricular geometry are associated with worsened status and prognosis [63, 64]. In 1998 Yettram et al. [65] proposed that the ratio of fiber to cross-fiber stiffness of the myocardium can be deduced in patients from the pattern of volume and shape changes during left ventricular filling. Left ventricle shape and shape changes have been studied extensively in large patients [66]. The ability to relate ventricular shape or shape changes to underlying myocardial mechanics could be of diagnostic value, because ventricular shape is easy to measure noninvasively and is known not only to change in some disease states but also to predict clinical prognosis [67].

LV geometry is an important determinant of the pumping performance of the left ventricle. Better correlation was expected between a morphological index and the degree of dysfunction, so a geometric index, the eccentricity ratio (ER), was proposed in 2006 [68] and was calculated as follows:

$$ER = \frac{\text{Eccentricity index (EI)}}{\text{Perimeter } (p)} \times 100$$

$$EI = \frac{\sqrt{L_1^2 - L_2^2}}{L_1}$$

where L_1 is the anterior to posterior length and L_2 is the septal to endocardial free wall length in the apical view.

The ER describes the geometry of LV well even when the regional and valvular dysfunction varies significantly between the two groups measured at end diastole and end systole ($p < 0.001\%$). In particular, the ER measured at end systole was found to be low for the dysfunction group (3.142 ± 0.7974) when compared with the other (5.48 ± 0.739, $p < 0.0001$). For patients having dysfunction, the systolic phase exhibits an irregular variation in ER that illustrates the improper contraction.

10.4.3 Region of Dysfunction Identification

The assessment of regional left ventricular function is an essential component in the evaluation of any patient with known or suspected heart disease. Although the title of the very first description of echocardiography by Edler and Hertz in 1954 [69] mentioned "recording of the movements of heart walls" as its goal, measuring wall motion objectively and quantitatively has remained difficult in clinical practice. Several methods have been developed to aid cardiologists in their clinical and research studies on wall motion. Some methods directly analyze image data, according to the optical flow concept [70], even if noise and temporal undersampling may limit accuracy. Other methods track markers that have been either surgically implanted in the LV wall [71] or noninvasively assigned to it by magnetic resonance tagging [72].

The most common methods [73, 74] analyze the LV contours as extracted from image sequences by means of manual or automated procedures. The use of contours makes the analysis independent of the imaging modalities, provided that the particular 2-D projection from which the heart is viewed is taken into account. In most cases, contour analysis considers two single contours taken at the end of the systolic (ES) and diastolic (ED) phases of a representative cardiac cycle. In a few cases, a complete contour sequence, spanning an entire cardiac cycle, is considered (frame-to-frame analysis).

The most successful current technique, tissue Doppler-based analysis of motion [75] and deformation [76], is hampered by the critical dependence of both measurements on the ultrasound beam direction, leading to signal dropout and velocity underestimation. To provide an objective framework, a quantitative motion description of wall segments is usually built by estimating the regional excursion of LV wall motion, and seldom by considering velocity and asynchrony indices. In the measurement of these parameters, geometric features of the LV, such as the LV long axis or centroid at ED or ES, or the centerline lying between the ED and ES contours [77], are usually exploited to establish a reference system for the correspondence of points belonging to subsequent frames. In recent years, new methods have been proposed that consider not only LV motion but also LV contour shape, which is independent from any reference system.

In diastole–systole analysis, a good discrimination capability of normal and abnormal contours has been provided through curvature-based descriptions of digitized contours [74]. In frame-to-frame analysis, some methods are based on a static approach, as shape descriptions in single images are used for the tracking of contour points or segments in an image sequence [78]; other computer vision methods are based on a dynamic approach in that they exploit spatiotemporal models for estimating nonrigid motion from 2-D contours [79, 80] or from 3-D data [81]. Another issue concerns the interpretation of results. In diastole–systole analysis, regional nonuniformity is commonly taken into account by assessing the variability range for the motion excursion of a number N of wall segments, in a group of normal subjects. Then, abnormalities are detected and evaluated by comparing N measured values of the patient against the normal range and by using statistical decision tests. The movement of any given ventricular segment is directly influenced by the adjacent segments, which means that any dysfunction can be masked by compensating motion from a neighboring segment.

Regional wall motion analysis can also be based on the wall thickening rather than endocardial motion. Wall thickening is insensitive to cardiac translation, is orientation model independent, and produces a single parameter of regional function. However, the thickness is difficult to measure on stop-frame 2-D images due to the reduction in endocardial and epicardial definition resulting from the loss of temporal continuity and the loss in spatiotemporal information provided by the preceding and succeeding frames [82]. Interpretation of the results of a frame-by-frame analysis is more difficult, because both spatial and temporal nonuniformity must be considered.

The qualitative evaluation of LV systolic function is based on the division of the LV into a number of segments, after which each segment is scored as normal, hypokinetic, akinetic, or dyskinetic. The main problems pertain to the distinction of hypokinesia from akinesia. Akinesia was considered to be present when endocardial excursion is 2 mm, and hypokinesia with endocardial excursion is 5 mm. However, movement may be passive, and thickening is the more reliable marker of contractility.

The standard 16-segment model of the ASE (septal, lateral, anterior, and inferior at the apex, with these segments as well as the anteroseptal and posterior segments at the base and midpapillary muscle level; Figure 10.15) is likely to remain in widespread use because the suggested 17-segment model (which includes a true apical segment) ignores the small but important detail that most echocardiograms fail to identify the true apex of the heart and so this can be modified. With this, a

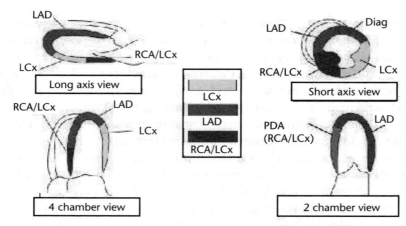

1 = normal; 2 = hypokinetic; 3 = akinetic; 4 = dyskinetic

Figure 10.15 The 16-segment ASE model for characterization of regional LV function, and the normal coronary artery distribution of the segments.

score of 1 is given for normal regions; with scores of 2, 3, and 4 for hypokinesis, akinesis, and dyskinesis, respectively; 5 for aneurysm; and 6 and 7 for akinesis or dyskinesis with thinning, respectively. The wall motion "score index" (obtained by averaging the scores of individual segments) gives a semiquantitative index of global systolic function, analogous to the ejection fraction and with similar prognostic significance [83].

Regional wall motion scoring is highly reproducible within individual sites, reflecting common reading styles. However, reproducibility of wall motion assessment between centers may be quite limited, especially during stress 2-D echo. Concordance may be improved with the use of standard reading criteria and harmonic imaging. Although it is unlikely that echocardiographers will stop visually assessing the LV, an objective measure that supplemented this assessment with objective criteria could act as a "common language" to reduce variation between readers.

A number of echo and Doppler modalities are able to offer quantitation of regional function, and most are outside the remit of this review (Table 10.1). The 2-D echo-based techniques include techniques for assessing radial displacement (using the centerline method or color kinesis) or thickening (anatomical M-mode).

10.4.3.1 Centerline Method

The centerline method is based on three steps: tracing the LV end-diastolic and end-systolic borders; superimposition of the traces with interpolation of the centerline; and measurement of the excursion from this line in a series of chords perpendicular to the centerline, which can be compared to a normal range of displacement. Each step poses potential pitfalls.

Tracing of contours, preferably in two orthogonal planes (usually apical four- and two-chamber views), is dependent on good quality border definition, and the reliance on apical views may compromise edge detection because of the parallel orientation of the echocardiographic beam with the endocardium.

Table 10.1 Alternative Techniques for the Quantification of Regional LV Function

	Radial	*Longitudinal*
Displacement	Center line (from 2-D echo) Color kinesis	Annular M mode Tissue tracking
Thickening	Anatomical M mode Integrated backscatter	
Velocity	Velocity from displacement Longitudinal velocity	Tissue Doppler velocity Strain
Timing	Time to peak systole Time to onset of diastole	Time to peak systole Time to onset of diastole

Although good border definition has become more available with the development of harmonic and contrast imaging, tracing the edge may still involve an element of guesswork. More than a single frame may need to be traced at both systole and diastole, making the procedure time consuming, although automated and semiautomated methods of tracking the wall have been developed (Figure 10.16). The superimposition of systole and diastole may have a critical effect on the measurement of excursion from the centerline. Either fixed or floating frames of reference can be used to compensate for rotational or translational movement of the heart. Failure to correct for such movements may cause false positives, but the use of correction may hinder the detection of milder abnormalities. Finally, different variations of the technique measure the chords relative to the centerline or relative to the center of LV mass.

10.4.3.2   Color Kinesis Method

The color kinesis method uses acoustic quantification to define the border, based on the difference in backscatter between the LV wall and cavity. This has the benefit of avoiding the onerous process of tracing the border in every frame, although the frame rate is somewhat limited, compared to standard 2-D imaging. The excursion of the myocardium from each frame to the next is filled with a different color, and the resulting display overlaid on the 2-D image ("color kinesis") [84]. The displacement is portrayed as segmental area shrinkage and is arranged in stacked histo-

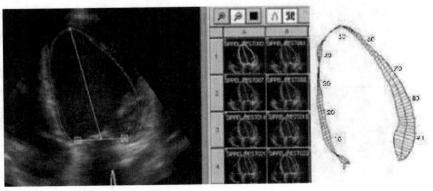

Figure 10.16   Centerline approach to quantification of regional LV function [73].

grams, as shown in Figure 10.17, which can be compared to normal ranges. This approach has been particularly applied during stress echo, where it correlates with expert wall motion analysis and may be of value to less expert readers. This technique is heavily dependent on image quality, and appears to be more feasible with the use of myocardial contrast for LV opacification. Measurements show a variation of 10% to 20%. As with any technique that measures endocardial motion, this is sensitive to extrinsic cardiac movement.

10.4.3.3 Anatomic M-Mode

Myocardial thickening is the optimal parameter for measurement because, unlike excursion, it is independent of cardiac rotation or translation. However, the measurement of thickening requires definition of both the endocardium and the epicardium—and the latter can pose a problem in the apical views. M-mode ultrasound has conventionally been used for gathering wall thickening data, but has been constrained by the angle dependence of standard M-mode imaging. Two-dimensional images at high temporal and spatial resolution have been used to reconstruct M-mode images in any plane (anatomic M-mode), although caution has to be applied with angle corrections of 60° to 70°. The results correlate well with visual assessments, but it has been difficult to designate a normal range, because of variations of baseline thickening, and the clinical benefit of this approach is not well defined.

Cardiologists refer to qualitative properties, such as late/early and inward/outward movements, rather than to numerical quantities as provided by most methods. This is the main reason why computer-aided analysis is seldom performed frame by frame. The model validation may have limitations due to the linearity assumption about deformation, which is, however, satisfied for small segments. On the other hand, validation with markers may also suffer from drawbacks, as the endocardial surface exhibits movements different from those of the midwall or epicardium, where markers are usually implanted. Nevertheless, both the radial method and the equidivision matching were validated with implanted markers. The centerline method has limited capabilities at basal regions, where the perpendicular chords do

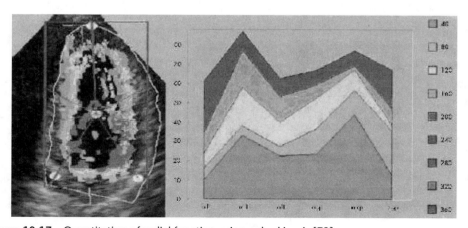

Figure 10.17   Quantitation of radial function using color kinesis [73].

not always match the first and last contour points. In fact, disagreements between different methods arise in borderline cases [85].

10.4.3.4   Radial Line Method

In Figure 10.18, radial lines have been drawn in each delineated LV from a reference point toward the endocardium. From the variations in the length of these lines over a cardiac cycle, the region of dysfunction can be identified. Sixteen lines (L1–L16) were drawn from the center of the delineated LV toward the endocardium, and its span of variation was measured between end of diastole and end of systole. If the muscles are active the variations will be greater. The number of lines drawn increases the efficiency of the diagnosis; normally, it is in between 8 and 24, with 32 being the maximum. This method was tested over two groups of subjects with and without regional dysfunctions, as shown in Figure 10.19.

In Figure 10.19, we can observe the difference in contractility of LV in terms of variation in radial line length between the two groups. In the dysfunctional group, the span of the lines is reduced when compared with the other group that shows the poor contractility of the muscles.

Figure 10.18   Radial line method.

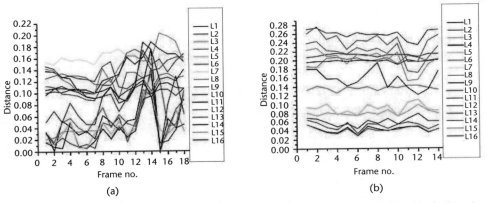

(a)   (b)

Figure 10.19   Radial line method: (a) normal (i.e., without dysfunction) and (b) with dysfunction.

10.4.3.5   Regional Area Method

In the regional area method, the LV is divided into different regional areas by drawing the radial lines from a fixed reference point (Figure 10.20). Then the area of each region was determined and normalized against the total endocardial length. The change in this area of each segment is directly related to the muscle contractility. Usually the number segments used for this analysis is 8 or 12.

The entire LV was divided in to eight segments (A1–A8) with respect to a fixed reference point. Then the areas were calculated for each segment and normalized against endocardial length, as shown for two cases in Figure 10.21. This figure shows that the LV muscles of the dysfunctions are not active as that of the other groups, because the spans have lesser heights of variation.

10.5   Real-Time 3-D Echocardiography

Routine clinical use of 3-D echocardiography has been hindered by the prolonged and tedious nature of data acquisition. Image processing is not only time consuming but also requires dedicated manpower to generate 3-D reconstructions and useful quantitative data. Accordingly, the recent introduction of real-time 3-D echocardiographic imaging techniques is of great interest since this methodology may circumvent many of the previously mentioned limitations.

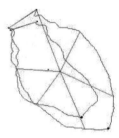

Figure 10.20   Regional area method.

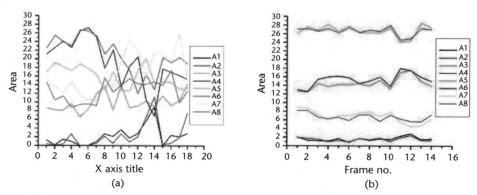

Figure 10.21   Regional area method: (a) normal (i.e., without dysfunction) and (b) with dysfunction.

10.5.1   Sparse Array Transducer

Real-time 3-D imaging was initially performed using a sparse array matrix transducer (2.5 or 3.5 MHz), which consisted of 256 nonsimultaneously firing elements. This transducer acquired a pyramidal volume dataset measuring 60°660° within a single heartbeat. Echocardiographic images were then displayed online using simultaneous orthogonal (B-scan images) as well as two to three parallel short-axis planes (C-scans) [45, 46]. Similar to other 3-D methods, the sparse array transducer resulted in accurate left ventricular volumes, ejection fraction, and mass measurements when compared to gold standard techniques, such as MRI and RNA. The sparse array transducer was also used advantageously during stress testing, because all postexercise images were simultaneously acquired during a single heartbeat, which was at higher peak stress heart rates compared to conventional stress tests.

Although 3-D real-time images were acquired using this novel transducer, it continued to have several disadvantages that precluded its routine clinical use. The ultrasound images were of relatively poor quality, frame rates were low, and the pyramidal volume had a relatively narrow sector angle of 60°, which resulted in the inability to accommodate larger ventricles. Moreover, the images obtained with this system were not volume rendered online; instead, they consisted of computer-generated, 2-D cut planes derived from the 3-D volume dataset.

10.5.2   Full Matrix Array

Recently, a full matrix array transducer (X4, Phillips Medical Systems, Andover, Massachusetts), which utilizes 3,000 elements in contrast to the 256 elements of the sparse matrix array probe, has been developed. This new development in transducer technology has resulted in: (1) improved sidelobe performance (contrast resolution), (2) higher sensitivity and penetration, and (3) harmonic capabilities that may be used for both grayscale and contrast imaging. In addition, this transducer displays online 3-D volume rendered images and is also capable of displaying two simultaneous orthogonal 2-D imaging planes (that is, biplane imaging).

10.5.3   Real-Time Volume Rendered 3-D Imaging

The full matrix array transducer has several modes of data acquisition (Figure 10.1): (1) narrow-angle acquisition, which consists of 60°630° pyramidal volumes displayed in a volume rendered manner in real time without the need for respiratory gating; (2) the "zoom mode," which allows a magnified view of a subsection of the pyramidal volume (30°630° sector in high resolution); and, last, (3) wide-angled acquisition, which is used to collect the entire left ventricular volume. In this acquisition mode, four wedges (15°660°) are obtained over eight consecutive cardiac cycles during a breath hold with ECG gating. The first two modes of data acquisition are predominantly used to visualize cardiac and valvar morphology. Images of heart valves acquired in this manner provide unique views, which are not always readily obtainable using conventional 2-D echocardiography. In contrast, the wide-angled acquisition mode is often used to acquire the entire left ventricular volume in order to perform detailed analysis of global and regional wall motion. This matrix array is capable of acquiring data at three levels of image resolution during both narrow- and wide-angled acquisition modes, which in turn has an impact on the size of the pyramidal volume. Data are stored digitally on CD-ROM and may be transferred to an off-line computer for quantitative purposes.

The left ventricle (apical four-, three-, and two-chamber views) is usually acquired from the apical window using wide-angled acquisition (Figure 10.22). Images may be displayed using either orthogonal long-axis views or multiple short-axis views, obtained at the level of the left ventricular apex, papillary muscles, and the base (Figure 10.22, bottom row). For example, when slicing the heart orthogonally in a four-chamber view, the septal and lateral walls together with the entire surface of the anterior or posterior wall are visualized (Figure 10.22, top row). Likewise, in a two-chamber view, the anterior and inferior walls are imaged in conjunction with the entire surface of the septum or lateral walls, depending on the orthogonal cut used (Figure 10.22, middle row). The right ventricle can also be assessed from a traditional four chamber. In contrast to 2-D echocardiography, 3-D echocardiography does not rely on geometric assumptions to calculate left ventricular volumes. This constitutes a real advantage in ventricles with odd shapes and wall motion abnormalities [86]. Similarly, the unique geometrical shape of the right ventricle has precluded accurate quantification using traditional echocardiographic methods.

Transthoracic 3-D echocardiography has the potential to overcome these limitations, resulting in accurate measurements of right ventricular size and function. Multiple studies have found 3-D calculated left ventricular volumes, ejection fractions, and left ventricular mass values to be comparable to those obtained with

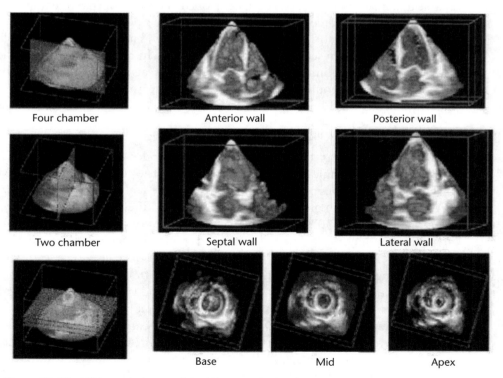

Figure 10.22 Wide-angled scan of the left ventricle sliced using multiple cut planes.

nuclear imaging and MRI. However, because gated acquisition methods including free-hand scanning are tedious, time consuming, and relatively nonportable, calculations of left ventricular ejection fractions from 3-D reconstructions have not been incorporated into routine echocardiographic studies. Additional disadvantages of the previously used 3-D methodologies include these: (1) datasets were acquired over multiple heart cycles; (2) data processing was slow because of the limitations of computer technology; (3) calculation of left ventricular volumes was performed on an off-line system, which requires tedious manual tracing of endocardial borders; and (4) the left ventricle was displayed using a static wire frame, which failed to provide anatomic information.

Quantification of left ventricular volumes and mass using real-time 3-D echocardiography is usually performed from using various apical wide-angled acquisition methods. Because a dataset comprises the entire left ventricular volume, multiple slices can be obtained from the base to the apex of the heart to evaluate wall motion. This acquisition can be combined with the use of an contrast infusion, particularly in patients who have a difficult acoustic window in whom it might be of benefit to improve the delineation of the endocardial border.

The 3-D volume dataset is then automatically divided into a number (that is, $n = 8$) of predetermined equiangled longitudinal slices through the apex (Figure 10.23,

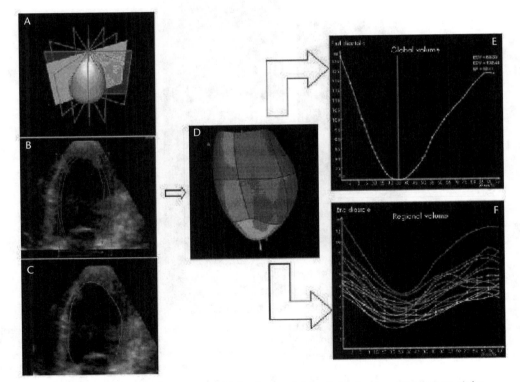

Figure 10.23   Panel A depicts the pyramidal volume of data divided automatically into eight equiangled longitudinal slices through the apex. Panels B and C depict the automated border detection algorithm used to track endocardial borders throughout the cardiac cycle. A dynamic ventricular cast is automatically displayed as a result of the endocardial borders and surface reconstruction (panel D). Global volumes and ejection fraction are displayed in panel E. Regional volumes of all 16 segments are shown in panel F.

panel A). The midpoint of the mitral valve annulus, apex, and aortic valve are identified as landmarks. In each of the longitudinal slices, an automated border detection algorithm is then used to track endocardial borders throughout the cardiac cycle (Figure 10.23, panels B and C) to obtain a dynamic cast of the left ventricle, as well as instantaneous global and regional left ventricular volumes versus time curves (Figure 10.23, panels D–F). This method of data analysis is semiautomated. In preliminary work, the volumes obtained with this method compare favorably with those obtained with cardiac MRI.

An alternative method of calculating ventricular volumes from a real-time 3-D cardiac volume dataset is to use the disc summation method, which has been well validated in the past (Figure 10.24). With this method, multiple short-axis cut planes are obtained from base to apex using a predefined distance interval. In each short-axis slice, endocardial borders at end systole and end diastole are traced and the summations of these volumes at end systole and end diastole are used to calculate left ventricular ejection fraction.

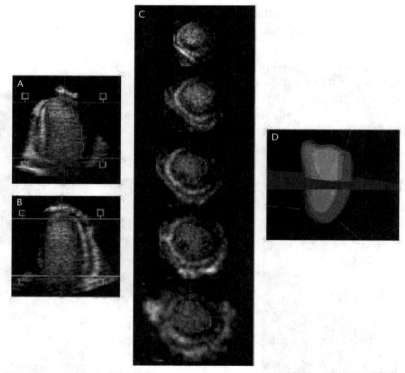

Figure 10.24 The disc summation method is an alternative method for calculating left ventricular volume and ejection fraction. The left ventricle is placed in a longitudinal position (panels A and B). With predefined distance intervals, multiple short-axis cut planes are derived and endocardial borders traced in all end-systolic and end-diastolic frames (panel C). The summation of the volumes of each slice results in the left ventricular volumes illustrated in panel D.

References

[1]   Yusuf, S., et al., "Global Burden of Cardiovascular Diseases: Part I: General Consider-ations, the Epidemiologic Transition, Risk Factors, and Impact of Urbanization," *Circulation*, Vol. 104, 2001, pp. 2746–2753.

[2]   Teerlink, J. R., S. Z. Goldhaber, and M. A. Pfeffer, "An Overview of Contemporary Etiolo-gies of Congestive Heart Failure," *Am. Heart J.*, Vol. 121, 1991, pp. 1852–1853.

[3]   Reichek, N., et al., "Noninvasive Determination of Left Ventricular End-Systolic Stress: Validation of the Method and Initial Application," *Circulation*, Vol. 65, 1982, pp. 99–108.

[4]   The CONSENSUS Trial Study Group, "Effects of Enalapril on Mortality in Severe Conges-tive Heart Failure. Results of the Cooperative North Scandinavian Enalapril Survival Study (CONSENSUS)," *N. Engl. J. Med.*, Vol. 316, 1987, pp. 1429–1435.

[5]   The SOLVD Investigators, "Effect of Enalapril on Survival in Patients with Reduced Left Ventricular Ejection Fractions and Congestive Heart Failure," *N. Engl. J. Med.*, Vol. 325, 1991, pp. 293–302.

[6]   The SOLVD Investigators, "Effect of Enalapril on Mortality and the Development of Heart Failure in Asymptomatic Patients with Reduced Left Ventricular Ejection Fractions," *N. Engl. J. Med.*, Vol. 327, 1992, pp. 685–691.

[7]   Paraskos, J. A., et al., "A Noninvasive Technique for the Determination of Velocity of Cir-cumferential Fiber Shortening in Man," *Circ. Res.*, Vol. 29, No. 6, 1971, pp. 610–615.

[8]   Tei, C., et al., "New Index of Combined Systolic and Diastolic Myocardial Performance: A Simple and Reproducible Measure of Cardiac Function—A Study in Normals and Dilated Cardiomyopathy," *J. Cardiol.*, Vol. 26, No. 6, 1995, pp. 357–366.

[9]   Tei, C., et al., "Noninvasive Doppler-Derived Myocardial Performance Index: Correlation with Simultaneous Measurements of Cardiac Catheterization Measurements," *J. Am. Soc. Echocardiogr.*, Vol. 10, No. 2, 1997, pp. 169–178.

[10] Bargiggia, G. S., et al., "A New Method for Estimating Left Ventricular dP/dt by Continu-ous Wave Doppler-Echocardiography. Validation Studies at Cardiac Catheterization," *Circulation*, Vol. 80, 1989, pp. 1287–1292.

[11] Grossman, W., et al., "Alterations in Preload and Myocardial Mechanics in the Dog and in Man," *Circ. Res.*, Vol. 31, No. 1, 1972, pp. 83–94.

[12] Brodie, B. R., et al., "Effects of Sodium Nitroprusside on Left Ventricular Diastolic Pres-sure-Volume Relations," *J. Clin. Invest.*, Vol. 59, No. 1, 1977, pp. 59–68.

[13] Grossman, W., et al., "Contractile State of the Left Ventricle in Man as Evaluated from End-Systolic Pressure-Volume Relations," *Circulation*, Vol. 56, No. 5, 1977, pp. 845–522.

[14] Sagawa, K., "The End-Systolic Pressure-Volume Relation of the Ventricle: Definition, Modifications and Clinical Use," *Circulation*, Vol. 63, 1981, pp. 1223–1227.

[15] Starling, M. R., et al., "Impaired Left Ventricular Contractile Function in Patients with Long-Term Mitral Regurgitation and Normal Ejection Fraction," *J. Am. Coll. Cardiol.*, Vol. 22, 1993, pp. 239–250.

[16] Gorcsan, J., et al., "Rapid Estimation of Left Ventricular Contractility from End-Systolic Relations by Echocardiographic Automated Border Detection and Femoral Arterial Pres-sure," *Anesthesiology*, Vol. 81, 1994, pp. 553–562; discussion, p. 27A.

[17] Hayward, C. S., et al., "Effect of Inhaled Nitric Oxide on Normal Human Left Ventricular Function," *J. Am. Coll. Cardiol.*, Vol. 30, No. 1, 1997, pp. 49–56.

[18] Glower, D. D., et al., "Linearity of the Frank-Starling Relationship in the Intact Heart: The Concept of Preload Recruitable Stroke Work," *Circulation*, Vol. 71, No. 5, 1985, pp. 994–1009.

[19] Weber, K. T., J. S. Janicki, and L. L. Hefner, "Left Ventricular Force-Length Relations of Isovolumic and Ejecting Contractions," *Am. J. Physiol.*, Vol. 231, No. 2, 1976, pp. 337–343.

[20] Lee, J. D., et al., "Application of End-Systolic Pressure-Volume and Pressure-Wall Thickness Relations in Conscious Dogs," *J. Am. Coll. Cardiol.,* 9, No. 1, 1987, pp. 136–146.

[21] Aroney, C. N., et al., "Linearity of the Left Ventricular End-Systolic Pressure-Volume Relation in Patients with Severe Heart Failure," *J. Am. Coll. Cardiol.,* Vol. 14, No. 1, 1989, pp. 127–134.

[22] McKay, R. G., et al., "Assessment of the End-Systolic Pressure-Volume Relationship in Human Beings with the Use of a Time-Varying Elastance Model," *Circulation,* Vol. 74, No. 1, 1986, pp. 97–104.

[23] Herrmann, H. C., et al., "Inotropic Effect of Enoximone in Patients with Severe Heart Failure: Demonstration by Left Ventricular End Systolic Pressure-Volume Analysis," *J. Am. Coll. Cardiol.,* Vol. 9, No. 5, 1987, pp. 1117–1123.

[24] McClain, L. C., et al., "Afterload Sensitivity of Nonlinear End-Systolic Pressure-Volume Relation vs. Preload Recruitable Stroke Work in Conscious Dogs," *J. Surg. Res.,* Vol. 75, No. 1, 1998, pp. 6–17.

[25] Burkhoff, D., et al., "Contractility-Dependent Curvilinearity of End Systolic Pressure-Volume Relations," *Am. J. Physiol.,* Vol. 252, No. 6, Pt. 2, 1987, pp. H1218–H1227.

[26] Little, W. C., et al., "Response of the Left Ventricular End-Systolic Pressure-Volume Relation in Conscious Dogs to a Wide Range of Contractile States," *Circulation,* Vol. 78, No. 3, 1988, pp. 736–745.

[27] Berko B, et al., "Disparity Between Ejection and End-Systolic Indexes of Left Ventricular Contractility in Mitral Regurgitation," *Circulation,* Vol. 75, 1987, pp. 1310–1319.

[28] Schiller, N. B., "Two-Dimensional Echocardiographic Determination of Left Ventricular Volume, Systolic Function, and Mass. Summary and Discussion of the 1989 Recommendations of the American Society of Echocardiography," *Circulation,* Vol. 84, Suppl. 3, 1991, pp. I280–I287.

[29] Davila-Roman, V. G., et al., "Quantification of Left Ventricular Dimensions on Line with Biplane Transesophageal Echocardiography and Lateral Gain Compensation," *Echocardiography,* Vol. 11, No. 2, 1994, pp. 119–125.

[30] Lindower, P. D., et al., "Quantification of Left Ventricular Function with an Automated Border Detection System and Comparison with Radionuclide Ventriculography," *Am. J. Cardiol.,* Vol. 73, No. 2, 1994, pp. 195–199.

[31] Vanoverschelde, J. L., et al., "On-Line Quantification of Ventricular Volumes and Ejection Fraction by Automated Backscatter Imaging-Assisted Boundary Detection: Comparison with Contrast Cineventriculography," *Am. J. Cardiol.,* Vol. 74, No. 6, 1994, pp. 633–635.

[32] Marcus, R. H., et al., "Ultrasonic Backscatter System for Automated On-Line Endocardial Boundary Detection: Evaluation by Ultrafast Computed Tomography," *J. Am. Coll. Cardiol.,* 22, No. 3, 1993, pp. 839–847.

[33] Stewart, W. J., et al., "Left Ventricular Volume Calculation with Integrated Backscatter from Echocardiography," *J. Am. Soc. Echocardiogr.* Vol. 6, No. 6, 1993, pp. 553–563.

[34] Gorcsan III, J., et al., "Online Estimation of Changes in Left Ventricular Stroke Volume by Transesophageal Echocardiographic Automated Border Detection in Patients Undergoing Coronary Artery Bypass Grafting," *Am. J. Cardiol.,* Vol. 72, No. 9, 1993, pp. 721–727.

[35] Gorcsan III, J., et al., "Left Ventricular Pressure-Volume Relations with Transesophageal Echocardiographic Automated Border Detection: Comparison with Conductance-Catheter Technique," *Am. Heart J.,* Vol. 131, No. 3, 1996, pp. 544–552.

[36] Glower, D. D., et al., "Linearity of the Frank-Starling Relationship in the Intact Heart: The Concept of Preload Recruitable Stroke Work," *Circulation,* Vol. 71, No. 5, 1985, pp. 994–1009.

[37] Little, W. C., et al., "Comparison of Measures of Left Ventricular Contractile Performance Derived from Pressure-Volume Loops in Conscious Dogs," *Circulation,* Vol. 80, No. 5, 1989, pp. 1378–1387.

[38] Takeuchi, M., et al., "Comparison Between Preload Recruitable Stroke Work and the End-Systolic Pressure-Volume Relationship in Man," *Eur. Heart J.*, Vol. 13, Suppl. E, 1992, pp. 80–84.

[39] Senzaki, H., C. H. Chen, and D. A. Kass, "Single-Beat Estimation of End-Systolic Pressure-Volume Relation in Humans: A New Method with the Potential for Noninvasive Application," *Circulation,* Vol. 94, No. 10, 1996, pp. 2497–2506.

[40] Karunanithi, M. K., and M. P. Feneley, "Single-Beat Determination of Preload Recruitable Stroke Work Relationship: Derivation and Evaluation in Conscious Dogs," *J. Am. Coll. Cardiol.,* Vol. 35, No. 2, 2000, pp. 502–513.

[41] Borow, K. M., et al., "Left Ventricular End-Systolic Stress Shortening and Stress-Length Relations in Human. Normal Values and Sensitivity to Inotropic State," *Am J. Cardiol.,* Vol. 50, No. 6, 1982, pp. 1301–1308.

[42] Schillar, N. B., et al., "Recommendations for Quantification of the Left Ventricle by Two Dimensional Echocardiography. American Society of Echocardiography Committee Standards, Subcommittee on Quantisation of Two Dimensional Echocardiograms," *J. Am. Soc. Echocardiogr.,* Vol. 2, 1989, pp. 358–367.

[43] Helmcke, F., et al., "Color Doppler Assessment of Mitral Regurgitation with Orthogonal Planes," *Circulation*, Vol. 75, 1987, pp. 175–183.

[44] Erbel, R., et al., "Comparison of Single-Plane and Biplane Volume Determination by Two-Dimensional Echocardiography. 1. Asymmetric Model Hearts," *Eur. Heart J.,* Vol. 3, 1982, pp. 469–480.

[45] Himelman, R. B., et al., "Reproducibility of Quantitative Two-Dimensional Echocardiography," *Am. Heart J.,* Vol. 115, No. 2, 1988, pp. 425–431.

[46] Rivera, J. M., et al., "Quantification of Mitral Regurgitation with the Proximal Flow Convergence Method: A Clinical Study," *Am. Heart J.,* Vol. 124, No. 5, 1992, pp. 1289–1296.

[47] Pu, M., et al., "Quantification of Mitral Regurgitation by the Proximal Convergence Method Using Transesophageal Echocardiography: Clinical Validation of a Geometric Correction for Proximal Flow Constraint," *Circulation,* Vol. 92, 1995, pp. 2169–2177.

[48] Blumlein, S., et al., "Quantitation of Mitral Regurgitation by Doppler Echocardiography," *Circulation*, Vol. 74, 1986, pp. 306–314.

[49] Hegger, J. J., A. E. Weyman, and L. S. Wann, "Cross Sectional Echocardiography in Acute Myocardial Infarction: Detection and Localization of Regional Left Ventricular Asynergy," *Circulation*, Vol. 60, 1979, pp. 531–538.

[50] Kisslo, J. A., et al., "A Comparison of Real-Time, Two Dimensional Echocardiography and Cineangiography in Detecting Left Ventricular Asynergy," *Circulation,* Vol. 55, 1977, pp. 134–141.

[51] Henry, W. L., et al., "Report of the American Society of Echocardiography Committee on Nomenclature and Standards in Two-Dimensional Echocardiography," *Circulation,* Vol. 62, 1980, pp. 212–217.

[52] Cerqueira, M. D., et al., "Standardized Myocardial Segmentation and Nomenclature for Tomographic Imaging of the Heart: A Statement for Healthcare Professionals from the Cardiac Imaging Committee of the Council on Clinical Cardiology of the American Heart Association," *Circulation,* Vol. 105, 2002, pp. 539–542.

[53] Gottdiener, J. S., et al., "American Society of Echocardiography Recommendations for Use of Echocardiography in Clinical Trials," *J. Am. Soc. Echocardiogr.,* Vol. 17, 2004, pp. 1086–1119.

[54] Berming, J., et al., "Rapid Estimation of Left Ventricular Ejection Fraction in Acute Myocardial Infarction by Echocardiographic Wall Motion Analysis," *Cardiology,* Vol. 80, 1992, pp. 257–266.

[55] McGowan, J. H., and J. G. Cleland, "Reliability of Reporting Left Ventricular Systolic Function by Echocardiography: A Systematic Review of 3 Methods," *Am. Heart J.,* Vol. 146, 2003, pp. 388–397.

[56] Galasko, G. I. W., et al., "A Prospective Comparison of Echocardiographic Wall Motion Score Index and Radionuclide Ejection Fraction in Predicting Outcome Following Acute Myocardial Infarction," *Heart,* Vol. 86, September 2001, pp. 271–276.

[57] Gustafsson, F., C. Torp-Pedersen, and B. Brendorp, "Long-Term Survival in Patients Hospitalized with Congestive Heart Failure: Relation to Preserved and Reduced Left Ventricular Systolic Function," *Eur. Heart J.,* Vol. 24, 2003, pp. 863–870.

[58] Ho, S. Y., M. Hocini, and T. Kawara, "Electrical Conduction in Canine Pulmonary Veins: Electrophysiological and Anatomic Correlations," *Circulation,* Vol. 105, 2002, pp. 2442–2448.

[59] Hedberg, P., "Left Ventricle Systolic Dysfunction in 75 Year Old Men and Women, A Community Based Prevalence, Screening and Mitral Annulus Motion for Diagnosis and Prognostics," Ph.D. Dissertation, Uppsala: Uppsala University, 2005.

[60] Willenheimer, I. R., et al., "Left Ventricular Atrioventricular Plane Displacement: An Echocardiographic Technique for Rapid Assessment of Prognosis in Heart Failure," *Heart,* Vol. 78, 1997, pp. 230–236.

[61] Jenson-Urstad K. I., et al., "Comparison of Different Echocardiographic Methods with Radionuclide Imaging for Measuring Left Ventricular Ejection Fraction During Acute Myocardial Infarction Treated by Thrombolytic Therapy," *The American Journal of Cardiology,* Vol. 81, No. 5, March 1, 1998, pp. 538–544.

[62] Manivannan, J., et al., "Endocardial Edge Detection by Fuzzy Inference System," *Proc. IEEE TENCON 2003,* Bangalore, India, October 2003, Vol. 1, pp. 3–5.

[63] Patel, A. R., et al., "Echocardiographic Analysis of Regional and Global Left Ventricular Shape in Chagas' Cardiomyopathy," *Am. J. Cardiol.,* Vol. 82, 1998, pp. 197–202.

[64] Douglas, P. S., et al., "Left Ventricular Shape, Afterload and Survival in Idiopathic Dilated Cardiomyopathy," *J. Am. Coll. Cardiol.,* Vol. 13, 1989, pp. 311–315.

[65] Yetram, A. L., and M. C. Beecham, "An Analytical Method for the Determination of Along-Fiber to Cross-Fiber Elastic Modulus Ratio in Ventricular Myocardium—A Feasibility Study," *Med. Eng. Phys.,* Vol. 20, 1998, pp. 103–108.

[66] Sundblad, P., and B. Wrane, "Influence of Posture on Left Ventricular Long and Short Axis Shortening," *Am. J. Physiol. Heart Circ. Physiol.,* Vol. 283, 2002, pp. 1302–1306.

[67] Doblas, G., et al., "Left Ventricular Geometry and Operative Mortality in Patients Undergoing Mitral Valve Replacement," *Clin. Cardiol.,* Vol. 24, 2001, pp. 717–722.

[68] Manivannan, J., et al., "Quantitative Evaluation of Left Ventricle Performance from Two Dimensional Echo Images," *Echocardiography,* Vol. 23, No. 2, pp. 87–92.

[69] Edler, I., and C. H. Hertz, "The Use of Ultrasonic Reflectoscope for the Continuous Recording of the Movements of Heart Walls," *Kungl. Fysiografiska sällskapets i Lund förhandlingar,* Vol. 24, 1954, pp. 1–19.

[70] Baraldi, P., et al., "Evaluation of Differential Optical Flow Techniques on Synthesized Echo Images," *IEEE Trans. on Biomedical Eng.,* Vol. 43, No. 3, 1996, pp. 259–272.

[71] Doss, J. K., et al., "A New Model for the Assessment of Regional LV Wall Motion," *Radiology,* Vol. 143, 1982, pp. 763–770.

[72] Buchalter, M. B., et al., "Noninvasive Quantification of LV Rotational Deformation in Normal Humans Using Magnetic Resonance Imaging Myocardial Tagging," *Circulation,* Vol. 81, 1990, pp. 1236–1244.

[73] Marwick, T. H., "Techniques for Comprehensive Two Dimensional Echocardiographic Assessment of Left Ventricular Systolic Function," *Heart,* Vol. 89, Suppl. III, 2003, III2–III8.

[74] Baroni, M., G. Barletta, and F. Fantini, "LV Wall Motion: From Quantitative Analysis to Knowledge-Based Understanding," *Proc. Computers in Cardiology,* September 19–22, 1989, 1990, pp. 483–486.

[75] Derumeaux, G., et al., "Doppler Tissue Imaging Quantitates Regional Wall Motion During Myocardial Ischemia and Reperfusion," *Circulation,* Vol. 97, 1998, pp. 1970–1977.

[76]   Uematsu, M., et al., "Myocardial Velocity Gradient as a New Indicator of Regional Left Ventricular Contraction: Detection by a Two-Dimensional Tissue Doppler Imaging Technique," *J. Am. Coll. Cardiol.,* Vol. 26, 1995, pp. 217–223.

[77]   Seehan, F. H., et al., "Advantages and Applications of the Centerline Method for Characterizing Regional Ventricular Function," *Circulation,* Vol. 74, 1986, pp. 293–305.

[78]   McEachen, J. C., and J. S. Duncan, "Shape-Based Tracking of Left Ventricular Wall Motion," *IEEE Trans. on Medical Imaging,* Vol. 16, 1997, pp. 270–283.

[79]   Metaxas, D., and D. Terzopoulos, "Shape and Nonrigid Motion Estimation Through Physics-Based Synthesis," *IEEE Trans. on Pattern Anal. Machine Intell.,* Vol. 15, 1993, pp. 580–591.

[80]   Coppini, G., et al., "Artificial Vision Approach to the Understanding of Heart Motion," *J. Biomed. Eng.,* Vol. 14, 1993, pp. 321–328.

[81]   Pentland, A., and B. Horowitz, "Recovery of Nonrigid Motion and Structure," *IEEE Trans. on Pattern Anal. Machine Intell.,* Vol. 13, 1991, pp. 730–742.

[82]   Song, I., et al., "Tracking Regional Left Ventricular Wall Movement Using 2-D Echocardiographic Cineloops in Concert with Synthetic M-Mode Images," *Computers in Cardiology,* September 8–11, 1996, pp. 269–272.

[83]   Nishimura, R. A., et al., "Prognostic Value of Predischarge 2-Dimensional Echocardiogram After Acute Myocardial Infarction," *Am. J. Cardiol.,* Vol. 53, 1984, pp. 429–432.

[84]   Bednarz, J., et al., "Color Kinesis: Principles of Operation and Technical Guidelines," *Echocardiography,* Vol. 15, 1998, pp. 21–34.

[85]   Baroni, M., "Computer Evaluation of Left Ventricular Wall Motion by Means of Shape Based Tracking and Symbolic Description," *Med. Eng. & Physics,* Vol. 21, 1999, pp. 73–85.

[86]   Lawson, M. A., et al., "Accuracy of Biplane Long-Axis Left Ventricular Volume Determined by Cine Magnetic Resonance Imaging in Patients with Regional and Global Dysfunction," *Am. J. Cardiol.,* Vol. 77, 1996, pp. 1098–1104.

CHAPTER 11

Diagnosis of Heart by Phonocardiography

Manivannan Jayapalan

11.1 Introduction

Blood flow through the human heart gives rise to vibrations and sounds that can be acquired from the surface of the chest. These sounds contain information regarding the condition of the heart. Two principal methods are used to investigate heart sounds: cardiac auscultation and phonocardiography. The act of listening to sounds produced by body vessels and organs is called auscultation. The recordings of these sounds produced in the heart are called phonocardiograms, and the device used to record then is called a phonocardiograph. A typical phonocardiograph and a recording are shown in Figure 11.1.

Software-aided heart sound analysis has several advantages compared to traditional cardiac auscultation using a stethoscope. The most obvious one is the improved reliability of diagnoses. In traditional cardiac auscultation, murmurs at very low frequencies may not be heard; in addition, classification of heart murmurs into innocent and pathological is strongly dependent on the listener's experience and training. In contrast, in a phonocardiogram, even the low-frequency sounds can be picked by using an improved performance sensor, and they can be better visualized in a spectrogram representation. Another significant advantage is that heart sounds can be saved as a part of the documentation, such as patient information and diagnosis. When necessary, this can easily be sent as an e-mail attachment to a cardiology specialist, who can analyze the results. Today and almost certainly even more so in the future, the ability to exchange documents electronically is a time and cost-efficient way to conduct a cardiologic consultation when necessary. Based on these strong advantages, this chapter first discusses the origin of heart sounds and the ways in which they can be captured. That discussion is followed by an explanation of the different possible techniques by which heart sounds can be analyzed to aid in diagnosis.

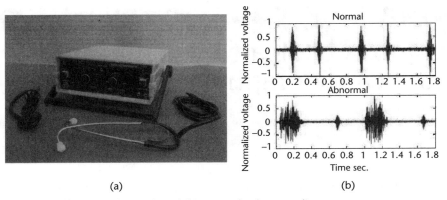

(a)   (b)

Figure 11.1 (a) Phonocardiograph and (b) a sample phonocardiogram.

11.2 Origin of Heart Sounds

The heart, although a single organ, can be considered as two pumps that propel blood through two different circuits. The right atrium receives venous blood from the head, chest, and arms via the large vein called the superior vena cava and receives blood from the abdomen, pelvic region, and legs via the inferior vena cava. Blood then passes through the tricuspid valve to the right ventricle, which propels it through the pulmonary artery to the lungs. In the lungs venous blood comes in contact with inhaled air, picks up oxygen, and loses carbon dioxide. Oxygenated blood is returned to the left atrium through the pulmonary veins. Valves in the heart allow blood to flow in one direction only and help maintain the pressure required to pump the blood. The cross-sectional view of a human heart is shown in Figure 11.2.

Alternating contractions and relaxation of the myocardium causes the heart to pump; the sound it makes is called the heartbeat. These contractions are stimulated by electrical impulses from a natural pacemaker, the sinoatrial (SA) node located in the muscle of the right atrium. An impulse from the SA node causes the two atria to contract, forcing blood into the ventricles. Contraction of the ventricles is controlled by impulses from the atrioventricular (AV) node located at the junction of the two atria. Following contraction, the ventricles relax, and pressure within them falls.

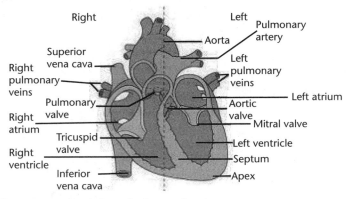

Figure 11.2 The cross-sectional view of a human heart.

Blood again flows into the atria and an impulse from the SA node starts the cycle over again. The conduction system of the human heart is shown in Figure 11.3. This process is called the cardiac cycle, which is comprised of a relaxation (diastole) and a contraction (systole).

Diastole is the longer of the two phases and it allows the heart to rest between contractions. The average adult rate is 70 beats per minute at rest. The rate increases temporarily during exercise, emotional excitement, and fever and decreases during sleep. The rhythmical noises accompanying a heartbeat are called heart sounds. Normally, two distinct sounds are heard through the stethoscope: a low, slightly prolonged sound that occurs at the beginning of ventricular contraction, or systole, and a sharper, higher-pitched "dup" (the second sound) at the end of systole. Occasionally audible in normal hearts is a third soft, low-pitched sound coinciding with early diastole. A fourth sound, also occurring during diastole, is revealed by graphic methods but is usually inaudible in normal subjects. Heart murmurs may be readily heard by a physician as soft swishing or hissing sounds that follow the normal sounds of heart action. Murmurs may indicate that blood is leaking through an imperfectly closed valve and may signal the presence of a serious heart problem.

11.2.1   Auscultatory Areas

Events related to the heart valves are best heard not over their anatomic locations, but in the auscultatory areas bearing their names. It is now commonly accepted that the externally recorded heart sounds are not caused by valve leaflet movements per se, as earlier believed, but by vibrations of the whole cardiovascular system triggered by pressure gradients. The cardiohemic system may be compared to a fluid-filled balloon, which, when stimulated at any location, vibrates as a whole. Externally, however, heart sound components are best heard at certain locations on the chest individually, and this localization has led to the concept of secondary

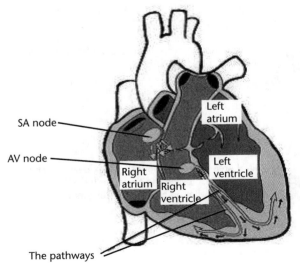

Figure 11.3   Conduction system of a human heart.

sources on the chest related to the well-known auscultatory areas: the mitral, aortic, pulmonary, and tricuspid areas. The standard auscultatory areas are shown in Figure 11.4. The mitral area is near the apex of the heart. The aortic area is to the right of the sternum, in the second right intercostal space. The tricuspid area is in the fourth intercostal space near the right sternal border. The pulmonary area lies at the left parasternal line in the second or third left intercostal space.

11.2.2   Rate and Rhythm

Identification of cardiac rate and rhythm is an important aspect of the cardiac examination. Depending on the patient's heart rate, normal is 60 to 100 beats per minute (bpm), bradycardia is <60 bpm, tachycardia is >100 bpm, and the rhythm is regular or irregular. If it is irregular, try to identify the pattern; for example, do early beats appear on a more or less regular basis (regularly irregular rhythm) or are they totally irregular (irregularly irregular rhythm)?

With the rate and rhythm known, you can narrow your differential diagnosis list considerably: A regular rhythm with tachycardia is consistent with sinus tachycar-

Aortic Area—located in the second intercostal space at the right sternal margin. The systolic murmurs of Aortic Stenosis and Increased Aortic Valve Flow.

Pulmonic Area—located in the second intercostal space at the left sternal border. The systolic murmur of Pulmonic Stenosis and the diastolic murmur of Pulmonic Regurgitation.

Tricuspid Area—located at the lower left sternal border (LLSB). The diastolic murmur of Tricuspid Stenosis.

Mitral Area—located about the apex beat, which is usually in the fifth intercostal space. It is also called the apical, or the left ventricular area. The systolic murmur of Mitral Regurgitation and the diastolic murmurs of Mitral Stenosis and Increased Valvular Flow.

Figure 11.4   Ausculatory areas.

dia, atrial or nodal (supraventricular) tachycardia, atrial flutter with a regular ventricular response, and ventricular tachycardia. A regular rhythm with a normal rate is consistent with either a normal sinus rhythm or atrial flutter with a regular ventricular response. A regular rhythm with bradycardia is often seen with sinus bradycardia, second-degree heart block, and complete heart block. A normal rate with a regularly irregular rhythm is consistent with sinus arrhythmia, ventricular premature contractions, and atrial or nodal (supraventricular) premature contractions. A normal rate with an irregularly irregular rhythm is seen in patients with atrial fibrillation and atrial flutter with varying block. Of course, almost any combination of heart rate and rhythm may occur in those patients with a malfunctioning electronic pacemaker. Don't forget to add this to your differential diagnosis list when appropriate.

11.2.3   Normal Heart Sounds

11.2.3.1   First Heart Sound (S1)

The normal first heart sound (S1) is traditionally thought to arise from mitral (M1) followed by tricuspid (T1) valve closure. Because the intensity of S1 depends on the velocity of the blood and resultant force of closure of the AV valves, factors that increase the force and velocity of ventricular pressure rise tend to increase the intensity of S1. The position of the AV valves at the onset of systole also affects S1 intensity. If ventricular contraction occurs against a wide-open valve, the LV leaflets attain a higher velocity (thus louder S1) than if the valve were partially closed.

S1 is normally loudest at the apex. Exercise and excitement increase the force and velocity of ventricular contraction and thus S1 intensity. Bradycardia is associated with an apparent softening of S1, since the AV valves are already closed at the onset of ventricular contraction. Splitting of S1, which may occur in normal patients, is best heard over the left lower sternal border.

11.2.3.2   Second Heart Sound (S2)

A normal second heart sound (S2) is produced by the pressure changes and vibration of valves and contiguous structures induced by motion of the aortic and pulmonic leaflets toward their respective ventricles. The normal first component is aortic and referred to as A2; the second is pulmonic (P2).

The A2 is heard best in the aortic area, but is also heard well in the mitral and pulmonic areas. P2 is usually heard only in the pulmonic area and at Erb's point. Normally, P2 is softer than the preceding A2, and this difference in intensity increases with age. Exercise increases the intensity of both A2 and P2. Deep inspiration tends to increase P2 intensity.

Inspiratory separation of S2 into two audible components, A2 and P2, is present throughout life, except in the newborn. Splitting is primarily due to delay in P2. During expiration, P2 occurs at nearly the same time as A2. Splitting of S2 is best heard in the third left intercostal space, or in the pulmonic area toward the end of inspiration, with the patient in the supine position.

Recognition of the first and second heart sounds can be made easier by following these hints: Because systole (the time of ventricular contraction) is usually shorter than diastole (the time of ventricular relaxation), there is a longer pause between S2 and S1 than between S1 and S2. S1 is of longer duration and of lower pitch; S2 is of shorter duration and of higher pitch. S1 is usually best heard at the apex; S2 is usually best appreciated in the aortic and pulmonic areas. The presence of tachycardia or bradycardia may change these relationships. In the patient, palpation of the apical impulse or the carotid pulse will assist in identification of S1, because S1 occurs slightly before these events.

11.2.3.3   Third Heart Sound (S3)

The physiologic third heart sound (S3) is a low-pitched vibration occurring in early diastole during the time of rapid ventricular filling. The S3 sound is produced by the abrupt transmission of forces to the chest wall when the blood mass enters the right ventricle. The S3 sound is commonly heard in children and adolescents and in some young adults. When heard after the age of 30, it is called a gallop sound, and is a sign of pathology, such as left ventricular failure.

The physiologic S3 occurs just after A2 and is best heard at the apex with a patient in the left semilateral position. S3 normally disappears completely when the patient sits or stands or performs any maneuver that lowers heart rate. Conversely, factors that increase heart rate, such as exercise, tend to accentuate a physiologic S3.

11.2.3.4   Fourth Heart Sound (S4)

The physiologic fourth heart sound (S4) is a very soft, low-pitched noise occurring in late diastole, just before S1. S4 generation is related to the ventricular filling by atrial systole. Vigorous atrial contraction produces rapid acceleration of blood mass. Associated with this event are vibrations in the left ventricle wall and mitral apparatus, which are heard as the S4.

A physiologic S4 may be heard in infants, small children, and adults over the age of 50. It is usually heard only at the apex with the patient placed in the left semilateral position. A physiologic S4 is poorly transmitted and is rarely accompanied by a shock (when the S4 can be felt as well as heard). Wide transmission of a loud S4 associated with a shock is pathologic and is referred to as an S4 gallop.

As in the case of S3, maneuvers that increase the force and frequency of ventricular contraction will accentuate S4. Conversely, maneuvers associated with cardiac slowing will diminish S4 intensity.

11.2.4   Abnormal Heart Sounds

11.2.4.1   Abnormal S1

An abnormally accentuated S1 is most often associated with mitral stenosis. Other conditions associated with a loud S1 are left-to-right shunts, hyperkinetic circulatory states (such as in anemia, thyrotoxicosis, and fever), accelerated AV conduction, and tricuspid stenosis.

An abnormally diminished S1 is associated with mitral and tricuspid stenosis, moderate or severe aortic regurgitation, delay in AV conduction, and diseases that reduce the force and velocity of ventricular contraction (acute myocardial infarction, cardiomyopathies, congestive heart failure, and acute myocarditis). A diminished S1 may also be found in patients with either pulmonary emphysema or very heavy upper body musculature.

A variable S1 is seen in those conditions that vary the velocity of AV valve closure and/or the relationship between atrial and ventricular systole: ventricular tachycardia, complete AV block, artificial ventricular pacemakers, junctional tachycardia with AV dissociation, and atrial fibrillation.

An abnormal S1 may be heard in the presence of complete right bundle branch block, premature ventricular contractions, Ebstein's anomaly (where tricuspid valve excursion is wide and closure is delayed), ventricular tachycardia, or with artificial pacing.

11.2.4.2   Abnormal S2

An accentuated S2 is seen with diastolic hypertension, systolic hypertension in the elderly, coarctation of the aorta, dilation of the aorta (as in syphilitic aortitis, Morphan's syndrome, and chronic severe hypertension), and advanced atherosclerosis of the aorta. Pulmonary hypertension (from left-to-right shunts, pulmonary thromboembolism, pulmonary emphysema, left ventricular failure, and mitral stenosis) is also associated with an accentuated S2.

A diminished S2 is seen in elderly patients with aortic stenosis, pulmonary emphysema, and pulmonic stenosis. Abnormal splitting of S2 (splitting other than that normally observed during inspiration) is associated with delayed electrical activation of the right ventricle (as in complete right bundle branch block and premature ventricular beats, and with left ventricular pacemakers), prolonged right ventricular or shortened left ventricular ejection time (as in valvular or infundibular pulmonic stenosis, mitral regurgitation, and ventricular septal defects), and altered impedance of the pulmonary vascular bed (massive pulmonary embolism).

Wide and fixed splitting of S2 (i.e., splitting of S2 that does not vary with respiration) is found in patients with large atrial septal defects, severe pulmonary stenosis, and right ventricular failure.

Paradoxical splitting of S2 (when splitting increases with expiration) is usually due to a delayed A2. This delay in A2 may be due to electrical conduction disorders (complete left bundle branch block, right ventricle premature contractions, and ventricular tachycardia) or mechanical disorders (severe valvular aortic stenosis, left ventricular outflow obstruction, hypertrophic cardiomyopathy, coronary artery disease, myocarditis, and congestive cardiomyopathy).

11.2.5   Murmur Classifications

Figure 11.5 discusses the various types of murmurs found in the heart.

11.2.5.1   Murmur Mechanisms

Figure 11.6 explains the mechanisms by which heart murmurs are caused.

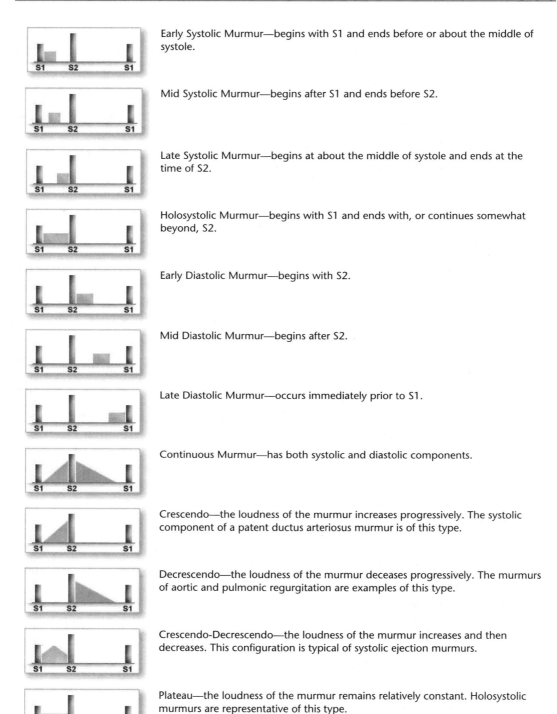

Early Systolic Murmur—begins with S1 and ends before or about the middle of systole.

Mid Systolic Murmur—begins after S1 and ends before S2.

Late Systolic Murmur—begins at about the middle of systole and ends at the time of S2.

Holosystolic Murmur—begins with S1 and ends with, or continues somewhat beyond, S2.

Early Diastolic Murmur—begins with S2.

Mid Diastolic Murmur—begins after S2.

Late Diastolic Murmur—occurs immediately prior to S1.

Continuous Murmur—has both systolic and diastolic components.

Crescendo—the loudness of the murmur increases progressively. The systolic component of a patent ductus arteriosus murmur is of this type.

Decrescendo—the loudness of the murmur deceases progressively. The murmurs of aortic and pulmonic regurgitation are examples of this type.

Crescendo-Decrescendo—the loudness of the murmur increases and then decreases. This configuration is typical of systolic ejection murmurs.

Plateau—the loudness of the murmur remains relatively constant. Holosystolic murmurs are representative of this type.

Figure 11.5   Murmurs in the heart.

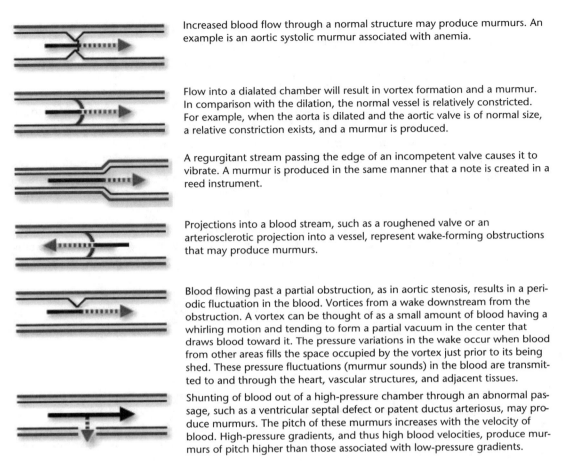

Increased blood flow through a normal structure may produce murmurs. An example is an aortic systolic murmur associated with anemia.

Flow into a dialated chamber will result in vortex formation and a murmur. In comparison with the dilation, the normal vessel is relatively constricted. For example, when the aorta is dilated and the aortic valve is of normal size, a relative constriction exists, and a murmur is produced.

A regurgitant stream passing the edge of an incompetent valve causes it to vibrate. A murmur is produced in the same manner that a note is created in a reed instrument.

Projections into a blood stream, such as a roughened valve or an arteriosclerotic projection into a vessel, represent wake-forming obstructions that may produce murmurs.

Blood flowing past a partial obstruction, as in aortic stenosis, results in a periodic fluctuation in the blood. Vortices from a wake downstream from the obstruction. A vortex can be thought of as a small amount of blood having a whirling motion and tending to form a partial vacuum in the center that draws blood toward it. The pressure variations in the wake occur when blood from other areas fills the space occupied by the vortex just prior to its being shed. These pressure fluctuations (murmur sounds) in the blood are transmitted to and through the heart, vascular structures, and adjacent tissues.

Shunting of blood out of a high-pressure chamber through an abnormal passage, such as a ventricular septal defect or patent ductus arteriosus, may produce murmurs. The pitch of these murmurs increases with the velocity of blood. High-pressure gradients, and thus high blood velocities, produce murmurs of pitch higher than those associated with low-pressure gradients.

Figure 11.6   Murmur mechanisms.

11.2.7   Gallop Sounds

The term *gallop sounds* refers to the cadence of S1 and S2 along with either an S3 or an S4, or their summation. This nomenclature arose because the trio of sounds simulates the cadence of a galloping horse, especially when the heart rate approaches 100 bpm. Gallop sounds are by definition diastolic events. The S3 or ventricular gallop occurs early in diastole, whereas the S4 or atrial gallop occurs in late diastole, just before the first heart sound (S1).

When both S3 and S4 gallops are present with a slow to moderate heart rate, one hears a quadruple rhythm. With faster heart rates, the S3 and S4 gallops become fused with each other in middiastole to produce a summation gallop. The term *gallop sounds* generally refers to the pathological third and fourth heart sounds, whereas the terms S3 and S4 usually refer to the normal, physiologic third and fourth heart sounds.

Repeating the word *Kentucky* can mimic the S3, or ventricular gallop, which occurs shortly after S2. Unlike the normal or physiologic S3, the S3 gallop may be palpable. In addition, standing diminishes the normal S3 intensity, but has no effect on an S3 gallop. The main causes of an S3 gallop include left and right ventricular diastolic overloading and reduced left and right ventricular compliance.

The left ventricular S3 gallop (LV S3G) is best heard at the cardiac apex, and with the patient in the left lateral recumbent position. Having the patient exercise and/or hold his or her breath in expiration also accentuates the LV S3G. The right ventricular S3 gallop (RV S3G) is best heard at the left lower sternal border, and is not heard at the apex. Inspiration tends to increase the intensity of an RV S3G.

Repeating the word *Tennessee* can mimic the S4, or atrial gallop, which occurs just prior to S1. A normal or physiologic S4 is rarely heard except in young children and the elderly, and is almost never palpable (as is an S4 gallop). Causes of the S4 gallop include reduced end-diastolic compliance, right ventricular systolic and diastolic overloading, and left ventricular diastolic overloading.

The left ventricular S4 gallop (LV S4G) is best heard at the cardiac apex, although it is often widely transmitted. The gallop may be accentuated by placing the patient in the left lateral recumbent position, or by having the patient exercise, or hold his or her breath in expiration.

The right ventricular S4 gallop (RV S4G) is best heard at the left lower sternal border, and is not heard at the apex. Rapid, deep inspiration and exercise tends to accentuate an RV S4G.

11.2.8 Clicks

Systolic clicks are usually referred to as either ejection or nonejection, depending mainly on when they occur during systole.

Ejection clicks occur early in systole at the time blood is being ejected from the left ventricle into the aorta or from the right ventricle into the pulmonary artery. Ejection clicks are usually associated with abnormalities of the semilunar valves or great vessels. When associated with either left or right ventricular outflow tract obstruction, ejection clicks indicate that the lesion is valvular in location.

Nonejection clicks occur in mid to late systole. They are usually the result of either mitral (mitral clicks) or tricuspid (tricuspid clicks) valve prolapse. By far, the most common cause of a nonejection click is mitral valve prolapse, especially of the posterior mitral leaflet.

Aortic ejection clicks are high-pitched sounds that occur in early systole. Intensity is not affected by respiration or patient position. Left-sided valvular ejection sounds are heard best in the aortic area and are not widely transmitted. The ejection click of aortic stenosis is heard best in the mitral area but is widely transmitted. Aortic ejection clicks are seen in congenital and rheumatic valvular aortic stenosis with a deformed but flexible aortic valve. Conditions associated with obstruction of the aorta, such as systemic hypertension and coarctation of the aorta, are also associated with aortic ejection clicks.

Pulmonic ejection clicks, like aortic ejection clicks, are high-pitched sounds of early systole. Unlike the aortic click, the pulmonic ejection click is heard best in the pulmonic area and is rarely transmitted. The pulmonic ejection click is the only right heart sound that decreases in intensity during inspiration. Pulmonic ejection clicks are often associated with mild to moderate valvular pulmonic stenosis, dilation of the pulmonary artery (as seen in pulmonary hypertension), and tetralogy of Fallot.

Nonejection clicks are high-frequency sounds of mid to late systole. Clicks associated with mitral valve prolapse (mitral clicks) are heard best in the mitral area and usually are not widely transmitted. Those associated with tricuspid valve prolapse (tricuspid clicks) are best appreciated along the left lower sternal border. Mitral and tricuspid clicks are sometimes only heard with the patient standing. In this position, the ventricles are smaller and the degree of prolapse of both mitral and tricuspid valves is increased. Having the patient exercise or move to a position other than standing usually diminishes the intensity of both mitral and tricuspid clicks.

11.2.9   Opening Snaps

11.2.9.1   Tricuspid Valve Opening Snap

In etiology an opening snap of the tricuspid valve is classically heard in cases of tricuspid stenosis and also has been reported in conditions associated with increased blood flow across the tricuspid valve, as in large atrial septal defects (ASDs).

The auscultatory characteristics of the tricuspid valve opening snap (TVOS) resemble those of the mitral valve opening snap (MVOS; discussed next). However, the TVOS is louder at the left lower sternal border or over the xiphoid area. Additionally, the loudness of the TVOS usually increases markedly during inspiration, whereas the MVOS is often louder on expiration. The interval between the first component of S2 (A2) and the TVOS tends to be longer than the interval between A2 and the MVOS.

11.2.9.2   Mitral Valve Opening Snap

In etiology when an elevated left atrial pressure forces the mitral valve to open rapidly to its point of maximal excursion, an opening snap is produced. The opening snap is related to the abrupt halting of the rapid opening motion of the AV valve, and to vibrations of the AV apparatus and the intracavitary blood mass at the end of isovolumetric relaxation of the ventricles.

Mitral stenosis is the most common cause of an opening snap of the mitral valve. Other, relatively rare, causes of an MVOS are endocardial fibroelastosis, ventricular septal defect (VSD), second- and third-degree AV block, tricuspid atresia associated with a large ASD, patent ductus arteriosus, and hyperthyroidism. The common element in these conditions is increased blood flow across the mitral valve.

With auscultatory characteristics the opening snap of the mitral valve has a quality similar to the normal heart sounds and is often confused with a splitting S2. The brief, sharp, rather snapping sound is heard shortly after the A2 component of S2.

The MVOS is heard best midway between the pulmonic and mitral areas. When loud, it is widely transmitted over the entire precordium. Optimum audibility is often achieved by turning the patient to the left lateral position. Standing tends to lower the left atrial pressure and thus increase the A2–OS interval.

A soft OS may be intensified after exercise that increases atrial pressure. The A2–OS interval is not altered during different phases of respiration; however, the MVOS is usually loudest on expiration.

11.2.9.3   Differential Diagnosis of MVOS Versus S3 Gallop

The OS is of higher pitch and occurs earlier in diastole than S3. In addition, the S3 gallop tends to be localized to the lower left sternal border (LSB) (RV S3G) or the cardiac apex (LV S3G), whereas the MVOS is usually widely transmitted.

In patients with significant isolated mitral valve stenosis, the MVOS is usually heard in association with the typical diastolic murmur of mitral stenosis. An S3 gallop is not heard because mitral stenosis prevents rapid ventricular filling and thus the generation of an S3. An S3 gallop in patients with mitral stenosis suggests that the mitral stenosis is trivial, the predominant lesion is mitral regurgitation, or the origin of the S3 gallop is from right ventricle dysfunction.

11.2.9.4   Differential Diagnosis of MVOS Versus Splitting of S2

The MVOS is of higher pitch and is usually louder than P2. In addition, the A2–OS interval is usually wider than the A2–P2 interval.

The OS is usually louder at the lower LSB than at the upper LSB, is heard well at the apex, and is often audible in the suprasternal notch. Splitting of S2 is best heard in the pulmonic area or at Erb's point, is not audible at the apex in the absence of pulmonary hypertension, and is not heard in the suprasternal notch.

The A2–OS interval usually widens in the upright position, whereas physiologic splitting of S2 narrows. Exercise-induced tachycardia decreases the A2–OS interval in patients with significant mitral stenosis, whereas the A2–P2 interval is usually unchanged.

11.2.10   Maneuvers

Heart sounds and murmurs can be altered by changes in venous return, vascular resistance, stroke volume, heart rate, and pressure gradients. Maneuvers (Figure 11.7) can be used alter these hemodynamic variables in a predictable manner.

11.3   Phonocardiograph

The technique of listening to the sounds produced by the organs and vessels of the body is called auscultation. In the early days of auscultation a physician listened to heart sounds by placing his ear on the chest of the patient, directly over the heart. The idea of transmitting the heart sounds from the patients chest's to the physician's ear via a section of cardboard tubing was then developed. This was the forerunner of the stethoscope, which has become a symbol of the medical profession. The stethoscope (from the Greek words *stehos*, meaning "chest," and *skopein*, meaning "to examine") is simply a device that carries sound energy from chest of the patient to the ear of the physician via a column of air. Because the system is strictly acoustical, there is no amplification of sound, except for any that might occur through resonance and other acoustical characteristics.

Instruments for graphically recording the heart sounds have been more successful. The device used to record the phonocardiogram of the heart sounds is called a phonocardiograph. It is true that a phonocardiogram reveals things that cannot be

Inspiration is associated with increased negative intrathoracic pressure, increased venous return to the heart, increased right ventricle stroke volume, and decreased pulmonary vascular resistance. These changes tend to increase the intensity of right-sided S3 and S4 gallops, mitral and tricuspid clicks, tricuspid and pulmonic stenosis, and tricuspid and pulmonic regurgitation murmurs. Splitting of S2 is also more apparent during inspiration when blood flow into the right side of the heart is increased and right ventricular ejection is prolonged.

Expiration is associated with decreased negative intrathoracic pressure and reduced venous return to the heart. Right-sided flow is decreased and splitting of S2 diminishes or disappears. However, murmurs and sounds originating on the left side of the heart (such as the left-sided S3 and S4, the mitral opening snap, and the murmurs of aortic stenosis and aortic regurgitation) are accentuated during expiration. In addition, the right-sided pulmonary ejection click may be appreciated best in expiration.

The Recumbent Position often accentuates the murmurs of mitral tricuspid stenosis.

The Left Semilateral Position can be used to accentuate the mitral opening snap and mitral regurgitation murmurs, in addition to the left-sided S3 and S4.

Exercise is associated with an increased heart rate, shorter diastole, elevated left atrial pressure, more abrupt closure of the heart valves, and increased blood velocity. These changes help account for the increased intensity of S1, S2, the mitral opening snap, the left-sided S3 and S4, the right-sided S4, and the murmurs of mitral regurgitation, mitral stenosis, and patent ductus arteriosus, noted in normal patients immediately following exercise.

Prompt Standing shifts blood to lower extremities and reduces left ventricle filling and size. This maneuver can often be used to accentuate the mitral and tricuspid clicks.

Sitting Up tends to accentuate the tricuspid valve opening snap.

Sitting Up, Leaning Forward tends to bring out the murmurs of aortic and pulmonic regurgitation, in addition to the murmur of aortic stenosis.

Figure 11.7 Maneuvers.

detected by a skilled auscultator. Phonocardiography has the ability to render susceptible to precise measurement time relationships, which hitherto could only be expressed in vague descriptive terms. It also produces a permanent record of what would otherwise be an ephemeral memory or an all-too-often inadequate note in words or diagrammatic form.

11.3.1 Instrument Description

The phonocardiograph consists of a transducer, filter, amplifier, computer equipped with a soundcard, and diagnosis software. The computer serves as a display for the graphical presentations, as a recorder for the phonocardiogram, and as a memory and disk storage for the documentation data. The computer also offers a way to send the auscultation data to other computers and printing devices. The phonocardiographic transducer (microphone or piezoelectric crystal) picks up the sounds within the patient's body and transforms them into electrical signals. The transducer is connected to an amplifier, which amplifies the signals to a suitable level and passes them to the filter. The filter allows only a useful range of frequencies for the soundcard. The soundcard transforms the signals into the digital domain for the diagnosis software. The general block diagram of the instrument is shown in Figure 11.8.

11.3.1.1 Transducer

A transducer is a device that converts a nonelectrical quantity into an electrical quantity. The purpose of the transducer in a phonocardiograph instrument is to convert the sound waves into electrical signals. Many sensing devices are commercially available for this purpose, including piezoelectric crystals and condensor microphones. Though the first offer good sensitivity, their stability is weak compared to the latter and they are also expensive. A good transducer is said to have the following properties:

- Extremely high impedance;
- Excellent frequency response;

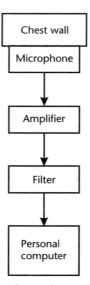

Figure 11.8 Block diagram of a phonocardiograph.

- Low distortion;
- Excellent transient response;
- Inexpensive.

11.3.2 Amplifier

The output voltage of the transducer is quite small and at a high impedance level; thus, an amplifier having high input impedance and gain is used at the condenser microphone. The amplifier can be a relatively simple ac amplifier, because a response to static or slowly varying voltages are not required. Condenser microphones are often used for the first stage of an FET input amplifier.

The output voltage is amplified from the millivolts range to a range in volts that is equivalent to an amplification factor of 1,000. To provide this gain, a two-stage noninverting operational amplifier can be used. A typical two-stage amplifier with an amplification factor of about 50 and 20 can be used. The output of the preamplifier will have the frequency response for the entire input. To obtain a frequency bandwidth that is limited to that of the heart sounds, the output is passed through a filter circuit.

11.3.3 Filter

An electrical filter is a frequency-selective circuit that passes a specified band of frequencies and blocks or attenuates signals of various frequencies outside this band. The requirement of a typical phonocardigraph filter is to attenuate any frequency below 10 Hz and above 1 kHz. Because the maximum cutoff frequency is only 1 kHz, active filters can be used. An active filter offers the following advantage over the passive filter:

1. *Gain and frequency adjustment flexibility:* Because the opamp is capable of providing a gain, the input signal is not attenuated as it is in a passive filter. In addition, the active filter is easier to tune or adjust.
2. *No loading problem:* Because of the high input resistance and low output resistance of the opamp, the active filter does not cause loading of the source or load.
3. *Cost:* Active filters are far more economical than passive filters. This is because of the variety of cheaper opamps and the absence of inductors.

11.3.4 Analysis

Many pathological conditions that occur in the cardiovascular system surface as murmurs and aberrations in a phonocardiogram long before they are reflected in other symptoms, such as changes in the electrocardiogram signal [1]. Furthermore, modern analytical methods provide new insights into heart sounds and murmurs and the precise timings of valve movements, firmly establishing auscultation as a cornerstone of the detection and evaluation of valvular disease and other cardiac disorders [2]. Auscultation has its demerits as well, because it is a highly subjective task and depends largely on the experience and training of the observer. Therefore, extensive training is required for a person to associate a particular heart sound with a particular diagnosis as well as classifying heart sounds into pathological or innocent sounds. Nonoptimal listening conditions (such as a busy hospital ward), the coexistence of other sounds or murmurs, a rapid heart rate, or the presence of noncardiac sounds such as chest wheeze lung sounds can make auscultation of little help in contributing to a diagnosis [3].

A phonocardiogram (PCG) together with an automatic algorithm assists a physician to localize and classify heart sounds and murmurs. Cardiac sounds have been the subject of many kinds of analyses. Frequency analyses of heart sounds have included application of the Fourier transform, short-time Fourier transform, and the Stransform [4]. Heart sounds have also been analyzed using parametric and nonparametric methods [5], as well as energy-based methods [6, 7]. Most of the current and past research focused on time segmenting the heart sound or finding features.

Heart sound segmenting [8] is useful for finding the traditional heart sounds, the major sounds S1 and S2, and the intervals in between—the systole and diastole—but does not give any clues as to the cause of the sounds. In the case of other advanced techniques, such as artificial neural networks, wavelet neural networks, and wavelet transforms, features are extracted from the heart sounds, and then these features are used to train a neural network to differentiate between pathological and physiologic sounds.

11.3.5 Segmentation of S1 and S2

Segmentation refers to partitioning of the PCG signal into cardiac cycles, and detection of the main events (S1, S2, murmur) and intervals (systole, diastole) in each cycle:

- The most common approach is to use other signals (mainly ECG) as an external reference to identify the beginning of S1. The CP notch may be used to identify the beginning of S2.
- Groch et al. [8] used the time-domain properties of the PCG signal (the systole S1–S2 is shorter and more constant than the diastole S2–S1).
- Iwata et al. [9] used the frequency-domain properties of the PCG: spectral tracking at designated frequencies of 100 and 150 Hz for S1 and S2, respectively, by a sliding 25-ms window, and peak detection by differentiating the tracking levels.
- Liang et al. [10] used the normalized average Shannon energy to compute the PCG envelope, and then applied a heuristic algorithm for peak selection and identification. They improved their method by discrete wavelet transform (DWT) decomposition and reconstruction.
- A more modern SP technique, matching pursuit (MP; a generalized wavelet transform), was applied by Sava et al. [11]. MP is a method proposed by Mallat and Zhang [12] for adaptive time–frequency decomposition. It decomposes the signal into waveforms selected from a dictionary of time–frequency atoms and selects the atoms that best match the local structure of the signal.

Two methods used for segmentation of PCG are explained in the following sections.

11.3.5.1 Blind Source Separation

Blind source separation (BSS) [13] denotes observing mixtures of independent sources, and making use of these mixture signals only—and nothing else—to recover the original signals. The basic BSS problem assumes instantaneous mixing of sources, and this is modeled by a linear relation between the observations x and sources s given by

$$x = As, \; x \in R'', S \in R^m, A \in R^{m \times n}$$

The components of s are assumed to be statistically independent and have probability distributions that are not Gaussian except for at most one component. To obtain a unique separation of sources given a set of mixtures, m, the number of sources is assumed to be less than or equal to n, the number of observations. The goal of BSS is to estimate a separation matrix $\mathbf{W}$ that satisfies

$$\mathbf{WA = PD, \; W} \in \mathbf{R}^{m \times n}, \mathbf{P} \in \mathbf{R}^{m \times n}, \mathbf{D} \in \mathbf{R}^{m \times m}$$

where $\mathbf{P}$ is a permutation matrix that has one large entry in each of its rows and columns and $\mathbf{D}$ is a diagonal matrix. With the separation matrix, the source can be reconstructed with

$$y = \mathbf{W}x$$

Learning $\mathbf{W}$ by observing x merely requires us to make use of higher order statistics (HOS). Typically, the separating matrix $\mathbf{W}$ is calculated iteratively by optimizing some cost function of the source estimates y [14]. Currently reported approaches to this problem can be divided into two categories. One category makes use of HOS explicitly [15, 16], and the other category makes use of HOS implicitly through the nonlinearity of neurons in a neural network [17–19]. Bell and Sejnowski [13] showed that another criterion for BSS can be the mutual information among the output y components.

11.3.5.2 Complexity-Based Gating

Simultaneous recordings of S2 from two acoustic sensors will contain other heart sounds, including S1 sounds. To extract S2 from two simultaneous PCG recordings, a simplicity-based gating scheme can be used. The simplicity-based gating scheme assumes that the human heart behaves like a hidden dynamic system whose evolution in time gives rise to the PCG. Because the complexity of the generated PCG is directly proportional to the complexity of the hidden dynamic system [20], major heart sounds such as S1 and S2, which are simpler in appearance, are generated when the hidden dynamic system exhibits a lower number of states (modes). By measuring the number of states (complexity) of the hidden dynamic system over time, the time intervals where S1 and S2 occur can be isolated, because during these intervals the PCG will exhibit lower complexity (higher simplicity). The details of the algorithm to measure simplicity of the PCG over time are given in Nigam and Priemer [20]. The simplicity profile of a PCG peaks at times when S1 and S2 occur and it is subsequently thresholded to identify these time intervals.

While comparing the simplicity-based gating scheme with a conventional gating scheme that uses Shannon energy, some noteworthy points were observed. The conventional energy-based gating method yields a time gate of duration, T_e, which gates only a portion of S2, the simplicity-based gating method gives a time gate of duration T_s, which encompasses most of the S2. This is because the energy-based gating method relies on the amplitude of the signal being gated and tends to gate only that portion of the signal that has significant amplitude. On the other hand, the simplicity-based gating method is invariant to amplitude fluctuations and depends only on the inherent simplicity of the signal (the number of modes of the hidden dynamic system).

Another advantage of using simplicity-based gating is that simplicity values of S1 and S2 are almost similar in all PCG recordings irrespective of their relative amplitudes, whereas their energy values depend on their absolute amplitudes. Therefore, when either of S1 or S2 is relatively weak in amplitude as compared to the other, simplicity-based gating will still gate both of them correctly, whereas energy-based gating might lose the sound with weaker amplitude.

11.3.6   Decomposition of PCG

A typical PCG can be decomposed by either parametric or nonparametric methods. Some of the parametric methods are autoregressive (AR), autoregressive moving average (ARMA), and adaptive spectrum analysis. Nonparametric method can be further classified into linear or quadratic where short-time Fourier transform (STTF) and continuous wavelet transform (CWT) come under the earlier one and Wigner-Ville distribution (WVD) and Choi-Williams distribution (CWD) are grouped under the later case.

In nonparametric methods, certain parameters in the analysis routine have to be set. For example, the spectrogram's window duration and degree of smoothing (cross-term reduce) in CWD. If a signal is a linear combination of some frequency components, its linear TFR is the same linear combination of each individual component. The STFT is a linear TFR that has been widely studied and used. A more recent linear TFR is the CWT. STFT suffers from a trade-off between time and frequency resolutions: The shorter analysis windows produce better temporal resolution, but degrade the spectral resolution.

Quadratic or bilinear TFRs can be expressed in a general form proposed by Cohen [21], where the particular distribution is defined by a *kernel* function of the frequency and the time lag. WVD might add artifact cross-terms to the TF representation of multicomponent signals, which can make the analysis difficult. CWD belongs to reduced interference distributions (RIDs), which reduces cross terms between components.

With parametric methods, the investigated signal is first modeled by an equivalent system, and then the parameters of the model are estimated from measurements of the signal made over a limited period of time. In the AR model, the current output of the system depends only on the values of previous outputs. In the ARMA model, the output depends on previous outputs and inputs. In adaptive signal decomposition, the largest amplitude component is identified (via STFT), reconstructed, and subtracted from the original signal to form a residual signal. The STFT is reapplied to the residual signal, and the procedure is repeated. MP is another adaptive spectrum analysis method that has been applied to heart sound decomposition.

11.3.7 Short-Term Fourier Transform

The FFT can provide a basic understanding of the frequency contents of the heart sounds. However, FFT analysis remains of limited value if the stationary assumption of the signal is violated. Because heart sounds exhibit marked changes with time and frequency, they are therefore classified as nonstationary signals. To understand the exact feature of such signals, it is thus important, to study their time–frequency characteristics. To gain information from both the time and frequency domains, Dennis Gabor [22] adapted the Fourier transform to analyze only a small section of the signal at a time. This adaptation is called the short-time Fourier transform. The STFT is formally defined by the integral transform:

$$X_T(f) = \int_{-T/2}^{T/2} x(t)\overline{w(t-\tau)}e^{-j2\pi ft}\ dt$$

where $x(t)$ is the signal, $w(t)$ is the conjugate of the window function, T is duration, and τ is the time location.

STFT maps a time signal into a 2-D time–frequency plane. The time and frequency information can only be obtained with limited precision, as determined by the size of the window. As the window function (time domain) gets narrower, the localization information in the frequency domain gets wider, and vice versa. In addition, once the window function is chosen, the STFT uses the fixed time–frequency window to analyze the whole signal (Figure 11.9), making it inaccurate for signals having relatively wide bandwidths that change rapidly with time. If a high degree of accuracy is required, the STFT must be repeatedly applied to the signal with a varying width each time. Due to the assumption of stationarity of the signal within the time interval of the window, the length of the stationary interval determines the time and frequency resolution.

In fact the spectrogram (STFT) cannot track very sensitive sudden changes in the time direction. To deal with these time changes properly, it is necessary to keep

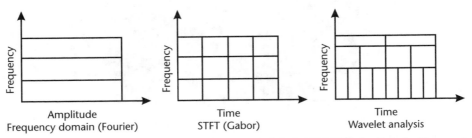

Figure 11.9 The windows of FT, STFT, and WT. (*From:* [7]. © 1998 IEEE. Reprinted with permission.)

the length of the time window as short as possible. This, however, will reduce the frequency resolution in the time–frequency plane. Hence, there is a trade-off between time and frequency resolutions [23]. A typical STFT of a normal PCG for a single cardiac cycle is shown in Figure 11.10.

11.3.8   Continuous Wavelet Transform

The CWT returns a coefficient that is a measure of the similarity between the wavelet at a particular scale and time point and the signal at that instance. The coefficient is calculated by scaling and shifting the basic wavelet Ψ to cover the whole signal:

$$C_{a,b} = \frac{1}{\sqrt{a}} \int_{-\infty}^{\infty} x(t) \, \overline{\Psi\left(\frac{t-b}{a}\right)} \, dt, \quad \text{for } a \neq 0$$

where $\overline{\Psi(t)}$ is the conjugate of the basic wavelet, a is the scale factor, b is the translation factor, and $x(t)$ is the signal to be analyzed.

The introduction of a scale parameter in CWT makes the time–frequency window flexible by allowing dilation and compression of the wavelet. This concept of scale gives CWT an optimized time–frequency resolution: good time resolution at high frequencies and good frequency resolution at low frequencies. This property makes wavelets useful in the detection of short-time transients.

Scale a is related to the frequency by the following expression:

$$F_a = \frac{F_c}{a\Delta}$$

where a is a scale, Δ is the sampling period, F_c is the center frequency of a wavelet in hertz, and F_a is the pseudofrequency corresponding to scale a, also in hertz.

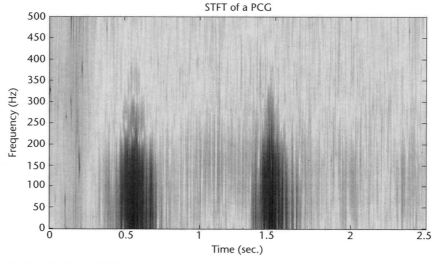

Figure 11.10   STFT of a PCG for one cardiac cycle.

A lower scale a results in a more compressed wavelet, which is used to analyze rapidly changing details. On the other hand, a higher-scale a results in a stretched wavelet, which is used to analyze coarse features.

Owing to the fact that windows are adapted to the transients of each scales, a wavelet lacks the requirement of stationarity. Hence, a WT is better suited to analyzing nonstationary signals than STFT. Because heart signals are transient signals, the WT is the technique chosen in this research.

11.3.8.1   Resolution

The resolution in time Δt and the resolution in frequency Δf for both the STFT and WT are bounded by Heisenberg's principle of uncertainty:

$$\Delta f \Delta t \geq \frac{1}{4\pi}$$

Only windows with a Gaussian shape satisfy the principle of uncertainty with an exact equality [24, 25].

11.3.9   Wigner Distribution

The subject of a time-frequency signal description is motivated by the need 5m of describing the fraction of total energy of a signal at a time t and frequency f. In light of this aim, the Wigner distribution was first proposed by Ville and hence is commonly referred to as the Wigner-Ville distribution [21]:

$$W(t, \omega) = \int_{-\infty}^{\infty} e^{-j\omega\tau} s^* (t - 1/2\tau) \, s(t + 1/2\tau) \; d\tau$$

The Wigner distribution is particularly useful because of its high resolution and its ability to satisfy a great number of desirable time-frequency properties. These clear advantages are offset by its poor performance when applied to multicomponent signals. Applying the Wigner distribution to a signal with two distinct components results in interference terms that appear in time and frequency where a true signal does not exist. Although it is generally accepted that these cross-terms contain signal information, their interpretation and significance are not as yet fully understood. The Wigner distribution of a typical PCG for a single cardiac cycle is shown in Figure 11.11.

From the WVD the instantaneous frequency of the signal can be derived that helps us classify heart sound and murmurs. Figure 11.12 shows the instantaneous frequency of a PCG for a single cardiac cycle.

11.3.10   The Choi-Williams Distribution

After the discovery of the Wigner distribution, a number of other distributions with various advantages were developed. These distributions showed common desirable properties, such as satisfying the margins, and all of them present a reasonable time–frequency (T-F) description of a signal. The appearance of this large array of

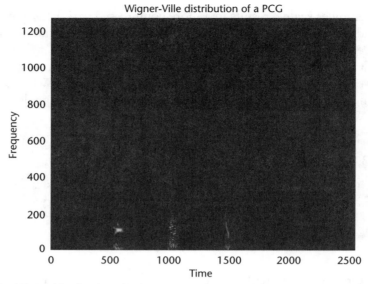

Figure 11.11   Wigner distribution of a PCG.

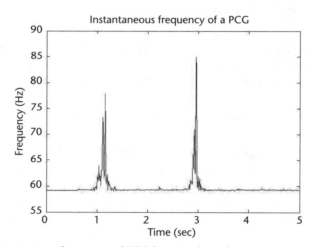

Figure 11.12   Instantaneous frequency of PCG for a single cycle.

seemingly valid transforms, all with alternative derivations and valid performances, created a confusing situation. This confusion was overcome when Cohen presented his unified theory for T-F distributions. Cohen proposed that there were in fact an infinite number of possible transforms that could be derived from the general formula:

$$P(t,\omega) = \frac{1}{4\pi^2} \iiint e^{-j\theta t - j\tau\omega + j\theta u} \, \phi(\theta,\tau) \, s^*(u - 1/2\theta) \, s(u + 1/2\tau) \; du \; d\tau \; d\theta$$

This general formula has the same bilinear structure $(s^*(u - 1/2\theta) \, s(u + 1/2\tau))$ as the Wigner distribution and in the general case suffers from the same cross-term problems. The important feature of the second equation is the term phi, which is commonly referred to as the *kernel function*. The advantage of this kernel notation is that properties of the kernel determine and in fact can be used to control the properties of the distribution. The satisfaction of desirable distribution properties has driven the development of various distributions. One in particular has been developed to address the cross-term problem. The distribution in question was first studied by Choi and Williams [26]. They named the new distribution the *exponential distribution* after its kernel function, but future work has referred to it as the Choi-Williams distribution. The CWD kernel is:

$$\phi(\theta,\tau) = e^{-\theta^2\tau^2/\sigma}$$

where σ is a constant. Substituting this into the generalized distribution we obtain the CWD:

$$CW(t,\omega) = \frac{1}{4\pi^{3/2}} \iint \frac{1}{\sqrt{\tau^2/\sigma}} e^{-[(u-t)^2/(4\tau^2/\sigma)]-j\tau\omega} s^*(u-1/2\theta)s(u+1/2\tau) \ du \ d\tau$$

The term σ controls the reduction of undesirable cross-terms; a low value for σ results in reduced cross-terms, and a high value for σ forces the distribution to behave as the Wigner distribution with large cross-terms. A side effect of reducing the cross-terms is that the distribution loses resolution. Hence a value for σ is used such that a balance is found where cross-term reduction occurs without a significant loss of resolution [27]. The CWD distribution of a typical PCG is shown in Figure 11.13.

11.3.11   Autoregressive Modeling

The quadrature Doppler signal can be modeled by a complex AR process. To compute the T-F domain by using AR modeling, an approach similar to spectrogram analysis is used. More precisely, the time-varying AR model is expressed as follows:

$$x(n) = \sum_{m=1}^{p} a(m,n)x(n-m) + e(n)$$

where $x(n)$ is the complex Doppler signal, p is the order of the model, $a(m,n)$ is the complex time-varying coefficients, and $e(n)$ is the modeling error. The complex coefficients $a(m,n)$ can be computed by using the Yule-Walker equations together with the Levinson-Durbin algorithm [28] for an efficient recursive solution. Once the AR coefficients are found, the power spectrum of each windowed signal segment is computed by using

$$P_n(\theta) = \frac{\delta_p^2(n)}{\left|1 + \sum_{m=1}^{p} a(m,n)\exp(-jm\theta)\right|^2}$$

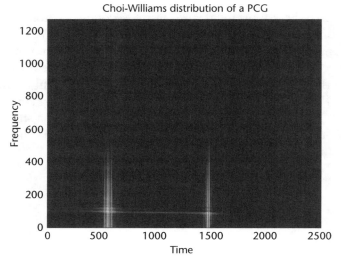

Figure 11.13   CWD distribution of a PCG.

where $\delta(n)$ is the variance of the modeling error signal corresponding to model order p. Thus, the T-F domain of the Doppler signal $x(n)$ is obtained by sliding the window with a prespecified time increment to compute the time-indexed spectra. For $0 = 2\prod k/N$, the T-F domain based on AR modeling is

$$AR_x(n,k) = P_n(\theta) = \frac{\delta_p^2(n)}{\left|1 + \sum_{m=1}^{P} a(m,n)\exp\left(-j\frac{2nk}{N}m\right)\right|^2}$$

11.3.12   Feature Extraction

Morphological feature extraction has been applied extensively to the analysis of heart sounds using spectrum analysis techniques [29–31]. Joo et al. [29] used the two most dominant second sound frequencies as diagnostic features. Stein et al. [30] used the dominant frequency of the second sound as a diagnostic feature, showing that there is clear correlation between this feature and the valve condition. Durand et al. [31] initially used nine diagnostic features, but subsequently [32] added more. These features have physical meaning, and they can be used to provide information about the patient's condition rather than provide a simple yes/no condition classification. Using T-F representations it is possible to extract time, frequency, and combined T-F features.

AR coefficients [33] modeled the heart sounds as a 12-order all-pole system, and used the 12 coefficients as features. Classification results using these features were better than with spectral features. DWT is a representation of a signal using an orthonormal basis of functions ("wavelets"). DWT-based features [34] compared the classification performance with nine morphological features from the time–frequency domain to the performance with a large set of features, extracted using the discrete wavelet transform (the coefficients of the transform). An optimal subset of features was selected using a search scheme. Results indicated that the DWT-based feature extraction technique yielded superior performance.

Muruganantham et al. [35] derived average power, total power, mean power frequency, median frequency, frequency variance, frequency skewness, frequency kurtosis, and jitter from the frequency domain. The frequency skewness and kurtosis are the third and fourth moment of the frequency spectrum. They used these features together with a standard deviation for the detailed coefficient of the wavelet transform. Also, from the FFT spectrum, they derived a Reimann sum, which involves averaging the FFT spectrum over a specified window length.

11.3.13 Classification

In general, the methods for feature extraction are crucial to a successful classification. A successful classification system depends not only on reliable, stable, and adaptable classifiers, but also on the strategies and methods used to extract and select features from the raw data. Classifiers perform better on feature vectors that can represent observed objects more efficiently such that the relevant information to patterns is emphasized the most.

The Bayes classifier is the most popular parametric technique of supervised pattern recognition. It is the optimal technique when the probability density functions of the patterns in the feature space are known. It is common to assume that a Gaussian PDF represents the distribution of the features for each class and to estimate the required mean and variance parameters from the training set.

When the probability density functions are difficult to estimate, a nonparametric approach such as the use of the neural network (NN) might be suitable. This method is based on distance measurements and, therefore, requires a distance metric. The nearest neighbors of a pattern of an unknown class are found by computing the distances between the pattern and a given number of patterns from the training set. The class of the input pattern is set to be the class of the K nearest neighbors.

The choice of distance metrics is important. In most cases, the Euclidian distance is used. But other measures, such as the Mahalanobis distance (the distance between two feature vectors, scaled by the statistical variation in each feature, that is, the covariance matrix) may produce better results.

ANN has been applied to classification of bioprosthetic valves and sounds classification of murmurs hidden Markov model (HMM), and fuzzy rule-based methods have been applied to ECG classification for detecting ectopic beats.

Muruganantham et al. developed NN, fuzzy, and wavelet-based classifiers for the classification of phonocardiograms and they were found to have better results with wavelet method, as discussed next.

11.3.13.1 Neural Networks

They designed a back propagation neural network with nine inputs and one output, which is either 1 for a normal case or 0 for an abnormal case. A single hidden layer with five neurons was implemented with a sigmoid transfer function at the input and output side. With the derived features the neural network is trained for about 10,000 epochs with performance criteria of 1.E–10. The mean square error was traced during each epoch. They found that during training MSE was considerably reduced, and at the end it reached a value of about 0.15. For testing the network, separate data were given to the network and the outputs were noted.

11.3.13.2   Fuzzy Systems

Muruganantham et al. signal was divided in to four beats and a FFT was performed on each of the four beats. The results were averaged and an averaged FFT resulted. Then a Reimann sum was performed on the averaged FFT data, on 50-point (Hz) intervals In essence, the data of 50-Hz intervals is binned and averaged into eight values (Figure 11.14). These eight values of all the signals of the same classes are grouped and averaged to get an 8×2 matrix data matrix, which is used to find the center values of the membership functions. The fuzzy system with eight inputs, each having two Gaussian membership functions with the known center value and two outputs corresponding to the two cases, was framed. Also the degree of fulfillment of each rule is needed, so two rules are written to determine at what extent the input at each bin of the sample corresponds to each heart condition.

11.3.13.3   Wavelet Method

Using "Daubechies 4" as the mother wavelet, the signals are decomposed in to six stages. The variances of the detailed coefficients at each stage are calculated for all signals and the average taken for both cases. The slopes of the curves for both cases are obtained by plotting log2 of these values against the log2 scales. While testing, this slope was calculated and compared with the standard from which the results were concluded.

They found that, via the wavelet method, 92% of signals were classified correctly, whereas a neural network correctly classifies 85% and a fuzzy system classifies 88%

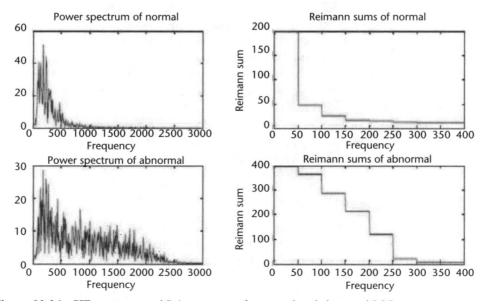

Figure 11.14   FFT spectrum and Reimann sum of a normal and abnormal PCG.

of the test signals. This is due to the fact that in a wavelet transform the signal is analyzed at a higher resolution in the frequency domain at varying window lengths, whereas in the neural network and fuzzy methods the parameters are extracted from the fixed-size window frequency representation, the Fourier transform.

Manivannan et al. [36] proposed three types of statistical classifier and compared their results. They proposed methods for classification of phonocardiograms using the Ho-Kashyap learning algorithm (HKL), Fisher discriminant analysis (FDA), and support vector machine (SVM). They found that both the FDA and HKL algorithm classifiers had almost equal efficiency of about 91% and 89.98%, respectively, whereas the efficiency of the SVM classifier was only 70%, and it can be increased by using larger amounts of data.

References

[1]   Liang, H., S. Lukkarinen, and I. Hartimo, "Heart Sound Segmentation Algorithm Based on Heart Sound Envelogram," *Comput. Cardiol.,* Vol. 24, 1997, pp. 105–108.

[2]   Hadjileontiadis, L. J., "Discrimination of Heart Sounds Using Higher Order Statistics," *Proc. 19th Int. IEEElEMBS Conf.,* 1997, pp. I 138–I 141.

[3]   Asir, K., and M. Abbasi, "Time-Frequency Analysis of Heart Sounds," *IEEE TENCON–DSP Applications,* 1996, pp. 2553–2558.

[4]   Livanos, G., N. Ranganathan, and J. Jiang, "Heart Sound Analysis Using the S-Transform," *Comput. Cardiol.,* Vol. 27, 2000, pp. 587–590.

[5]   Haghighi-Mood, T., *Application of Advanced Signal Processing Techniques in Analysis of Heart Sound,* London, U.K.: Institution of Electrical Engineers, 1995.

[6]   Sharif, Z. M. S., A. Z. Sha'ameri, and S. H. S. Salleh, "Analysis and Classification of Heart Sounds and Murmurs Based on the Instantaneous Energy and Frequency Estimations," *Proc. IEEE TENCON 2000,* Vol. 2, pp. 130–134.

[7]   Liang, H., "A Heart Sound Feature Extraction Algorithm Based on Wavelet Decomposition and Reconstruction," *Proc. 20th Annual Int. Conf. IEEE Engineering in Medicine and Biology Society,* Vol. 20, No. 3, 1998, pp. 1539–1542.

[8]   Groch, M. W., J. R. Domnanovich, and W. D. Erwin, "A New Heart-Sounds Gating Device for Medical Imaging," *IEEE Trans. on Biomedical Engineering,* Vol. 39, March 1992, pp. 307–310.

[9]   Iwata, I., et al., "Algorithm for Detecting the First and the Second Heart Schools by Spectral Tracking," *Med. Biol. Eng. Comput.,* Vol. 18, 1980, pp. 13–26.

[10] Liang, H., L. Sakari, and H. Iiro, "A Heart Sound Segmentation Algorithm Using Wavelet Decomposition and Reconstruction," *Proc. 19th Intl. Conf. IEEE IMBS,* 1997, pp. 1630–1633.

[11] Sava, H., P. Pibarot, and L. G. Durand, "Application of the Matching Pursuit Method for Structural Decomposition and Averaging of Photocardiographic Signals," *Med. Biol. Eng. Comput.,* Vol. 36, 1998, pp. 302–308.

[12] Mallat, S. G., and Z. Zhang, "Matching Pursuit with Time-Frequency Dictionaries," *IEEE Trans. on Signal Processing,* Vol. 41, No. 12, 1993, pp. 3387–3415.

[13] Bell, A. J., and T. Sejnowski, "An Information-Maximization Approach to Blind Separation and Blind Deconvolution," *Neural Computation,* Vol. 7, No. 6, 1995, pp. 1129–1159.

[14] Hyvarinen, A., "Survey on Independent Component Analysis," *Neural Comput. Surveys,* Vol. 2, 1998, pp. 94–128.

[15] Cardoso, J.-F., "Source Separation Using Higher Order Moments," *Proc. IEEE ICASSP,* 1989, pp. 2109–2112.

[16]   Yellin, D., and E. Weinstein, "Criteria for Multichannel Signal Separation," *IEEE Trans. on Signal Process.*, Vol. 42, 1994, pp. 2158–2168.

[17]   Jutten, C., and J. Herault, "Blind Separation of Sources: Part 1, An Adaptive Algorithm Based on Neuromimetic Architecture Signal Process," *Signal Processing*, Vol. 24, 1991, pp. 1–10.

[18]   Burel, G., "Blind Separation of Sources: A Nonlinear Neural Algorithm," *Neural Networks*, Vol. 5, 1992, pp. 937–947.

[19]   Cardoso, J.-F., A. Belouchrani, and B. Laheld, "A New Composite Criterion for Adaptive and Iterative Blind Source Separation," *Proc. IEEE ICASSP*, Vol. 4, 1994, pp. 273–276.

[20]   Nigam, V., and R. Priemer, "Accessing Heart Dynamics to Estimate Durations of Heart Sounds," *Physiol. Meas.*, Vol. 26, 2005, pp. 1005–1018.

[21]   Cohen, L., "Time-Frequency Distributions—A Review," *Proc. IEEE*, Vol. 77, 1989, pp. 941–981.

[22]   Gabor, D., "Theory of Communication," *J. Inst. Electrical Engineers*, Vol. 93, 1946, pp. 429–549.

[23]   Obaidat, M. S., "Phonocardiogram Signal Analysis: Techniques and Performance Comparison," *J. Medical Engineering Technol.*, Vol. 17, No. 6, November–December 1993, pp. 221–227.

[24]   Sarkar, T. K., M. S. Palma, and M. C. Wicks, *Wavelet Applications in Engineering Electromagnetics*, Norwood, MA: Artech House, 2002.

[25]   Chui, K. C., *An Introduction to Wavelets,* London: Academic Press Limited, 1992.

[26]   Choi, H. I., and W. J. Williams. "Improved Tune-Representation of Multicomponent Signals Using Exponential Kernels," *IEEE Trans. on Acoustics, Speech and Signal Processing*, Vo. 37, No. 6, 1989, pp. 862–871.

[27]   Jeong, J., and W. J. W. M., "Kernel Design for Red & Interference Distributions," *IEEE Trans. on Signal Processing*, Vol. 40, No. 2, 1992, pp. 452–412.

[28]   Kay, S. M., and S. L. Marple, "Spectrum Analysis—A Modem Perspective," *Proc. IEEE,* Vol. 69, No. 11, November 1981, pp. 1380–1419.

[29]   Joo, T. H., et al., "Pole-Zero Modeling and Classification of Phonocardiograms," *IEEE Trans. on Biomed. Eng.,* Vol. BME-30, 1983, pp. 110–117.

[30]   Stein, P., et al., "Frequency Spectra of First Heart Sound and of Aortic Component of the Second Heart Sound in Patients with Degenerated Porcine Bioprosthetic Valves," *Am. J. Cardiol.*, Vol. 53, 1984, pp. 557–561.

[31]   Durand, L.-G., et al., "A Bayes Model for Automatic Detection and Quantification of Bioprosthetic Valve Degeneration," *Mathematical. Comput. Modeling*, Vol. 11, 1988, pp. 158–163.

[32]   Durand, L-G., et al., "Comparison of Spectral Techniques for Computer-Assisted Classification of Spectra of Heart Sounds in Patients with Porcine Bioprosthetic Valve," *Med. Biolog. Eng. Computing*, Vol. 31, No. 3, 1993, pp. 229–236.

[33]   Guo, Z., et al., "Cardiac Doppler Blood Flow Signal Analysis, Part II: The Time-Frequency Distribution by Using Autoregressive Modeling," *Med. Bio. Eng. Comput.*, Vol. 31, No. 3, 1993, pp. 242–248.

[34]   Bentley, P. M., P. M. Grant, and J. T. E. McDonnell, "Time-Frequency and Time-Scale Techniques for the Classification of Native and Bioprosthetic Heart Valve Sounds," *IEEE Trans. on Biomedical Engineering*, Vol. 45, No. 1, January 1998, pp. 125–128.

[35]   Muruganantham, J., et al., "Methods for Classification of Phonocardiograms," *Proc. TENCON 2003, Conference on Convergent Technologies Asia-Pacific Region*, Vol. 4, 2003, pp. 1514–1515.

[36]   Manivannan, J., B. Jaganatha Pandian, and J. Muruganantham, "Heart Dysfunction Identification from Phonocardiography," *ICAIET 2006*, Sabah, Malaysis, November 22–24, 2006.

PART V

3-D Prostate Ultrasound Imaging

CHAPTER 12

Prostate Analysis Using Ultrasound Imaging

Fuxing Yang, Jasjit Suri, and Aaron Fenster

Prostate cancer is the most commonly diagnosed malignancy in men over the age of 50 [1]. It is the second leading cause of death due to cancer in men worldwide [2]. Ultrasound is a widely used imaging modality for prostate biopsy. The accurate detection of the prostate boundary in ultrasound images is of great importance for volume measurement and disease monitoring. However, in ultrasound images, the contrast is usually low and missing or the image has diffuse boundaries. These are very challenging problems for accurate segmentation of the prostate from the background.

Currently, many methods have been introduced to facilitate more accurate segmentation of the prostate boundaries from ultrasound images. Many of the methods are based on edge information. Aarnink et al. [3] proposed a method for determining of the contour of the prostate in ultrasound images via an edge detection technique based on nonlinear Laplace filtering. The authors combined the information about edge location and strength to construct an edge intensity image. Then edges representing a boundary are selected and linked to build the final outline. Abolmaesumi and Sirouspour [4] used a Sticks filter to reduce the speckle, and the problem is then discretized by projecting equispaced radii from an arbitrary seed point inside the prostate cavity toward its boundary. Candidate edge points obtained along each radius include the measurement points and some false returns. Pathak et al. [5], proposed an algorithm for guided edge delineation, which provided automatic prostate edge detection as a visual guide to the user. The edge detection contained contrast enhancement and an anisotropic diffusion filter for smoothing the images. However, the method needed a manual linking procedure on the detected edges.

Another group of methods require human interactivity for initialization. For example, Ladak et al. [6] proposed an algorithm, that used model-based initialization and a discrete dynamic contour. The user selected four points around the prostate. Then the outline of the prostate was estimated using cubic interpolation functions and shape information, and the estimated contour was deformed automatically to better fit the prostate. Wang et al. [7] used a similar 2-D method. However, the authors segmented the remaining slices by iteratively propagating the result to the other slices and implementing the refinement. Hu et al. [8] used a model-based initialization and mesh refinement with deformable models. Six points were required to initialize the outline of the prostate using shape information. The initial outline was then automatically deformed to better fit the prostate boundary. Chiu et al. [9] introduced an algorithm based on the dyadic wavelet transform and the discrete dynamic contours. A spline interpolation was used to determine the initial contour based on four user-defined initial points. Then the discrete dynamic contour refined the initial contour based on the approximate coefficients and the wavelet coefficients generated using the dyadic wavelet transform. Ghanei et al. [10] proposed a three-dimensional deformable surface model for prostate segmentation based on a discrete structure, which was made from a set of vertices in the 3-D space as triangle facets. But this model also needed manual initialization from a few polygons drawn on different slices.

Pattern recognition—based methods have also been widely used. Richard and Keen [11] presented a texture-based algorithm, which segments a set of parallel 2-D images of the prostate into prostate and nonprostate regions. The algorithm utilizes a pixel classifier based on four texture energy measures associated with each pixel in the image. Clustering techniques are used to label each pixel in the image with the label of the most probable class. Prater and Richard [12] use feed-forward neural networks for segmentation of the prostate in transrectal ultrasound images. These networks are trained using a small portion of a training image segmented by an expert and then applied to the entire training image.

Due to the weak boundary information, some researchers exploit statistical shape knowledge to help more accurate segmentation. Shen et al. [13] introduced a statistical shape model to segment the prostate in transrectal ultrasound images. The authors used a Gabor filter bank in both multiple scales and multiple orientations to characterize the prostate boundaries. Then, a hierarchical deformation strategy was used. The model focused on the similarity of different Gabor features at different deformation stages using a multiresolution technique. Although the approach had very high segmentation accuracy, the complexity and the computational cost were big barriers. Gong et al. [14] presented an approach based on deformable superellipses. Model initialization and constraining model evolution were based on prior knowledge about the prostate shape.

Generally in ultrasound images, prostate segmentation methods based on edge information have limitations when the image contains shadows with similar gray levels and textures as the prostate region, or missing boundary segments. The similar pattern inside and outside the prostate presents more difficulties for the pattern recognition-based method. Algorithms based on active contours depend on user interaction to determine the seed points. Although statistical shape information has been successfully applied, the convergence criterion based on the edge information makes the approach a nonrobust solution.

In recent years, region based active contour techniques [15–18] have been widely used in medical image analysis. The method can detect contours both with or without a gradient, for instance, objects with very smooth boundaries or even with discontinuous boundaries. The model has a level set formulation with a simple solution. The initialization of the contour also becomes simpler because the energy function defined by the prior shape and intensity prior, extracted from the training data set, will drive the contour to the location close to the object. Compared with edge-based methods, this approach is robust and does not need extra human intervention. In this chapter, we introduce a system for fully automated 3-D prostate segmentation from transrectal ultrasound images, that uses shape and intensity knowledge in a level set framework. Readers will gain an understanding about the regional-based active contour in a level set framework; how it solves the segmentation problem without using image gradient information, which is a major control factor for other approaches; and how it can be implemented via the level set method. In addition, readers will be shown how statistic shape information can be extracted from a training procedure and combined with regional information and intensity prior to segmentation.

In Section 12.1, a short introduction of the level set method and the theoretical background is given. Section 12.2 introduces the relationship between active methods and level set methods. Also, the active contours using level set frameworks are discussed in more detail, including the mathematical background. Section 12.3 describes a practical application: a system for fully automated 3-D prostate segmentation from transrectal ultrasound (TRUS) images, which uses shape and intensity knowledge in a level set framework. Finally, Section 12.6 gives a brief conclusion and a discussion of the pros and cons of the method.

12.1   Level Set Method

12.1.1   Front Evolution

Front evolution is a useful technique in image analysis for object extraction, object tracking, and so on. The basic idea behind the method is to evolve a curve toward the lowest potential of a cost function, where its definition reflects the task to be addressed and imposes certain smoothness constraints. Propagating interfaces occur in a wide variety of settings, and include ocean waves, burning flames, and material boundaries. Less obvious boundaries are equally important and include shapes against backgrounds, handwritten characters, and isointensity contours in images. Furthermore, there are applications not commonly thought of as moving interface problems, such as optimal path planning and construction of shortest geodesic paths on surfaces, which can be recast as front propagation problems with significant advantages [19].

Consider a boundary, either a curve in 2-D or a surface in 3-D, separating one region from another. Assume that this curve/surface moves in a direction normal to itself with a known speed function F. The goal is to track the motion of this interface as it evolves. Here, the motions of the interface in its tangential directions are ignored. At a specific moment, the speed function $F(L, G, I)$ describes the motion of the interface in the normal direction.

1. Speed factor L depends on local geometric information (e.g., curvature and normal direction).
2. Speed factor G depends on the shape and position of the front (e.g., integrals along the front, heat diffusion).
3. Speed factor I does not depend on the shape of the front (e.g., an underlying fluid velocity that passively transports the front).

First, Lagragian techniques are based on parameterizing the contour according to some sampling strategy and then evolving each element according to the speed function. Although such a technique can be very efficient, it suffers from various limitations such as deciding on the sampling strategy, estimating the internal geometric properties of the curve, changing its topology, addressing problems in higher dimensions, and so on [20].

The level set method was initially proposed to track a moving front by Osher and Sethian in 1988 [21] and has been widely applied across various imaging domains in the late 1990s. They can be used to efficiently address the problem of curve or surface propagation in an implicit manner. The central idea is to represent the evolving contour using a signed function, which is in a higher dimensional space, where its zero level corresponds to the actual contour. Then, according to the motion equation of the contour, one can easily derive a similar flow for the implicit surface that when applied to the zero level will reflect the propagation of the contour. Basically, there are two kinds of formulations regarding the level set method: boundary value formulation and initial value formulation.

12.1.2   Boundary Value Formulation

Assume for the moment that $F > 0$ and the front is moving all the way outward. We can define the arrival time $T(x, y)$ of the front as it crosses each (x, y). Based on the fact, *distance = rate × time*, we have the following expression in 1-D situation:

$$1 = F \frac{dT}{dx} \tag{12.1}$$

∇T is orthogonal to the level sets of T, and its magnitude is inversely proportional to the speed. So

$$|\nabla T|F = 1, \quad T = 0 \text{ on } \Gamma \tag{12.2}$$

where Γ is the initial location of the interface. Then, the front motion is characterized as the solution of a boundary value problem. If the speed F depends only on position, then the equation reduces to what is known as the *Eikonal equation*.

The efficient way to solve the Eikonal equation is to use the fast marching method. For readers who are interested in the details of the method, please refer to the literature [22].

12.1.3   Initial Value Formulation

Assume that the front moves with a speed that is neither strictly positive nor negative, so the front can move inward and outward and a point (x, y) can be crossed several times. $T(x, y)$ is not a single-value function. The solution is to treat the initial position of the front as the zero level set of a higher dimensional function Φ (Figure 12.1) The evolution of this function will give the propagation of the front through a time-dependent initial value problem, and at any time, the front is given by the zero level set of time-dependent level set function Φ. Taking the 1-D situation as example, with $x(t)$ as the front, we have the following definition of the zero-level set:

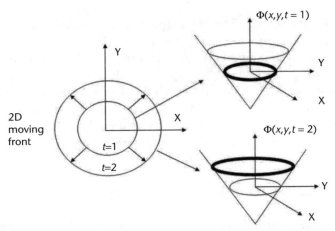

Figure 12.1 The zero level set of a higher-dimensional function is taken as the moving front.

$$\Phi\big(x(t),t\big) = 0 \tag{12.3}$$

Taking the partial differentials on the preceding equation and we will have the following equation:

$$\Phi_t + \nabla\Phi\big(x(t),t\big) \cdot x'(t) = 0 \tag{12.4}$$

$$F = x'(t) \cdot n, \quad \text{where } n = \frac{\nabla\Phi}{|\nabla\Phi|} \tag{12.5}$$

$$\Phi_t + F|\nabla\Phi| = 0, \quad \text{given } \Phi\big(x,t = 0\big) \tag{12.6}$$

So, from (12.6) the time evolution of the level set function Φ can be described such that the zero-level set of this evolving function is always identified with the propagating interface.

There is also an efficient way to solve the initial value problem we just introduced, the narrowband method. For readers who are interested in the details of the method, please refer to the literature [23].

12.1.4 Applications of Level Set Method

Nowadays, level set methods have been widely used in medical image processing. The diversity of applications of level sets has reached into [24] geometry [25–27], grid generation [22], fluid mechanics [28, 29], combustion [30], solidification [31], device fabrication, [32], morphing [33], object tracking/image sequence analysis in images [34–36], stereo vision [37], shape from shading [38, 39], mathematical morphology [40, 41], color image segmentation [42], 3-D reconstruction and modeling [43], surfaces and level sets [26, 44], topological evaluations [45], inpainting, and 2-D and 3-D (even 4-D) medical image segmentation. All of these give just a glimpse of the powerful level set method.

Among the applications just mentioned, segmentation has always been a critical component in 2-D and 3-D medical imagery since it assists largely in medical therapy [24]. The applications of shape recovery have been increasing since scanning methods became faster, more accurate and less artifacted [47]. The recovery of shapes of the human body is more difficult compared to other imaging fields. This is primarily due to the large variability in shapes, complexity of medical structures, several kinds of artifacts and restrictive body scanning methods. In the following sections, we focus on a family of the most advanced medical image segmentation methods in level set frameworks and also introduce one important application based on the method of 3-D prostate segmentation using TRUS images.

12.2 Active Contours in Level Set Framework

The level set method is closely connected with the active contour or deformable model method, which is a powerful tool for image analysis. Active contours are object-delineating curves or surfaces that move within 2-D or 3-D digital images under the influence of internal and external forces and user-defined constraints. Since their introduction by Kass et al. [48], these algorithms have been at the core of one of the most active and successful research areas in edge detection, image segmentation, shape modeling, and visual tracking. The first kind of active contour models is represented explicitly as parameterized contours (i.e., curves or surfaces) in a Lagrangian framework, which are called parametric active contours. Due to the special properties of the level set method, active contours can be represented implicitly as level set functions that evolve according to a Eulerian formulation; they are called geometric active contours. Although it is only a simple difference of representation for an active contour, the level set method–based geometric active contours have many advantages over parametric active contours:

1. They are completely intrinsic and therefore are independent of the parameterization of the evolving contour. In fact, the model is generally not parameterized until evolution of the level set function is complete. Thus, there is no need to add or remove nodes from an initial parameterization or adjust the spacing of the nodes as in parametric models.
2. The intrinsic geometric properties of the contour such as the unit normal vector and the curvature can be easily computed from the level set function, contrast to the parametric case, where inaccuracies in the calculations of normals and curvatures result from the discrete nature of the contour parameterization.
3. The propagating contour can automatically change topology in geometric models (e.g., merge or split) without requiring an elaborate mechanism to handle such changes as in parametric models.
4. The resulting contours do not contain self-intersections, which are computationally costly to prevent in parametric deformable models.

Since its introduction, the concept of using active contours for image segmentation defined in a level set framework has motivated the development of several families of methods including the front evolving geometric model, geodesic active contours, and region-based level set active contours.

12.2.1   Front Evolving Geometric Models of Active Contours

Front evolving geometric models of active contours [49, 50] are based on the theory of curve evolution, implemented via level set algorithms. They can automatically handle changes in topology. Hence, without resorting to dedicated contour tracking, unknown numbers of multiple objects can be detected simultaneously. Evolving curve C in the normal direction with speed F amounts to solving the differential equation:

$$\frac{\partial \Phi}{\partial t} = |\nabla \Phi| g(|\nabla u_0|) \left(\text{div} \left(\frac{\nabla \Phi}{|\nabla \Phi|} \right) + \gamma \right) \tag{12.7}$$

where Φ is the level set function corresponding to the current curve C, $g()$ is the gradient function on the image, and γ is constant to regulate the evolving speed.

12.2.2   Geodesic Active Contour Models

The geodesic model was proposed in 1997 by Caselles et al. [51]. This is a problem of geodesic computation in a Riemannian space, according to a metric induced by the image. Solving the minimization problem consists of finaling the path of minimal new length in that metric:

$$J(C) = 2\int_0^1 |C'(s)| \cdot g\left(|\nabla u_0(C(s))|\right) \ ds \tag{12.8}$$

where the minimizer C will be obtained when $g(|\nabla u_0(C(s)|)$ vanishes, that is, when the curve is on the boundary of the object. The geodesic active contour model also has a level set formulation as follows:

$$\frac{\partial \Phi}{\partial t} = |\nabla \Phi| \left(\text{div}(g|\nabla u_0|) \frac{\nabla \Phi}{|\nabla \Phi|} \right) + vg(|\nabla u_0|) \tag{12.9}$$

The geodesic active contour model is based on the relation between active contours and the computation of geodesics or minimal distance curves. The minimal distance curve lies in a Riemannian space whose metric is defined by the image content. This geodesic approach for object segmentation allows classical "snakes" to be connected based on energy minimization and geometric active contours based on the theory of curve evolution. Previous models of geometric active contours are improved, allowing stable boundary detection when their gradients suffer from large variations.

12.2.3   Tuning Geometric Active Contour with Regularizers

The main problem of boundary-based level set segmentation methods is related to contour leakage at locations of weak or missing boundary data information. One approach can be followed to solve these limitations, to fuse regularizer terms in the speed function.

Suri et al. [24] reviewed recent works on the fusion of classical geometric and geodesic deformable models speed terms with regularizers, that is, regional statistics information from the image. Regularization of the level set speed term is desirable to add prior information on the object to the segment and prevent segmentation errors when using only gradient-based information in the definition of the speed. Four main kinds of regularizers were identified by the authors of the review:

1. Clustering-based regularizers;
2. Bayesian-based regularizers;
3. Shape-based regularizers;
4. Coupling-surfaces regularizers.

12.2.3.1   Clustering-Based Regularizers

Suri proposed in [52] the following energy functional for level set segmentation:

$$\frac{\partial \Phi}{\partial t} = \left(\varepsilon k + F_p\right)|\nabla \Phi| - F_{ext}\nabla \Phi \tag{12.10}$$

where F_p is a regional force term expressed as a combination of the inside and outside regional area of the propagating curve. The term is proportional to a region indicator taking a value between 0 and 1, derived from a fuzzy membership measure. The second part of the classical energy model constituted the external force given by F_{ext}. This external energy term depended on image forces that were a function of image gradient.

12.2.3.2   Bayesian-Based Regularizers

Baillard et al. [53] proposed an approach that is similar to the previous one where the level set energy functional is expressed as:

$$\frac{\partial \Phi}{\partial t} = g\left(|\nabla I|\right)\left(k + F_0\right)|\nabla \Phi| \tag{12.11}$$

It uses a modified propagation term F_0 as a local force term. This term was derived from the probability density functions inside and outside the structure to the segment.

12.2.3.3   Shape-Based Regularizers

Another application of the fusion of Bayesian statistics into a geometric boundary/ surface to model the shape in the level set framework was done by Leventon et al. [54]. The authors introduced shape-based regularizers where curvature profiles act as boundary regularization terms more specific to the shape to extract than standard curvature terms. A shape model is built from a set of segmented exemplars using principal component analysis applied to the signed-distance level set functions derived from the training shapes (analogous to Cootes et al.'s [55] technique). The principal modes of variation around a mean shape are computed. Projection coefficients of a shape on the identified principal vectors are referred to as shape parameters. Rigid transformation parameters aligning the evolving curve and the shape model are referred to as pose parameters. To be able to include a global shape constraint in the level set speed term, shape and pose parameters of the final curve $\Phi^*(t)$ are estimated using maximum a posteriori estimation. The new functional with a solution for the evolving surface is expressed as:

$$\Phi(t+1) = \Phi(t) + \lambda_1 \left(g(|\nabla I|)(k+c)|\nabla\Phi| + \nabla g(|\nabla I|) \cdot \nabla\Phi \right) + \lambda_2 \left(\Phi^*(t) - \Phi(t) \right) \quad (12.12)$$

where λ_1 and λ_2 are two parameters that balance the influence of the gradient curvature term and the shape model term.

In [56], Leventon at al. proposed further refinements of their method by introducing prior intensity and curvature models using statistical image surface relationships in the regularizer terms. Some clinical validations have been reported to show efficient and robust performance of the method.

12.2.3.4   Coupling-Surfaces Regularizers

This kind of regularizer was motivated by segmentation of embedded organs such as the cortical gray matter in the brain. The application tries to use a level set segmentation framework to perform simultaneous segmentation of the inner and outer organ surfaces with coupled level set functions. This method was proposed by Zeng et al. [57]. In this framework, segmentation is performed with the following system of equations:

$$\begin{cases} \Phi_{in} + F_{in}|\nabla\Phi_{in}| = 0 \\ \Phi_{out} + F_{out}|\nabla\Phi_{out}| = 0 \end{cases} \quad (12.13)$$

where the terms F_{in} and F_{out} are functions of the surface normal direction (e.g., curvature), image-derived information and the distance between the two surfaces. When this distance is within the desired range, the two surfaces propagate according to the first two terms of the speed term. When the distance is out of the desired range, the speed term based on the distance controls the deformation to correct for the surface positions.

12.2.4   Region-Based Active Contour Models

These classical snakes and active contour models rely on the edge function, depending on the image gradient, to stop the curve evolution, these models can detect only objects with edges defined by gradient. Some of the typical edge functions are illustrated in Figure 12.2. In practice, the discrete gradients are bounded and then the stopping function is never zero on the edges, and the curve may pass through the boundary. If the image is very noisy, the isotropic smoothing Gaussian has to be strong, which will smooth the edges too. This region-based active contour method is a different active contour model, without a stopping edge-function, that is, a model which is not based on the gradient of the image for the stopping process.

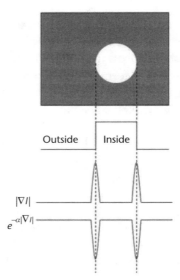

Figure 12.2 Typical definition of the edge-functions to control the propagation of the moving front.

12.2.4.1   Mumford-Shah (MS) Function

One kind of stopping term is based on Mumford-Shah [58] segmentation techniques. In this way, the model can detect contours both with or without gradients, for instance, objects with very smooth boundaries or even with discontinuous boundaries. In addition, the model has a level set formulation, interior contours are automatically detected, and the initial curve can be anywhere in the image.

The original MS function is defined as:

$$F^{MS}(u,C) = \mu Length(C) +$$
$$\lambda \int_{\Omega} |u_0(x,y) - u(x,y)|^2 \, dx \, dy + \lambda \int_{\Omega/C} |\nabla u(x,y)|^2 \, dx \, dy \qquad (12.14)$$

where u_0 approximates original image u, u_0 does not vary too much on each segmented region and the boundary C is becoming as short as possible. In fact, u_0 is simply a cartoon version of the original image u. Basically u_0 is a new image with edges drawn sharply. The objects are drawn smoothly without texture. Such cartoons are perceived correctly as representing a simpler version of the original scene containing most of its essential features. In Chan and Vese's approach [59], a level set approach is proposed to solve the modified Mumford-Shah functional.

12.2.4.2   The Level Set Approach to Solving the Mumford-Shah Function

To explain the model clearly, let us first define the evolving curve C in Ω as the boundary of an open subset w of Ω (i.e., $w \subset \Omega$, and $C = \partial w$). In what follows, inside (C) denotes region w, and outside (C) denotes region $\Omega / \overline{w}$. The method is the minimization of an energy-based segmentation. Assume that image u_0 is formed by two regions of approximatively piecewise-constant intensities, of distinct values

u_0^i and u_0^o. Assume further that the object to be detected is represented by the region with the value u_0^i. Let denote its boundary by C_0. Then we have $u_0 \approx u_0^i$ inside the object [or inside (C_0)], and $u_0 \approx u_0^o$ outside the object [or outside (C_0)]. The fitting term is defined as:

$$F_1(C) + F_2(C) =$$
$$\int_{\text{inside}(C)} |u_0(x,y) - c_1|^2 \, dx \, dy + \int_{\text{outside}(C)} |u_0(x,y) - c_2|^2 \, dx \, dy \qquad (12.15)$$

where C is any other variable curve, and the constants c_1, c_2, depending on C, are the averages of u_0 inside C and outside C, respectively. In this simple case, it is obvious that C_0, the boundary of the object, is the minimizer of the fitting term:

$$\inf_C \{F_1(C) + F_2(C)\} \approx 0 \approx F_1(C_0) + F_2(C_0) \qquad (12.16)$$

If curve C is outside the object, then $F_1(C) > 0$ and $F_2(C) \approx 0$. If curve C is inside the object, then $F_1(C) \approx 0$ and $F_2(C) > 0$. If curve C is both inside and outside the object, then $F_1(C) > 0$ and $F_2(C) > 0$. The fitting term is minimized when $C = C_0$, that is, curve C is on the boundary of the object (Figure 12.3).

To solve more complicated segmentation, we regularize terms, such as the length of curve C, or the area of the region inside C. A new energy functional $F(c_1, c_2, C)$ is defined as:

$$F(c_1, c_2 C) = \mu \, \text{Length}(C) + v \, \text{Area}(\text{inside}(C))$$
$$+ \lambda_1 \int_{\text{inside}(C)} |u_0(x,y) - c_1|^2 \, dx \, dy \qquad (12.17)$$
$$+ \lambda_2 \int_{\text{outside}(C)} |u_0(x,y) - c_2|^2 \, dx \, dy$$

where $\mu \geq 0$, $v \geq 0$, $\lambda_1, \lambda_2 \geq 0$. (In Chan-Vese's approach, $\lambda_1 = \lambda_2 = 1$ and $v = 0$.) Correspondingly, the level set based on C is defined as:

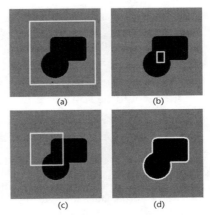

(a)   (b)

(c)   (d)

Figure 12.3 Consider all possible cases in the position of the curve. The fitting term is minimized only in the case when the curve is on the boundary of the object (a) The curve C is outside the object, and $F_1(C) > 0$ and $F_2(C) \approx 0$. (b) The curve C is inside the object, then $F_1(C) \approx 0$ and $F_2(C) > 0$. (c) The curve C is both inside and outside the object, then $F_1(C) > 0$ and $F_2(C) > 0$. (d) The fitting term is minimized when the curve C is on the boundary of the object

$$\begin{bmatrix} C = \partial w = \{(x,y) \in \Omega : \Phi(x,y) = 0\} \\ \text{inside}(C) = w = \{(x,y) \in \Omega : \Phi(x,y) > 0\} \\ \text{outside}(C) = \Omega/\overline{w} = \{(x,y) \in \Omega : \Phi(x,y) < 0\} \end{bmatrix} \qquad (12.18)$$

So, the unknown C could be replaced by Φ. We use Heaviside function H and Dirac measure δ_0, which are defined by:

$$H(z) = \begin{cases} 1, (z \geq 0) \\ 0, (z < 0) \end{cases}, \quad \delta_0 = \frac{dH(z)}{dz} \qquad (12.19)$$

Now, the energy function $F(c_1, c_2, C)$ could be rewritten as:

$$\begin{aligned} F(c_1, c_2, \Phi) = \mu \int_\Omega \delta(\Phi(x,y)) |\nabla\Phi(x,y)| dx\ dy + v \int_\Omega H(\Phi(x,y)) dx\ dy \\ + \lambda_1 \int_{\text{inside}(C)} |u_0(x,y) - c_1|^2 H(\Phi(x,y)) dx\ dy \\ + \lambda_2 \int_{\text{outside}(C)} |u_0(x,y) - c_2|^2 (1 - H(\Phi(x,y))) dx\ dy \end{aligned} \qquad (12.20)$$

where $c_1(\Phi)$ and $c_2(\Phi)$ are defined as:

$$c_1(\Phi) = \frac{\int_\Omega u_0(x,y) H(\Phi(x,y)) dx\ dy}{\int_\Omega H(\Phi(x,y)) dx\ dy} \qquad (12.21)$$

$$c_2(\Phi) = \frac{\int_\Omega u_0(x,y)(1 - H(\Phi(x,y))) dx\ dy}{\int_\Omega (1 - H(\Phi(x,y))) dx\ dy} \qquad (12.22)$$

Finally, the corresponding level set equation which is minimizing the energy, can be solved by the following equation:

$$\frac{\partial\Phi}{\partial t} = \delta(\Phi)\left[\mu\ \text{div}\left(\frac{\nabla\Phi}{|\nabla\Phi|}\right) - v - \lambda_1 (u_0 - c_1)^2 + \lambda_2 (u_0 - c_2)^2 \right] \qquad (12.23)$$

The level set equation could be solved iteratively using time step Δt. However, there are inherent time-step requirements to ensure stability of the numerical scheme via the CFL[1] condition. In Chan and Vese's approach, the time step could be set based on the following equation:

---

1. The domain of dependence of a hyperbolic partial differential equation (PDE) for a given point in the problem domain is that portion of the problem domain that influences the value of the solution at the given point. Similarly, the domain of dependence of an explicit finite difference scheme for a given mesh point is the set of mesh points that affect the value of the approximate solution at the given mesh point. The CFL condition, named for its originators Courant, Friedrichs, and Lewy, requires that the domain of dependence of the PDE must lie within the domain of dependence of the finite difference scheme for each mesh point of an explicit finite difference scheme for a hyperbolic PDE. Any explicit finite difference scheme that violates the CFL condition is necessarily unstable, but satisfying the CFL condition does not necessarily guarantee stability.

$$\Delta t \le \frac{\min(\Delta x, \Delta y, \Delta z)}{\left(|\mu| + |v| + |\lambda_0 + \lambda_1|\right)} \qquad (12.24)$$

12.3   Fully Automated 3-D Prostate Segmentation from Transrectal Ultrasound (TRUS) Images

The problem of missing or diffuse boundaries is a very challenging one for medical image processing. Missing or diffuse boundaries can be caused by patient movements, by a low SNR of the acquisition apparatus, or by being blended with similar surrounding tissues. Under such conditions, without a prior model to constrain the segmentation, most algorithms (including intensity- and curve-based techniques) fail—mostly due to the underdetermined nature of the segmentation process. Similar problems arise in other imaging applications as well and they also hinder the segmentation of the image. These image segmentation problems demand the incorporation of as much prior information as possible to help the segmentation algorithms extract the tissue of interest.

A number of model-based image segmentation algorithms in the literature were used to deal with cases when boundaries in medical images are smeared or missing. In 1995, Cootes et al. [55] developed a parametric point distribution model (PDM) for describing the segmenting curve that uses linear combinations of the eigenvectors to reflect variations from the mean shape. The shape and pose parameters of this PDM are determined to match the points to strong image gradients. Wang and Staib [60] developed a statistical point model in 1998 for the segmenting curve by applying principal component analysis (PCA) to the covariance matrices that capture the statistical variations of the landmark points. They formulated their edge detection and correspondence determination problem in a maximum a posteriori Bayesian framework. An image gradient is used within that framework to calculate the pose and shape parameters that describes their segmenting curve.

Leventon et al. [56] proposed a less restrictive model-based segmenter. They incorporated shape information as a prior model to restrict the flow of the geodesic active contour. Their prior parametric shape model is derived by performing PCA on a collection of signed distance maps of the training shape. The segmenting curve then evolves according to the gradient force of the image and the force exerted by the estimated shape. In 2002, Mitchell et al. [61] developed a model-based method for fixedmage segmentation. Comprehensive design of a 3-D active appearance model (AAM) is reported for the first time as an involved extension of the AAM framework introduced by Cootes et al. [62]. The model's behavior is learned from manually traced segmentation examples during an automated training stage. Information about the shape and image appearance of the cardiac structures is contained in a single model. The clinical potential of the 3-D AAM is demonstrated in short-axis cardiac MR images and four-chamber echocardiographic sequences. The AAM method showed good agreement with the independent standard using quantitative indexes of border positioning errors, endocardial and epicardial volumes, and left ventricular mass. The AAM method shows high promise for successful application to MR and echocardiographic image analysis in a clinical setting. The reported method combined the appearance feature with the shape knowledge and it provided robust matching criterion for segmentation. However, this method needs to set up point correspondences, which makes the procedure complicated.

In [15], the authors adopted implicit representation [54] of the proposed segmenting curve and calculated the parameters of the implicit model to minimize the region-based energy based on the Mumford-Shah functional for image segmentation [59]. In [16], the authors combined intensity prior with the shape knowledge and transferred the segmentation problem in a Bayesian interface. Efficient kernel functions were utilized, which made the method more robust and more efficient. Motivated by these state-of-the-art medical imaging techniques, we designed and implemented a system for fully automated 3-D prostate segmentation from TRUS images that uses shape and intensity knowledge in a level set framework. The method used in this system includes fixed important steps:

1. 3-D shape prior extraction;
2. Intensity prior extraction;
3. Segmentation in Bayesian interface;

A brief flow chart of the method is shown in Figure 12.4.

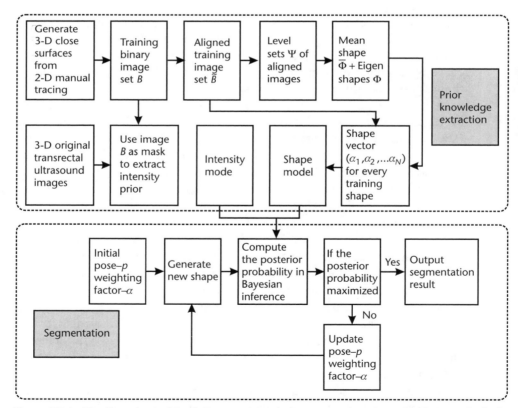

Figure 12.4 The flow chart of the 3-D prostate analysis system using ultrasound images. Basically there are two stages. The first stage is system training to extract shape and intensity prior and the second is to segment image using the statistical information from first step.

12.3.1   3-D Shape Prior Extraction

Extraction of 3-D shape priors is more complicated than in 2-D space. Consider the 3-D close surface from 2-D contours method: A set of binary 3-D images $\{B_1, B_2, ..., B_n\}$, has 1 as an object and 0 as the background. The 3-D binary shape is from manual segmentation in each 2-D slice. The first step toward 3-D shape model creation is to stack the 2-D segmentation. However, the stacked version has a zigzagged surface and so its accuracy is not guaranteed. We apply the Min/Max flow [63] method to smooth the 3-D surface and increase the accuracy between every two slices. Min/Max flow uses the curvature-based force to fix the surface and the patch on the surface with large positive or small negative curvatures will be smoothed, which can not be done by simply creating meshes based on adjacent contours.

To show the effect of using curvature as the driving force in the level set method, we give the following 2-D example. In (12.6), we define the F as:

$$
\begin{aligned}
F_{\min} &= \min(k, 0.0) \\
F_{\max} &= \max(k, 0.0)
\end{aligned}
\tag{12.25}
$$

Here, we have chosen the negative of the signed distance in the interior, and the positive sign in the exterior region. As shown in Figure 12.5, the effect of flow under $F_{\min}$ is to allow the inward concave fingers to grow outwards, while suppressing the motion of the outward convex regions. Thus, the motion halts as soon as the convex hull is obtained. Conversely, the effect of flow under $F_{\max}$ is to allow the outward

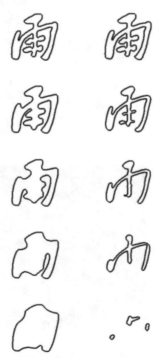

Figure 12.5   Motion of Curve under Min/Max Flow. The first column are the five intermediate results using $F_{\min} = \min(k, 0.0)$. The second column are the five intermediate results using $F_{\max} = \max(k, 0.0)$.

regions to grow inward while suppressing the motion of the inward concave regions. However, once the shape becomes fully convex, the curvature is always positive and, hence, the flow becomes the same as regular curvature flow; hence, the shape collapses to a point.

We can summarize the preceding paragraphs by saying that, for the preceding case, flow under F_{min} preserves some of the structure of the curve, whereas flow under F_{max} completely diffuses away all of the information. Similar effects can be achieved in 3-D space. By using Min/Max flow in a small number of iterations, the 3-D surface will be smoothed. In Figure 12.6, one example is shown of how Min/Max flow smooths the 3-D surface for 3-D prostate shape extraction.

After using the Min/Max flow method for each 3-D surface, in order to extract accurate shape information, alignment has to be applied. Alignment is a task to calculate pose parameter p, which includes translation in the x, y, and z directions, rotation around x, y, and z axis and scale in x, y, and z directions. In Figure 12.7, we listed the original 3-D shape after Min/Max flow. The example includes only eight data sets for illustration.

The strategy to compute the pose parameters for n binary images is to use the gradient descent method to minimize the specially designed energy functional E^j_{align} align for each binary image corresponding to the fixed one, say, the first binary image B_1. The energy is defined by the following equation:

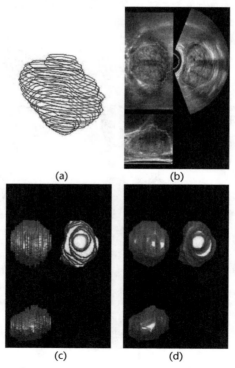

(a)   (b)   (c)   (d)

Figure 12.6 Using Min/Max flow to smooth 3-D surface. (a) The image is the 2-D contours from manual tracing in 3-D space. (b) Three views of after overlay the stacked 2-D contours onto the original 3-D ultrasound image data. (c) Three views of the surface rendering of the stacked 2-D contours. (d) Three views of the surface rendering of after Min/Max flow.

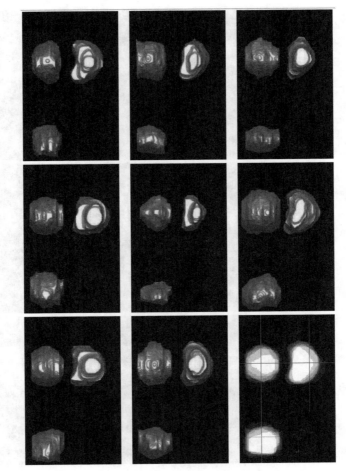

Figure 12.7   The three views of the surface rendering of the 3-D surface after Min/Max flow for the original 3-D shapes. The last image is the overlapping of the eight images.

$$E_{\text{align}}^{j} = \frac{\iint_{\Omega} \left(\tilde{B}_{j} - B_{1} \right)^{2} dA}{\iint_{\Omega} \left(\tilde{B}_{j} + B_{1} \right)^{2} dA} \tag{12.26}$$

where Ω denotes the image domain, $\tilde{B}_{j}$ denotes the transformed image of B_{j} based on the pose parameters p. We term the aligned 3-D shapes as $\{\tilde{B}_{1}, \tilde{B}_{2}, ..., \tilde{B}_{n}\}$. Minimizing this energy is equivalent to minimizing the difference between current binary image B_{1} and the fixed image in the training database. The normalization term in the denominator is employed to prevent the images from shrinking to improve the cost function. The hill climbing or Rprop [64] method could be applied for the gradient descent. Figure 12.8, shows the 3-D shapes after 3-D alignment. The example includes only eight data sets for illustration and they correspond to the original shapes shown in Figure 12.7.

Based on the 3-D binary images from the training datasets, we collected a set of 3-D training shapes encoded by their signed distance functions. The popular and natural approach to represent shapes is via point models where a set of marker

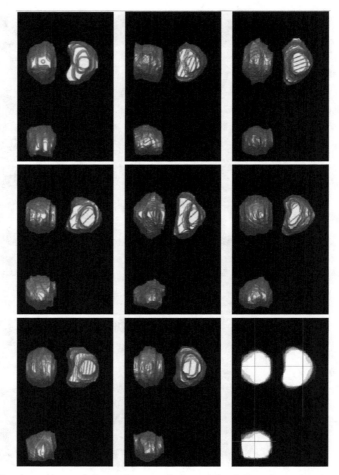

Figure 12.8   The three views of the surface rendering of the 3-D surface after 3-D alignment. The last image is the overlapping of the eight images after alignment.

points is often used to describe the boundaries of the shape. This approach suffers from problems such as numerical instability, inability to accurately capture high curvature locations, difficulty in handling topological changes, and the need for point correspondences. To overcome these problems, Eulerian approach to shape representation based on the level set methods [21] of Osher and Sethian can be utilized.

The signed distance function is chosen as the representation for shape. In particular, the boundaries of each of the aligned shapes in $\{\tilde{B}_1, \tilde{B}_2, ..., \tilde{B}_n\}$, are embedded as the zero-level set of separate signed distance functions $\{\Psi_1, \Psi_2, ..., \Psi_n\}$ with negative distances assigned to the inside and positive distances assigned to the outside of the object. The mean level set function of the shape database as the average of these signed distance functions can be computed as

$$\overline{\Phi} = \frac{1}{n}\sum_{i=1}^{n} \Psi_i$$

To extract the shape variabilities, $\overline{\Phi}$ is subtracted from each of the n signed distance functions to create n mean-offset functions $\{\widetilde{\Psi}_1, \widetilde{\Psi}_2, ..., \widetilde{\Psi}_n\}$. These mean-offset functions are analyzed and then used to capture the variabilities of the training shapes.

Specifically, n column vectors are created, $\widetilde{\psi}_i$, from each $\widetilde{\Psi}_1$. A natural strategy is to utilize the $N_1 \times N_2 \times N_3$ grid of the training images to generate $N = N_1 \times N_2 \times N_3$ lexicographically ordered samples (where the columns of the image grid are sequentially stacked on top of one other to form one large column). Next, define the shape-variability matrix S as: $[\widetilde{\psi}_1, \widetilde{\psi}_2, ..., \widetilde{\psi}_n]$. The procedure for creation of the matrix S is shown in Figure 12.9.

An eigenvalue decomposition is employed:

$$\frac{1}{n}SS^T = U\Sigma U^T \tag{12.27}$$

where U is an $N = N \times n$ matrix whose columns represent the orthogonal modes of variation in the shape and Σ is an $n \times n$ diagonal matrix whose diagonal elements represent the corresponding nonzero eigenvalues. The N elements of the ith column of U, denoted by U_i, are arranged back into the structure of the $N = N_1 \times N_2 \times N_3$ image grid (by undoing the earlier lexicographical concatenation of the grid columns) to yield Φ_i, the ith principal mode or eigenshape. Based on this approach, a maximum of n different eigenshapes $\{\Phi_1, \Phi_2, ..., \Phi_n\}$ are generated. In most cases, the dimension of the matrix $\frac{1}{n}SS^T$ is large so the calculation of the eigenvectors and eigenvalues of this matrix $\frac{1}{n}SS^T$ is computationally expensive. In practice, the eigenvectors and eigenvalues of can be efficiently computed from a much smaller $n \times n$ matrix W given by $\frac{1}{n}S^TS$. It is straightforward to show that if d is an eigenvector

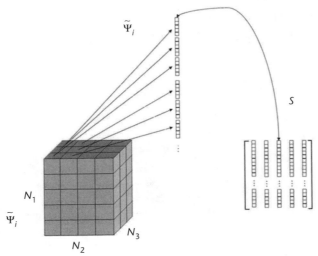

Figure 12.9   Create the shape-variability matrix S from 3-D mean-offset functions $\widetilde{\Psi}_i$.

of W with corresponding eigenvalue λ, then Sd is an eigenvector of $\dfrac{1}{n} SS^T$ with eigenvalue λ.

For segmentation, it is not necessary to use all of the shape variabilities after the preceding procedure. Let $k \leq n$, which is selected prior to segmentation, be the number of modes to consider. In general, k should be chosen large enough to be able to capture the main shape variations present in the training set. One way to choose the value of k is by examining the eigenvalues of the corresponding eigenvectors. In some sense, the size of each eigenvalue indicates the amount of influence or importance its corresponding eigenvector has in determining the shape. By looking at a histogram of the eigenvalues, one can estimate the threshold for determining the value of k. However, an automated algorithm is hard to implement as the threshold value for k varies for each different application. It is hard to define a universal k that can be set. The histogram of the eigenvalues for the preceding examples are shown in the Figure 12.10. In total, there are eight eigenshapes, corresponding to the eight eigenvalues. However, only the first seven are large enough to be used to represent all of the shape variations.

The preceding processing method is based on work from Tsai et al. [15], who proposed an efficient method to reduce the segmentation problem to one of finite-dimensional optimization by constraining the optimization problem to the finite-dimensional subspace spanned by the training shapes. Then a new level set function is defined to represent the 3-D shape as:

$$\Phi[\alpha] = \overline{\Phi} + \sum_{i=1}^{k} a_i \Phi_i \qquad (12.28)$$

where $\alpha = \{a_1, a_2, ..., a_k\}$ are the weights for the k eigenshapes. Now we can use this newly constructed level set function Φ as the implicit representation of shape. specif-

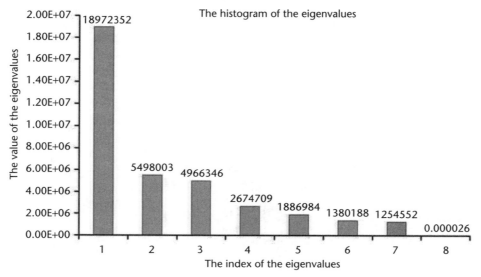

Figure 12.10   The histogram of the eight eigenvalues after PCA analysis on the eight 3-D prostate shapes.

ically, the zero-level set of Φ describes the shape, with the shape's variability directly linked to the variability of the level set function. Therefore, by varying α, Φ will be changed, which indirectly varies the shape. The average shape and several shape variances are illustrated in Figure 12.11.

12.3.2   Intensity Prior Extraction

From the original energy definition for segmentation in (12.17), we found that the intensity of each pixel inside the image is taken with the same probability. In other words, the energy function never uses the intensity prior based on the shape model. Although for certain imaging modalities, such as ultrasound or CT, the structures of interest do not differ much from their background, the intensity information in the image can be utilized by probabilistic intensity models. This extra

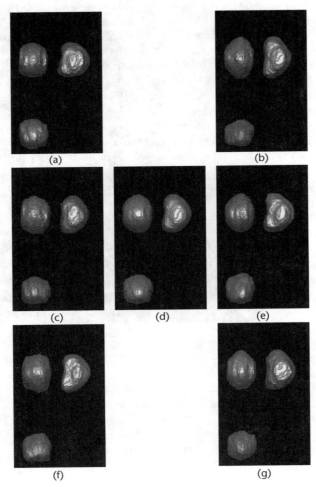

Figure 12.11   Three views and surface rendering of the average shape and several shape variances. (a) The mean shape + 0.5 of the first eigenshape. (b) The mean shape −0.5 of the first eigenshape. (c) The mean shape + 0.5 of the second eigenshape. (d) The mean shape. (e) The mean shape −0.5 of the second eigenshape. (f) The mean shape + 0.5 of the third eigenshape. (g) The mean shape −0.5 of the third eigenshape.

information will increase the accuracy of the energy function definition. Because we are using a statistic shape model, from the training dataset, we can classify the image as foreground and background. The corresponding histograms provide the conditional probability for different gray levels inside and outside the segmentation. This information can be efficiently combined with the shape prior for final segmentation (Figure 12.12).

12.3.3   Segmentation in Bayesian Interface

Now that we have collected the statistical shape priors, we need to combine with the intensity knowledge too. However, how to accurately use this information in segmentation is still an unsolved problem. From (12.24), readers have already seen how

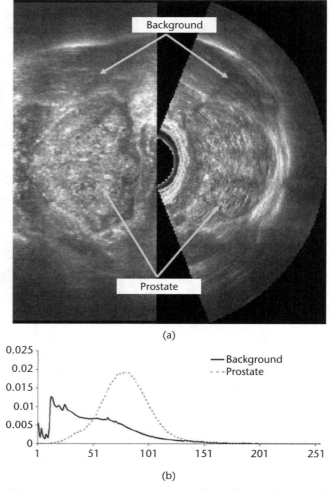

(a)

(b)

Figure 12.12   In this prostate segmentation project, we have the two intensity distributions corresponding to the empirical probability of intensities inside and outside the prostate. (a) One example of the total training ultrasound image data sets. Sagittal and Coronal views are shown, with background and object (prostate) labeled. (b) The probability of intensities inside and outside the prostate from all the training data sets.

to solve the region-based energy function for segmentation in a level set framework, without using any statistical shape and intensity information.

Recently, many researchers have tried to utilize the level set method with statistical shape priors. Given a set of training shapes, one can impose information about which segmentations are a priori more or less likely. Such prior shape information can improve segmentation results in the presence of noise or occlusion from different imaging modalities [15, 54, 65]. However, most of these approaches are based on the assumption that the training shapes, encoded by their signed distance function, form a uniform distribution [15] or Gaussian distribution [54]. As introduced in [16], these assumptions have been proven with the following problems:

1. The space of signed distance functions is not a linear space, therefore, the mean shape and linear combinations of eigenmodes are typically no longer signed distance functions.
2. Even if the space were a linear space, it is not clear why the given set of sample shapes should be distributed according to a Gaussian density. In fact, they are generally not Gaussian distributed.

Cremers et al. [65] proposed the use of nonparametric density estimation in the space of level set functions in order to model nonlinear distributions of training shapes. Although this resolves the above-mentioned problems, one sacrifices the efficiency of working in a low-dimensional subspace (formed by the first few eigenmodes) to a problem of infinite-dimensional optimization. This method is an efficient solution of the two problems just listed, and it defines kernel density estimation in a linear subspace that is sufficiently large to embed all training data. We therefore choose this approach to use for the shape and intensity priors.

Segmentation in the level set framework can be formulated as the estimation of the optimal embedding function $\Phi: \Omega \to \Re$ given an image $I: \Omega \to \Re$ [16]. In the Bayesian framework, this can be computed by maximizing the posterior distribution

$$P(\Phi|I) \propto P(I|\Phi)P(\Phi) \qquad (12.29)$$

Based on the result from (12.28), the new shape used for segmentation is determined by parameters α. The orientation, location and scale of segmentation are determined by pose parameter p. Combined with both factors, the Bayesian interface problem stated in (12.29) therefore can be solved in a problem as the following conditional probability:

$$P(\alpha, p|I) \propto P(I|\alpha, p)P(\alpha, p) \qquad (12.30)$$

which is optimized with respect to the weight a and the pose parameter p. Maximization of the posterior probability or equivalently minimization of its negative logarithm is able to create the most probable segmentation of a given image. Combined with the shape and intensity priors we collected from the previous two sections, a new energy can be defined as

$$E(\alpha, p) = -\log P(I|\alpha, p) - \log P(\alpha, p) \qquad (12.31)$$

In [16], the energy is defined as the following equations in more details:

$$E(\alpha, p) = -\int_{\Omega} H(\Phi) \log p_{\text{in}}(I) + (1 - H(\Phi)) \log p_{\text{out}}(I) \, dx$$
$$- \log\left(\frac{1}{N_\sigma} \sum_{i=1}^{N} \text{Kernel}\left(\frac{\alpha - \alpha_i}{\sigma}\right)\right) \tag{12.32}$$

where $H()$ function is the Heaviside function we defined in (12.18), p_{in} and p_{out} are the kernel density estimation of the intensity distributions of object and background which are given by the corresponding smoothed intensity histograms as shown in Figure 12.12. The Kernel() function is the special invention proposed in [16]. Before we describe the kernel function and kernel density estimation in more detail, we define the following notation. Given a set of aligned training shapes represented by signed distance $\{\Psi_1, \Psi_2, ..., \Psi_N\}$, we can represent each of them by their corresponding shape vector $\{\alpha_1, \alpha_2, ..., \alpha_N\}$. These shape vectors can be easily extracted by projection to the basis formed by the eigenmodes.

Inferring a statistical distribution from the training shape, in other words, the shape vectors, is done differently in the work by Tsai et al. [15], Leventon et al. [54], and Rousson and Cremers [16]. In Tsai's work, all of the shapes were regarded equally and, correspondingly, a uniform distribution was utilized. In Leventon's work, the authors used a Gaussian distribution from the training shape. However, in Rousson's work, the authors made use of a nonparametric density estimation to approximate the shape distribution within the linear subspace spanned by the training shapes. The kernel function and the corresponding kernel density estimate are defined as:

$$P(\alpha) - \frac{1}{N_\sigma} \sum_{i=1}^{N} \text{Kernel}\left(\frac{\alpha - \alpha_i}{\sigma}\right)$$
$$\text{Kernel}(x) = \frac{1}{\sqrt{2\pi}} e^{\left(\frac{x^2}{2}\right)} \tag{12.33}$$
$$\sigma^2 = \frac{1}{N} \sum_{i=1}^{N} \min_{j \neq i} (\alpha_i - \alpha_j)^2$$

Eventually, we are able to compute all of the elements defined in (12.32). The segmentation task is transformed into a parameter adjustment problem in the Bayesian interface. Many methods can be utilized to adjust parameters for the model based segmentation. Artificial neural networks such as multilayer perceptrons (MLPs) have become standard tools for regression. In general, an MLP with fixed structure defines a differentiable mapping from a parameter space to the space of functions. In the case where an MLP has a fixed structure, the adaption of the network to the sample data becomes a regression problem on the parameters of the network itself. This procedure is often referred to as learning. Because MLPs are differentiable, gradient-based adaptation techniques are typically applied to determine the weights. The earliest and most straightforward adaptation rule, ordinary gradient descent, adapts weights proportional to the partial derivatives of the error functional. Several improvements of this basic adaptation rule have been proposed, some of them based on elaborated heuristics, others on theoretical reconsideration of gradient-based learning. Resilient backpropagation (Rprop) is a well-established modification of the ordinary gradient descent method. The basic idea is to adjust an individual step size for each parameter to be optimized. These step sizes are not proportional to the partial derivatives but are themselves adapted based on some heuristics. Ordinary gradient descent computes the direction of steepest descent by implicitly assuming a Euclidean metric on the weight space.

The Rprop algorithms are among the best performing first-order batch learning methods. They are many advantages for Rprop methods:

1. Fast and accurate;
2. Robust;
3. First-order methods, therefore time and space complexity scales linearly with the number of parameters to be optimized;
4. Only dependent on the sign of the partial derivatives of the objective function and not on their amount, therefore they are suitable for applications where the gradient is numerically estimated or the objective function is noisy;
5. Easy to implement and not very sensitive to numerical problems.

The Rprop algorithms are iterative optimization methods. Let t denote the current iteration (epoch). In epoch t, each weight is changed according to:

$$w^i(t+1) = w^i(t) - \text{sign}\left(\frac{\partial E(t)}{\partial w^i}\right) \cdot \Delta^i(t) \tag{12.34}$$

The direction of the change depends on the sign of the partial derivative, but is independent of its amount. The individual step sizes $\Delta^i(t)$ are adapted based on changes of sign of the partial derivatives of $E(w)$ with regard to the corresponding weight:

- If $\dfrac{\partial E(t-1)}{\partial w^i} \cdot \dfrac{\partial E(t)}{\partial w^i} > 0$ then $\Delta^i(t)$ is increased by a factor $\eta^+ > 1$.

- If $\dfrac{\partial E(t-1)}{\partial w^i} \cdot \dfrac{\partial E(t)}{\partial w^i} \leq 0$ then $\Delta^i(t)$ is increased by a factor $\eta^- \in [0, 1]$.

Additionally, some Rprop methods implement a weight-backtracking heuristic. That is, they partially retract "unfavorable" previous steps. Whether a weight change was "unfavorable" is decided by a heuristic. In the following an improved version [66] of the original algorithm is described in pseudocode. The difference compared to the original Rprop method is that the weight-backtracking heuristic considers both the evolution of the partial derivatives and the overall error.

Step 1: Enter the iteration n.

Step 2: For each weighting factor w^i, if $\dfrac{\partial E(t-1)}{\partial w^i} \cdot \dfrac{\partial E(t)}{\partial w^i} > 0$, then:

$$\min\left(\Delta^i(t-1)\cdot\eta^+,\,\Delta_{max}\right)$$
$$w^i(t+1) = w^i(t) - \text{sign}\left(\frac{\partial E(t)}{\partial w^i}\right)\cdot\Delta^i(t) \tag{12.35}$$

Step 3: Else if $\dfrac{\partial E(t-1)}{\partial w^i}\cdot\dfrac{\partial E(t)}{\partial w^i} < 0$, then: $\min(\Delta^i(t-1)\cdot\eta^-,\Delta_{min})$ if $E(t) > E(t-1)$,

then $w^i(t+1) = w^i(t-1)$, $\dfrac{\partial E(t)}{\partial w^i} = 0$

Step 4: Else if $\dfrac{\partial E(t-1)}{\partial w^i}\cdot\dfrac{\partial E(t)}{\partial w^i} = 0$, then:

$$w^i(t+1) = w^i(t) - \text{sign}\left(\frac{\partial E(t)}{\partial w^i}\right)\cdot\Delta^i(t) \tag{12.36}$$

Step 5: If w^i still remain to be updated, go to step 2, otherwise go to Step 6.
Step 6: If no more iteration is needed, exit. Otherwise, $n = n + 1$; go to step 2.

12.4 Evaluation of the System Performance

12.4.1 Transrectal Ultrasound Images

The 3-D TRUS images were obtained from the Imaging Research Laboratories of the Robarts Research Institute, using an Aloka 2000 US machine (Aloka, CN) with a biplane side-firing TRUS transducer mounted on a motorized scanning mechanism. The transducer rotated about its axis through a scanning angle, while a series of 120 2-D B mode images were acquired with an 8-bit video frame grabber at 15 Hz. For each 3-D scan, the acquired 2-D images were reconstructed into a 3-D image, using reconstruction software developed in the laboratory, with the z-axis parallel to the transducer axis. The sizes of the images are $860 \times 427 \times 348$ and the voxel size is mm^3 0.154 by 0.154 by 0.154 mm.

12.4.2 Validation Method

Eleven manually extracted prostates were collected from 11 different patients. We employed a jackknifing[2] strategy by removing the image of interest from the training phase.

We used the outlines from manual tracing as an independent standard, which was done by an experienced radiologist. In Section 12.3, we have collected the 3-D close surface using the 2-D manual tracing and used these smoothed 3-D surfaces as the final ground truth.

---

2. Cross-validation is a method for estimating generalization error based on "resampling" [67–70]. In k-fold cross-validation, the data are divided into k subsets of equal size. The system is trained k times, each time leaving out one of the subsets from training, but using only the omitted subset to compute error. If k equals the sample size, this is called "leave-one-out" cross-validation. Sometimes, the "leave-one-out" method is also called *jackknifing*.

The validation indices are defined as the following two parameters: correct segmentation rate (CSR) and incorrect segmentation rate (ISR). CSR is defined as the ratio of correct segmentation voxel number and the total voxel number of the ground truth. ISR is defined as the ratio of the incorrect (the nonprostate voxel is classified as prostate voxel) segmentation voxel number and the total voxel number of the ground truth.

12.5   Results

Our segmentation method successfully determined the prostate in all 20 ultrasound images. As introduced in the previous section, we used the jackknifing method for validation. For every ultrasound image, we use the remaining 19 to extract the shape and intensity priors. Using the computed shape model and intensity knowledge, we initialized the mean shape at the center of the image and started the gradient descent method to final the best pose parameters and weight factors. On

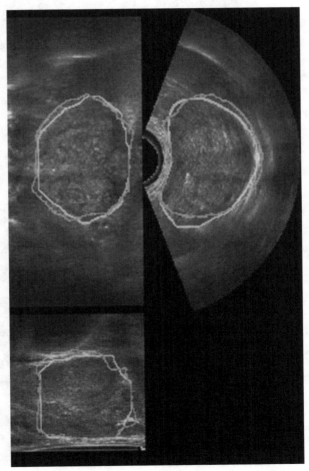

Figure 12.13   Sagittal, coronal, and transversal views of the prostate segmentation, a patient, and the overlaid manual segmentation.

average, the CSR was 0.84 and the ISR was 0.15. Also, the standard deviation of the CSR and ISR were 0.06 and 0.08, respectively. Figure 12.13 shows a detailed result after using this method. The manual and automated segmentation results are both overlapped onto the original prostate ultrasound image data. The CSR was 0.89 and the ISR was 0.15. Figures 12.14 and 12.15 show the surface rendering of the 2-D and 3-D imagery, respectively, of the same image sequence as in Figure 12.13.

12.6   Conclusions and Future Work

The segmentation of prostate from ultrasound images is an important and challenging problem. In the context of ultrasound imagery, the main challenges include these: The edge information of the objects is too weak; speckle noise degrades the images. Existing algorithms that use edge information depend on extra image enhancement methods and even need user interaction for initialization or editing.

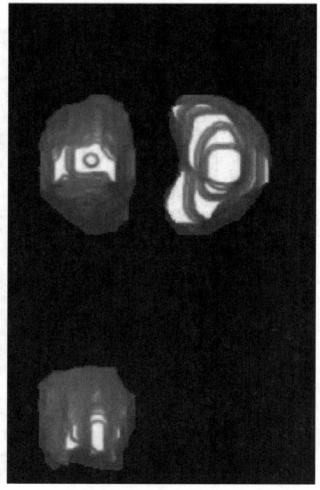

Figure 12.14   The surface rendering of the manual segmentation from 2-D tracing, corresponding to the close surface in Figure 12.13.

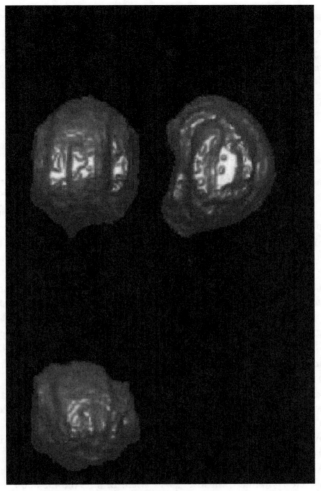

Figure 12.15   The surface rendering of the automated segmentation, corresponding to the close surface in Figure 12.13.

In this chapter, we reported on a system, that incorporates learned shape models and intensity prior for the objects of interest using level set methods. The algorithm allows the shape to evolve until the optimal segmentation is found by minimizing a region-based energy factor without special initialization. The quantitative analysis proved that the method is a robust solution for fully automated prostate segmentation tasks in ultrasound images.

The 2-D manual tracing for training was not created at every slice from the original ultrasound image. Although we employed the Min/Max flow method to make a smooth interpolation, the resulting 3-D surface was still not accurate enough. Adding manual tracing in more slices is a possible solution.

Although there are many advantages to using the proceeding methods in a level set framework, as a fully automated 3-D approach, the computational cost is still a big barrier. In this chapter, we exploited kernel density estimation [16] to increase the efficiency. The total convergence time is still more than 70 seconds. (The hardware environment is a PC with a Pentium 4, 2.80-GHz CPU and 1.00 GB of RAM.)

Optimization work will be done in the future to make the method more practical for prostate segmentation from ultrasound images.

We provided a 3-D validation result in this chapter too. For readers who are still interested in this performance of the method compared with other methods, segmentation accuracy and computational and memory cost evaluations can be made through your own attempts.

References

[1]   Garfinkel, L., and M. Mushinski, "Cancer Incidence, Mortality, and Survival Trends in Four Leading Sites," *Stat. Bull. Metrop. Insur. Co.*, Vol. 75, 1994, pp. 19–27.

[2]   Silverberg, E. C. C., Boring, and T. S. Squires, "Cancer Statistics," *CA Cancer J. Clin.*, Vol. 40, 1990, pp. 9–26.

[3]   Aarnink, R., et al., "A Practical Clinical Method for Contour Determination in Ultrasonographic Prostate Images," *Ultrasound in Medicine and Biology,* Vol. 20, 1994, pp. 705–717.

[4]   Abolmaesumi, P., and M. Sirouspour, "Segmentation of Prostate Contours from Ultrasound Images," *Proc. of IEEE Inter. Conf. on Acoustics, Speech and Signal Processing*, Vol. 3, 2004, pp. 517–520.

[5]   Pathak, S. V., et al., "Edge-Guided Boundary Delineation in Prostate Ultrasound Images," *IEEE Trans. on Medical Imaging*, Vol. 19, 2000, pp. 1211–1219.

[6]   Ladak, H., et al., "Prostate Boundary Segmentation from 2D Ultrasound Images," *Medical Physics,* Vol. 27, 2000, pp. 1777–1788.

[7]   Wang, Y., et al., "Semiautomatic Fixed-Dimensional Segmentation of the Prostate Using Two-Dimensional Ultrasound Images," *Medical Physics,* Vol. 30, September 2003, pp. 887–897.

[8]   Hu, N., et al., "Prostate Boundary Segmentation from 3D Ultrasound Images," *Medical Physics,* Vol. 30, 2003, pp. 1648–1659.

[9]   Chiu, B. et al., "Prostate Segmentation Algorithm Using Dyadic Wavelet Transform and Discrete Dynamic Contour," *Phys. Med. Biol.,* Vol. 49, 2004, pp. 4943–4960.

[10] Ghanei, A., et al., "A Fixed-Dimensional Deformable Model for Segmentation of Human Prostate from Ultrasound Image," *Medical Physics,* Vol. 28, 2001, pp. 2147–2153.

[11] Richard, W., and C. Keen, "Automated Texture-Based Segmentation of Ultrasound Images of the Prostate," *Computerized Medical Imaging and Graphics,* Vol. 20, 1996, pp. 131–140.

[12] Prater, J. S., and W. Richard, "Segmenting Ultrasound Images of the Prostate Using Neural Networks," *Ultrasound Imaging,* Vol. 14, 1992, pp. 159–185.

[13] Shen, D., Y. Zhan, and C. Davatzikos, "Segmentation of Prostate Boundaries from Ultrasound Images Using Statistical Shape Model," *IEEE Trans. on Medical Imaging,* Vol. 22, No. 4, 2003, pp. 539–551.

[14] Gong, L., et al., "Parametric Shape Modeling Using Deformable Superellipses for Prostate Segmentation," *Medical Physics,* Vol. 23, 2004, pp. 340–349.

[15] Tsai, A., et al., "A Shape-Based Approach to the Segmentation of Medical Imagery Using Level Sets," *IEEE Trans. on Medical Imaging,* Vol. 22, 2003, pp. 137–154.

[16] Rousson, M., and D. Cremers, "Efficient Kernel Density Estimation of Shape and Intensity Priors for Level Set Segmentation," *Proc. of Inter. Conf. on Medical Image Computing and Computer Assisted Intervention,* Vol. 1, 2005, pp. 757–764.

[17] Yang, F., J. S. Suri, and M. Sonka, "Volume Segmentation Using Active Shape Models in Level Set Framework," in *Parametric and Geometric Deformable Models: An Application in Biomaterials and Medical Imagery,* J. S. Suri and A. Farag, (eds.), Berlin: Springer, 2006.

[18] Suri, J. S., D. L. Wilson, and S. Laxminarayan, *Handbook of Medical Image Analysis,* Berlin: Springer, 2006.

[19] Sethian, J. A., *Level Set Methods and Fast Marching Methods Evolving Interfaces in Computational Geometry, Fluid Mechanics, Computer Vision, and Materials Science,* Cambridge: Cambridge University Press, 1999.

[20] Osher, S., and N. Paragios, *Geometric Level Set Methods in Imaging Vision and Graphics,* Berlin: Springer, 2003.

[21] Osher, S., and J. A. Sethian, "Fronts Propagating with Curvature-Dependent Speed: Algorithms Based on Hamilton-Jacobi Formulation," *Comput. Phys.,* Vol. 79, 1988, pp. 12–49.

[22] Sethian, J. A., and R. Malladi, "An O(N log N) Algorithm for Shape Modeling," *Proc. of the National Academy of Sciences,* Vol. 93, 1996, pp. 9389–9392.

[23] Adalsteinsson, D., and J. A. Sethian, "A Fast Level Set Method for Propagating Interfaces," *Journal of Computational Physics,* Vol. 118, 1995, pp. 269–277.

[24] Suri, J. S., et al., "Shape Recovery Algorithms Using Level Sets in 2-D/3-D Medical Imagery: A State-of-the-Art Review," *IEEE Trans. on Information Technology in Biomedicine,* Vol. 6, 2002, pp. 8–28.

[25] Angenent, S., D. Chopp, and T. Ilmanen, "On the Singularities of Cones Evolving by Mean Curvature," *Commun. Partial Differential Equations,* Vol. 20, 1995, pp. 1937–1958.

[26] Chopp, D. L., "Computing Minimal Surfaces Via Level Set Curvature Flow," *Journal of Computational Physics,* Vol. 106, 1993, pp. 77–91.

[27] Sethian, J. A., "Numerical Algorithms for Propagating Interfaces: Hamilton Jacobi Equations and Conservation Laws," *J. Differential Geometry,* Vol. 31, 1990, pp. 131–161.

[28] Sethian, J. A., "Algorithms for Tracking Interfaces in CFD and Material Science," *Annu. Rev. Comput. Fluid Mechanics,* 1995.

[29] Sussman, M., P. Smereka, and S. J. Osher, "A Level Set Method for Computing Solutions to Incompressible Two-Phase Flow," *J. Comput. Phys.,* Vol. 114, 1994, pp. 146–159.

[30] Rhee, C., L. Talbot, and J. A. Sethian, "Dynamical Study of a Premixed V-Flame," *J. Fluid Mechanics,* Vol. 300, 1995, pp. 87–115.

[31] Sethian J. A., and J. D. Strain, "Crystal Growth and Dentritic Solidification," *J. Comput. Phys.,* Vol. 98, 1992, pp. 231–253.

[32] Adalsteinsson, D., and J. A. Sethian, "A Unified Level Set Approach to Etching, Deposition and Lithography I: Algorithms and Two-Dimensional Simulations," *J. Comput. Physics,* Vol. 120, 1995, pp. 128–144.

[33] Breen, D. E., and R. T. Whitaker, "A Level-Set Approach for the Metamorphosis of Solid Models," *IEEE Trans. on Visualization and Computer Graphics,* Vol. 7, April 2001, pp. 173–192.

[34] Mansouri, A. R., and J. Konrad, "Motion Segmentation with Level Sets," *IEEE Int. Conf. Image Processing,* Vol. 2, October 1999, pp. 126–130.

[35] Paragios, N., and R. Deriche, "Geodesic Active Contours and Level Sets for the Detection and Tracking of Moving Objects," *IEEE Trans. on Pattern Analysis and Machine Intelligence,* Vol. 22, No. 3, 2000, pp. 266–280.

[36] Kornprobst, P., R. Deriche, and G. Aubert, "Image Sequence Analysis Via Partial Differential Equations," *J. Math. Imaging Vision,* Vol. 11, 1999, pp. 5–26.

[37] Faugeras, O., and R. Keriven, "Variational Principles, Surface Evolution, PDEs Level Set Methods and the Stereo Problem," *IEEE Trans. on Image Processing,* Vol. 7, May 1998, pp. 336–344.

[38] Kimmel, R., "Tracking Level Sets by Level Sets: A Method for Solving the Shape from Shading Problem," *Comput. Vision Image Understanding,* Vol. 62, 1995, pp. 47–58.

[39] Kimmel, R., and A. M. Bruckstein, "Global Shape from Shading," *Comput. Vision Image Understanding,* Vol. 62, 1995, pp. 360–369.

[40] Sapiro, G., et al., "Implementing Continuous-Scale Morphology Via Curve Evolution," *Pattern Recognition,* Vol. 26, 1997, pp. 1363–1372.

[41] Sochen, N., R. Kimmel, and R. Malladi, "A Geometrical Framework for Low Level Vision," *IEEE Trans. Image Processing,* Vol. 7, 1998, pp. 310–318.

[42] Sapiro, G., "Color Snakes," *Computer Vision and Image Understanding,* Vol. 68, 1997, pp. 247–253.

[43] Caselles, V., et al., "Fixed Dimensional Object Modeling Via Minimal Surfaces," *Proc. European Conf. Comput. Vision,* 1996, pp. 97–106.

[44] Kimmel, R., A. Amir, and A. M. Bruckstein, "Finding Shortest Paths on Surfaces Using Level Sets Propagation," *IEEE Trans. on Pattern Anal. Machine Intell.,* Vol. 17, 1995, pp. 635–640.

[45] DeCarlo D., and J. Gallier, "Topological Evolution of Surfaces," *Graphics Interface,* 1996, pp. 194–203.

[46] Telea, A., "An Image Inpainting Technique Based on the Fast Marching Method," *Graphics Tools,* Vol. 9, 2004, pp. 23–24.

[47] Suri, J. S., "Two Dimensional Fast MR Brain Segmentation Using a Region Based Level Set Approach," *Int. J. Eng. Medicine Biol.,* Vol. 20, 2001.

[48] Kass, M., A. Witkin, and D. Terzopoulos, "Snakes: Active Contour Models," *Inter. J. Computer Vision,* 1988, pp. 321–331.

[49] Caselles, V., et al., "A Geometric Model for Active Contours in Image Processing," *Numer. Math.,* Vol. 66, 1993, pp. 1–31.

[50] Malladi, R., J. A. Sethian, and B. C. Vemuri, "Evolutionary Fronts for Topology-Independent Shape Modeling and Recovery," *Third European Conf. on Computer Vision, Lecture Notes in Computer Science,* Vol. 800, 1994, pp. 3–13.

[51] Caselles, V., R. Kimmel, and G. Sapiro, "Geodesic Active Contours," *International Journal of Computer Vision,* Vol. 22, 1997, pp. 61–79.

[52] Suri, J. S., "Leaking Prevention in Fast Level Sets Using Fuzzy Models: An Application in Mr Brain," *Proc. Int. Conf. Inform. Technol. Biomedicine,* 2000, pp. 220–226.

[53] Baillard, C., P. Hellier, and C. Barillot, "Segmentation of 3-D Brain Structures Using Level Sets," *IRISA, Res. Rep.,* November 2000.

[54] Leventon, M. E., W. L. Grimson, and O. Faugeras, "Statistical Shape Influence in Geodesic Active Contours," *Proc. of the IEEE Conference Computer Vision and Pattern Recognition,* Vol. 1, 2000, pp. 316–323.

[55] Cootes, T. F., et al., "Active Shape Models—Their Training and Application," *Comput. Vision Image Understand.,* Vol. 61, 1995, pp. 38–59.

[56] Leventon, M. E., et al., "Level Set Based Segmentation with Intensity and Curvature Priors," *Proc. IEEE Workshop on Mathematical Methods in Biomedical Image Analysis,* July 2000, pp. 1121–1124.

[57] Zeng, X., et al., "Segmentation and Measurement of the Cortex from 3-D Mr Images Using Coupled Surfaces Propagation," *IEEE Trans. on Medical Imaging,* Vol. 18, September 1999, pp. 927–937.

[58] Mumford, D., and J. Shah, "Optimal Approximations by Piecewise Smooth Functions and Associated Variational Problems," *Communications on Pure and Applied Mathematics,* Vol. XLII, pp. 577–685, 1989.

[59] Chan, T. F., and L. A. Vese, "Active Contours Without Edges," *IEEE Trans. on Image Processing,* Vol. 10, 2001, pp. 266–277.

[60] Wang, Y., and L. H. Staib, "Boundary Finding with Correspondence Using Statistical Shape Models," *Computer Vision and Pattern Recognition,* 1998, pp. 338–345.

[61] S. C. Mitchell, S. C., et al., "3-D Active Appearance Models: Segmentation of Cardiac MR and Ultrasound Images," *IEEE Trans. on Medical Imaging,* Vol. 21, No. 9, 2002, pp. 1167–1178.

[62] Cootes, T. F., et al., "Active Appearance Models," *IEEE Trans. on Pattern Analysis and Machine Intelligence,* Vol. 23, 2000, pp. 681–685.

[63] Malladi, R., and J. A. Sethian, "Image Processing Via Level Set Curvature Flow," *Proc. National Academy of Sciences*, Vol. 92, July 1995, pp. 7046–7050.

[64] Riedmiller, M., and H. Braun, "A Direct Adaptive Method for Faster Backpropagation Learning: The Rprop Algorithm," *Proc. IEEE Inter. Conf. on Neural Networks*, 1993, pp. 586–591.

[65] Cremers, D., S. J. Osher, and S. Soatto, "Kernel Density Estimation and Intrinsic Alignment for Knowledge-Driven Segmentation: Teaching Level Sets to Walk," *Pattern Recognition*, Vol. 3175, 2004, pp. 36–44.

[66] Igel, C., M. Toussaint, and W. Weishui, "Rprop Using the Natural Gradient Compared to Levenberg-Marquardt Optimization," *4th Inter. Meeting on Constructive Approximation (IDoMAT 2004)*, 2004.

[67] Weiss, S. M., and C. A. Kulikowski, *Computer Systems That Learn: Classification and Prediction Methods for Statistics, Neural Nets, Machine Learning, and Expert Systems*, San Francisco, CA: Morgan Kaufmann, 1991.

[68] Efron, B., and R. J. Tibshirani, *An Introduction to the Bootstrap*, London, U.K.: Chapman & Hall, 1993.

[69] Plutowski, M., S. Sakata, and H. White, "Cross-Validation Estimates Integrated Mean Squared Error," in *Advances in Neural Information Processing Systems*, J. Cowain, G. Tesaulo, and J. Alspector, (eds.), San Francisco, CA: Morgan Kaufmann, 1994, pp. 1135–1142.

[70] Shao, J., and D. Tu, *The Jackknife and Bootstrap*, New York: Springer-Verlag, 1995.

PART VI

Application of 3-D Therapeutic and Image Guidance Ultrasound

CHAPTER 13

HIFU Therapy Planning Using Pretreatment Imaging and Simulation

Viren R. Amin

13.1 Introduction

High-intensity focused ultrasound (HIFU) is being studied by many research groups as a potential noninvasive therapy for many clinical problems that otherwise require surgery or other approaches. The HIFU technology delivers high levels of ultrasound energy (from 1,000 to 10,000 W/cm^2 at the focal spot compared to less than 0.1 W/cm^2 for diagnostic ultrasound), focused to a tissue of interest, raising the tissue temperature to greater than 50°C within seconds. Maintaining this temperature for 1 to 3 seconds results in cell death, and a single exposure can destroy up to 0.5 mL volume of tissue at focus without significantly damaging the surrounding tissues. The effects on the tissue at focus for varying energy levels and exposure times include protein denaturation, coagulation and hemostasis, cavitation, and tissue ablation. The HIFU works similar to using a magnifying glass to concentrate the sun's energy to burn a hole in a dry leaf. Just as the hole in the leaf occurs only when it is at the focus of the magnifying glass, the HIFU also delivers maximum concentration of energy at its focus without significant damage to the overlaying tissues. Thus, HIFU is emerging as an exciting new therapy option that is trackless, bloodless, and portable without the need for general anesthesia. It can be used for many clinical applications including cancer tumors, hemostasis, and cardiology that otherwise require surgery or other approaches. Figure 13.1 shows examples of HIFU lesions created in a phantom, chicken breast, and pork chop.

Although promising, HIFU is currently limited by issues such as lesion size, its single exposure, the time required between exposures to allow local tissue cooling, the inability to treat through air or bone, difficulty monitoring the therapy in real time, and the inability to plan the dosimetry before therapy. Technological advances, such as array beam generators and image-based guidance, promise to overcome many of these problems. The HIFU dosimetry and pretherapy planning require tools to define the HIFU beam through multiple layers of inhomogeneous tissues and tissue lesion generation. Such tools are prerequisites to sophisticated treatment planning using pretherapy imaging (e.g., CT, MRI, or US), similar to computational dosimetry and pretreatment planning for radiation therapy.

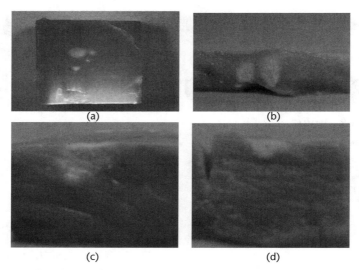

(a)   (b)

(c)   (d)

Figure 13.1   Examples of HIFU effects on tissue-mimicking phantom, chicken breast, and pork chop. (a) Localized HIFU lesion inside a translucent tissue-mimicking phantom. (b) An example of HIFU lesions in chicken breast using a 1.561-MHz HIFU transducer driven with continuous sinusoidal waves for durations of 60 (left lesion) and 45 (right lesion) seconds. (c) An example of HIFU lesion in pork chop using 60 seconds of HIFU exposure. (d) Same as (c) but with 30-second duration.

HIFU application to tissue ablation requires high precision and tools for dosimetry therapy planning, and real-time feedback of the intended and actual target locations. Pretreatment planning is an important step for a successful HIFU therapy outcome. Typically, the therapy planning approach involves the use of pretreatment imaging data, defining the target and surrounding tissues by manual or semiautomatic segmentation, development of a 3-D anatomy model of the region of interest from segmentation or registration with a reference dataset, simulation of the HIFU beam and thermal dosimetry around the target tissue, display and 3-D visualization of imaging and simulation data, and review of the treatment plan options. Recent developments in therapy planning using imaging are targeted for specific applications such as prostate cancer using 3-D ultrasound images and uterine fibroids using MRI. However, significant developments have been accomplished in image guidance and feedback during the delivery of HIFU treatments.

This chapter reviews recent work toward therapy planning and presents approaches for developing strategies for HIFU therapy. It describes general and target-specific techniques and software tools for HIFU treatment planning using pretherapy imaging (e.g., MRI, CT, or US) similar to dosimetry and planning for radiation therapy. An understanding of this subject matter aids development of optimized, high-precision HIFU dosimetry and patient-specific planning strategies for applications such as tumor ablation. It also potentially reduces the guesswork on dosage parameters and thereby reduces the overall treatment duration and exposure to nontarget tissues.

13.2   Recent Developments in HIFU and Therapy Planning

The progress on HIFU and current research focus can be summarized into five categories: (1) in vitro and in vivo study of HIFU applications to specific problems and organs, (2) therapy guidance technology development (e.g., US and MRI guidance of HIFU therapy), (3) HIFU instrumentation (transducer and equipment), (4) theoretical and experimental study of mechanism of tissue effects by HIFU, and (5) therapy planning.

13.2.1   HIFU Applications

Many HIFU applications have been studied and reported in the recent literature. Some examples include the following:

- Ablation of cancer tumors in the prostate [1, 2], uterine fibroids [3], breast [4], liver [5], and rectum [6]. An overview of 5-year clinical trials in 1,038 patients with solid carcinoma in China is reported in [7].
- Brain tumor ablation via the trans-skull approach [8, 9] and via the craniotomy approach [10, 11] as an aid to surgery or as a less invasive surgery.
- An overview of the scientific and engineering advances that are allowing the growth in clinical focused ultrasound applications is presented by Clement [12].
- Homeostasis of blood vessels [13], parenchyma of liver [14], and spleen [15].
- Heart diseases such as revascularization of myocardium [16], hypertrophic cardiomyopathy [17], and creating communication in the atria septum [18].
- Blood–brain barrier opening [19] and fetal tissues [20].

13.2.2   HIFU Therapy Guidance

Image-based guidance and monitoring of HIFU treatments has been addressed significantly in recent reports. Techniques based on real-time MRI, US, and CT have been developed to monitor tissue and temperature changes during HIFU application. The MRI thermometry relies on proton resonance frequency shift methods to characterize tissue damage and temperature around the HIFU focus [4, 21]. A general method for temperature control based on Fourier transformation of the bioheat equation is proposed that takes into account heat diffusion, energy absorption, and temperature distribution derived from rapid continuous MRI temperature mapping [22]. McDannold et al. [23] demonstrated the need and usefulness of quantitative temperature mapping during HIFU surgery by analyzing MRI-based temperature imaging data from the treatments of 62 fibroids in 50 women to test the ability of thermal dose maps acquired during treatment and to predict the resulting nonperfused area in contrast-enhanced images acquired immediately after treatment.

Examples of ultrasound-based HIFU monitoring techniques include real-time visualization using gated US imaging [24], elastography for better detection of coagulated tissue damage [2], US-stimulated acoustic emission or vibroacoustography for localizing temperature elevation [25], differential attenuation imaging to characterize HIFU-induced lesions [26], US phase-contrast transmission imaging [27, 28], and radiation-force imaging by means of echo-shift estimations [29].

Additionally, CT imaging has also been proposed to monitor and guide HIFU treatment [30]; however, MRI and US have been increasingly used for monitoring HIFU therapy in research and early clinical trials. When possible, MRI thermometry is considered the gold standard for quantifying temperature changes during HIFU exposure in clinical settings.

13.2.3 HIFU Transducers and Instrumentation

Recent developments in HIFU transducers and instrumentation include an array transducer [31], solid aluminum cone surgical applicator [32], split-focus transducer [33], parabolic reflection device for mechanically variable focal lengths [34], HIFU synchronization with ultrasound imaging [24], and robotics for HIFU delivery [35, 36]. Similarly, different tissue effects are observed by different shapes and duty cycles of the signal driving the transducers. Such developments allow for various applications of HIFU but create challenges when modeling the HIFU beam and its effects on tissues.

13.2.4 Theoretical and Experimental Study of the HIFU Mechanism

The biological effects of HIFU have been studied since the 1950s [37]. HIFU tissue damage and ablation occurs primarily via two mechanisms: thermal energy absorption and cavitation and its related phenomena. Traditionally, the thermal mode has been used in understanding lesion size and shape predicted by models. With cavitation, a series of bioeffects can occur. These include shock wave propagation, bubble generation, higher absorption and a higher bulk temperature near the focus point, higher scattering that blocks ultrasound propagation, and subsequent power deposition.

Nonlinear effects in focused ultrasound beams are of importance for the intensities used in therapeutic applications [38]. However, linear wave propagation models are commonly applied to study the effects of overlying tissues (in which the field intensities are substantially lower than those in the focal zone) on the HIFU focus. Such models help determine the probability density function for the distribution of absorbed energy and the resulting temperature rise in the target tissue at the focus point. Experimental and computational studies to examine the impact of phase aberration through overlying inhomogeneous tissues on ultrasound beam intensity distribution have been reported [39]. The effect of inhomogeneous overlying tissues on the heating patterns and steady-state temperature distributions have been studied analytically using a random-phase screen model [40]. These model findings suggested the presence of many hot and cold spots throughout the medium.

Recent work has focused on understanding this complex interacting nonlinear phenomena at the very high intensities present at the focus point, evolution of cavitations and bubbles, increased focal heating from nonlinearity, and tissue lesion formation [41–43]. Curiel et al. [44] adapted the bioheat transfer equation (BHTE) to take into account the activity of cavitation bubbles generated during HIFU exposure. This modeling was used to predict the lesions produced by three different transducer geometries, and such predictions matched experimental results significantly better than the BHTE without bubble effects [45].

Several researchers have worked on the problem of computationally modeling the nonlinearly coupled equation set describing the interaction of tissue with HIFU. Curra et al. [38] provided a summary of these modeling efforts. This is a challenging computational problem complicated by the inability to directly measure the impact of the acoustic wave on tissue during exposure. Because all direct measurement techniques (e.g., thermocouples) alter the impact of the acoustic wave, measurements are done outside the area of greatest interest or are implied after the fact by observing the tissue damage done. In addition, the speckled distribution of acoustic energy, the formation of bubbles, and the energy absorption rate are nonlinearly coupled together [38]. One group of researchers has taken the first steps toward addressing these phenomena and used the results to "optimize the firing session parameters for dynamically focused HIFU treatment" [46, 47]. The researchers noted several weaknesses in the models relating to nonlinear coupling and temperature effects, and the density of the bubbles. In spite of this, reasonable estimates of tissue were made, and the researchers felt that additional work was warranted to develop a model-based approach to HIFU treatments.

Deposition of energy within tissue and the subsequent heat transfer within the tissue are complex and time-consuming computational problems. A complete model requires coupling of the equation set for the sound propagation model with the heat transfer equations within the tissue and blood. Simultaneous determination of multiple parameters is a challenging inverse problem because of a discontinuous search space and the nonlinear nature of the problem. Mahoney et al. [45] proposed a method that uses actual measured ultrasound field distributions in combination with backward projection to calculate the temperature elevation and potentially improve treatment planning.

13.2.5   HIFU Dosimetry and Therapy Planning

A systematic approach for developing pretreatment planning and HIFU dose calculations for specific target locations using simulations and experimental validation is important. A thorough understanding of the HIFU lesion mechanism, target identification, tools for imaging and visualization, and validated modeling tools are prerequisites to sophisticated treatment planning, similar to dosimetry and planning for radiation therapy. This approach will help HIFU technology to achieve its full potential and high precision for emerging complex applications.

Huber et al. [4] reported possibly the first MRI-guided HIFU in a malignant tumor in humans. The main objective of the study was to integrate MRI thermometry (phase imaging to estimate the temperature-dependent, proton-resonant frequency shift) to guide the HIFU treatment (as a neoadjuvant approach preceding open surgical removal) of a well-circumscribed invasive ductal carcinoma of the breast. Although the therapy regimen used was sufficient to homogeneously induce tumor cell death, they concluded that a tighter scanning pattern of the ultrasound foci, higher intensity, or longer sonication per pulse might enhance the instantaneous antitumor effect in breast cancer. They suggested that the issue of optimal dosing (e.g., acoustic power, duration, duty cycle, and total number of pulses) had to be addressed further. This work clearly demonstrated a need for high-precision dosimetry planning tools based on validated simulation for deep targets in inhomogeneous tissues.

Some dosimetry work has been done using empirical experimental data analysis on different tissues. For example, Wang et al. [48] demonstrated, with in vitro and in vivo experiments on pig liver, that the HIFU lesion occurs at a so-called "biological focal region" as a function of acoustic focal region (as measured in water), acoustic intensity, exposure time, irradiation depth within the tissue or overlying tissues, and the tissue structure and its functional status.

HIFU applications to brain tumors are especially challenging due to the skull bone. Recent work on HIFU applications for brain tumor therapy has focused mainly on trans-skull approaches [8, 9], but HIFU delivery via craniotomy has also been suggested as an aid to the surgical approach or as a less invasive therapy [10, 11]. In addition to MRI imaging for tumor identification, therapy planning for the trans-skull approach also involves evaluation of skull bone thickness via CT imaging and calculation of phase correction terms for an array of beams for accurate dose delivery at the target. Additionally, ultrasound imaging is increasingly used in brain tumor surgeries for intraoperative guidance, and HIFU could be envisioned to treat the tumor cells remaining in the surgical bed (rather than the applied chemotherapy).

In summary, recent reports in the literature have addressed HIFU dosimetry and therapy planning issues. Most advances have been made for HIFU applications to easily accessible tissues and lesions (e.g., breast and prostate). As discussed earlier, a significant amount of experimental work has been done in vitro on tissues or in vivo on exposed organs. However, additional potential for HIFU can be realized with further development of our understanding of the mechanism of various tissue effects of HIFU, modeling those effects when HIFU is delivered in vivo through overlying tissues with blood flow, the availability of such modeling tools, and visualization of simulation based on patient-specific anatomies. Such an approach will aid in the development of optimized high-precision HIFU dosimetry and planning strategies for complex and sensitive applications in cancer, cardiology, and the brain, as well as emerging areas of targeted delivery or activation of molecules. It also potentially optimizes HIFU therapy by reducing the guesswork on dosage parameters and thereby reducing the overall treatment duration and reduced exposure to nontarget tissues.

13.3 Generalized Strategies for HIFU Therapy Planning

This section describes some generalized strategies and issues to consider when developing a HIFU therapy planning system. It also presents some approaches to an integrated software platform where diverse imaging, segmentation, HIFU simulation, visualization, and delivery tools can be implemented and studied further. This approach is generally similar to that of a sophisticated radiation therapy planning system (e.g., the Eclipse system from Varian Medical Systems [49]) and would be significantly beneficial to the HIFU research, industrial, and clinical communities.

13.3.1   Overview of Generalized Strategy

A complete therapy planning procedure consists of multiple steps. The first step is pretreatment imaging (MRI, CT, or US). The imaging data are then processed for segmenting the target and surrounding anatomy (manual or semiautomatic or using coregistration with a reference dataset) to develop 3-D anatomic models of target and overlying tissues (voxel and surface models). This information is then used for ultrasound simulation of beam intensity distribution and HIFU temperature profiles. Finally, integrated visualization of 3-D anatomy, ultrasound beam simulation, and HIFU delivery parameters (e.g., power, duty cycle, and path for multiple foci) provides an interactive strategy planning tool for studying the approaches of HIFU delivery. Recent reports describe the use of such an approach for general targets using MRI images [11] and for specific targets, for example, for prostate cancer using 3-D US images [50]. Figure 13.2 illustrates an overall approach to HIFU therapy planning simulation for an example case of a through-craniotomy brain tumor application. (Note that this example is intended to illustrate the steps involved in generalized HIFU therapy planning system.)

13.3.2   Pretreatment Imaging

The primary purpose of pretreatment imaging is to define the target volume and surrounding anatomy, especially the overlying tissue layers and critical tissues such as blood vessel areas to protect from exposure. As discussed in the previous section, different imaging modalities have been used for different organs and regions. MRI is used frequently for diagnosis and full assessment before surgery or treatment. Often, CT or 3-D US is used for certain applications. Imaging provides a 3-D dataset with voxel intensities that define tissue properties. It is generally assumed that the voxel intensities relate to the tissue density and tissue type so that they can

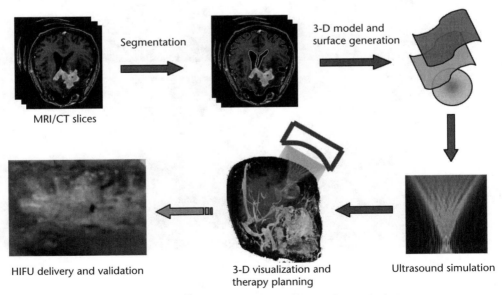

Figure 13.2   Illustration of overall approach to HIFU therapy planning simulation.

be used for simulation of ultrasound propagation through varying tissue densities or ultrasound velocities. Imaging techniques such as Doppler, diffusion MRI, and contrast imaging can help identify blood vessels, tissue perfusion, and targeted tumor cells in the region of interest.

13.3.3 Defining the Target and Surrounding Tissues

From each slice of the image dataset, an expert can trace the borders of a target organ to create a 3-D organ definition. Often, relevant surrounding and overlying tissues need to be identified and labeled. A manual segmentation approach is very time consuming, and variability between experts is reported to be up to 15%. A large amount of literature about the segmentation of the anatomy by different imaging modalities has been published. The challenges in automatic segmentation include effective representation of the area to be segmented, difficulty in incorporating variable domain knowledge (underlying anatomy, image processes, and so forth), and overlapping gray levels and regions. A wide variety of segmentation techniques has been proposed, however, no one standard segmentation technique exists that can produce satisfactory results for all imaging applications. For an overview of the fundamentals of medical image segmentation and recent level set–based techniques, refer to [51, 52]. Accurate segmentation is used to derive 3-D models for the target and relevant surrounding tissues and overlying surface geometry.

Figure 13.3 illustrates an example of the semiautomatic segmentation process for brain tumor MRI images using a level set–based technique for geometric active contour algorithms. A user segments an initial slice, and the successive slices are segmented automatically by using parameters calculated from the previous slice. Fully automatic segmentation is often challenging due to the large variation of the target

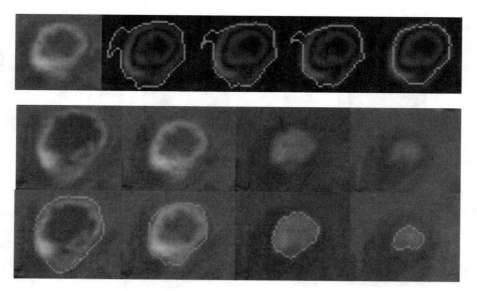

Figure 13.3 Illustration of an example of the semiautomatic segmentation process for brain tumor MRI images using a level set–based technique for geometric active contour algorithms. The top panel shows screen shots of the iterative process of active contour evolution and the bottom panel shows segmentation of successive slices.

shape, intensity, continuity, location, and size and due to the complex anatomic structures in the neighborhood.

13.3.4   Simulation of HIFU

Information from the 3-D anatomic models of target and overlying tissues (voxel and surface models) is used for ultrasound simulation of beam intensity distribution. To quantitatively connect a spatial variation of wave velocity and the resulting beam amplitude distortion, computational models are employed, as discussed in the previous section. Figures 13.4 and 13.5 show examples of computed and experimentally measured wavefield distortion due to tissue inhomogeneity. The ultrasound beam in a homogeneous medium is shown. An explicit random specification of acoustic velocity, in which increasing pixel brightness indicates increasing velocity, is also shown. Velocity values are randomly specified to vary 4% about the mean. Examples of ultrasound beam distortion due to this and similar realizations of 3-D random media are shown. These distortions are seen to be substantial, but their significance for high-intensity focused beams remains to be seen. The simulation tools allow for multilayered 3-D tissue configurations with inhomogeneity in velocity values and could be used for more realistic scenarios derived from imaging of the structure of the tissues.

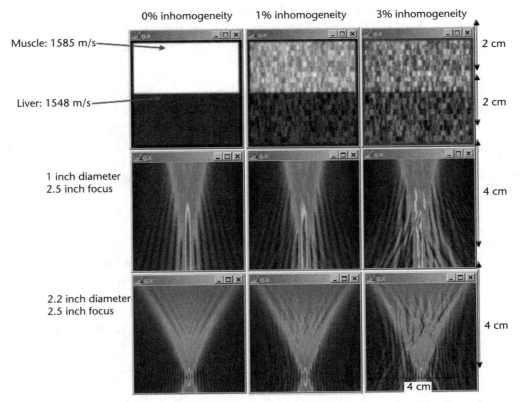

Figure 13.4   An example of computational results of the impact on phase aberration through overlying inhomogeneous tissue layers. The impact increases as the variation within tissues increases, which results in blurry and shifted energy focus. (*From:* [39]. © 2005 American Institute of Physics. Reprinted with permission.)

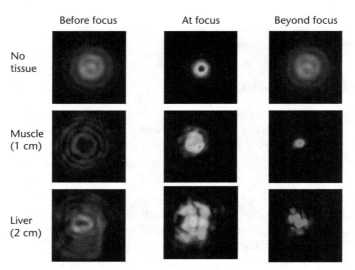

Figure 13.5 Experimental results of mapping ultrasound beam cross sections in low intensity in water with and without intervening tissue layers (1-cm layer of bovine muscle and 2-cm layer of bovine liver). The significant effect of phase aberration, caused by tissue inhomogeneity, is observed as predicted by the simulation. (*From:* [39]. © 2005 American Instititue of Physics. Reprinted with permission.)

From the segmented target volume and ultrasound beam intensity distributions, the BHTE is applied with given tissue parameters to calculate the temperature rise for a given HIFU exposure duration. The rate of acoustic energy absorption within tissue is tightly coupled to the temperature rise and development of oscillating microbubbles that form within the tissue [46]. Computing the impact of HIFU is further confounded by phenomena resulting from 3-D structures within the tissue such as acoustic streaming in the blood and back scattering [38]. Additionally, tissue changes (e.g., boiling) alter the characteristics of the energy absorption. Recent efforts attempt to account for the added terms of the BHTE including the volumetric heat rate generated by the presence of microbubbles [41] and the blood perfusion losses [53]. Basic principles of cavitation in therapeutic ultrasound have recently been actively studied [54, 55].

13.3.5 Integration and Visualization

Recent advances in computer graphics capabilities allow for decent visualization of 2-D and 3-D datasets on a desktop computer. A variety of computational visualization methods are used in medical applications; [51] provides an overview of many such techniques. For visualization, ray casting [56] is most commonly used in volume rendering, but a 3-D texture mapping technique makes the rendering real time, which is usually impossible with ray casting. Advanced computer graphics techniques are often used to take full advantage of a modern video card's high speed and programmable capability. For example, a preintegrated volume rendering technique [57] improves the rendering quality without sacrificing the rendering performance. Also, issues such as rendering multivolume data (e.g., voxels from both imaging and simulation data) need to be considered carefully. The visualization is not limited to volume data themselves. Many other 2-D or 3-D features can be rendered simultaneously, and that works well along with the partial exposure techniques [58]. During implementation, special care is taken to set rendering states and decide the rendering order for different features.

Figures 13.6, 13.7, 13.8, and 13.9 show some examples of visualization techniques. Using such visualization tools, the user can register and overlay the HIFU simulation results on top of the 3-D medical images, and interactively change the parameters and see the results.

13.3.6   Planning of HIFU Delivery and Control

HIFU treatment options can be explored interactively once the target geometry and HIFU-related simulation data are precalculated for given ranges of transducer position and orientation, HIFU durations, frequency, and power. It may be beneficial if simulation of an imaging modality (e.g., MRI or ultrasound technique to be used for guidance during HIFU delivery) can be developed and incorporated in the planning tool. If accurate, this can provide training for delivery control and guidance.

13.3.7   Overall Software Design and Integration of Tools

Figure 13.10 shows the overall approach and system architecture for the HIFU therapy planning software. Each module surrounding the control center incorporates one or more individual tools and communicates with the control center. This approach allows implementation of any individual tool in any programming language that can be plugged in dynamically. This software is being developed using Visual Studio.Net, OpenGL, ITK, and MATLAB programming environments on a Windows platform.

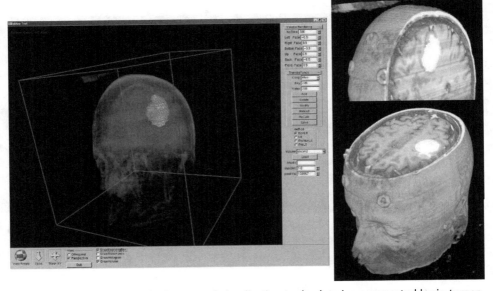

Figure 13.6   Screen capture for integrated visualization tools, showing segmented brain tumor inside a rendered volume. (*From:* [11]. © 2005 American Institute of Physics. Reprinted with permission.)

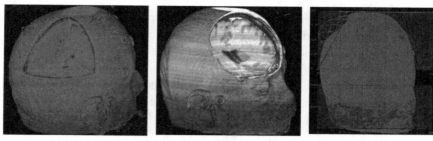

Figure 13.7 Examples of advanced visualization techniques. Traditional cube-shaped cutout for visualizing (left panel), sphere-shaped cutaway view similar to a large craniotomy (middle panel), and simultaneous rendering of multiple features including volume, cutaway object profile, and transfer functions (right panel).

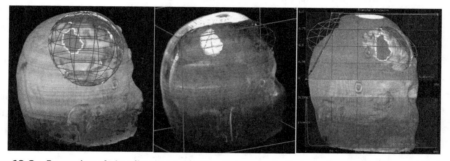

Figure 13.8 Examples of visualization with simultaneous rendering of multiple features. Volume and cutaway object profile (left panel); volume, cutaway object profile, and bounding box (middle panel); volume, cutaway object profile, transfer functions, and histogram (right panel). (*From:* [58]. © 2006 SPIE. Reprinted with permission.)

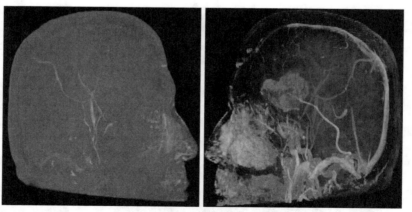

Figure 13.9 Examples of advanced visualization techniques. Real-time, view-oriented 3-D texture-based volume rendering using graphics Shader Language.

As with all software-implemented clinical techniques, the user interface design is crucial to successful implementation of HIFU planning tools. Using a Windows platform and Visual Studio.Net technologies, an integrated environment is developed

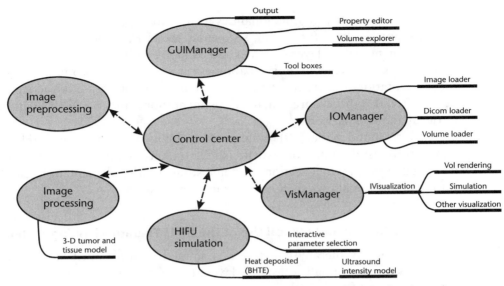

Figure 13.10 Overview of architecture for general-purpose HIFU therapy planning software.

that has a very flexible and user-friendly graphics user interface (Figure 13.11). Different panels/windows can be freely rearranged according to the user's preference. Multiple visualization windows can be opened simultaneously, and various 2-D or 3-D visualization windows can be arranged side by side for better comparison. The integrated system also makes communication among different tools smoother, user friendly, and resource efficient.

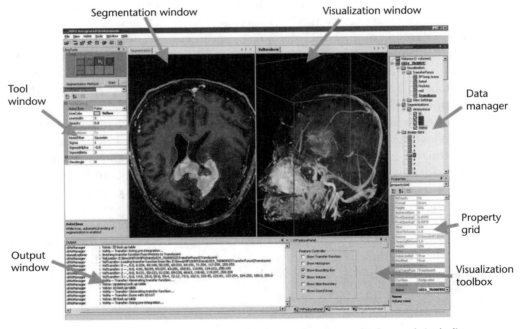

Figure 13.11 Screen capture of the integrated software showing multiple panels including segmentation and visualization.

The current software functionalities include loading and viewing DICOM files and relevant header information, segmentation, HIFU simulation using an ultrasound beam model for layered tissues and coupling with BHTE, and visualization of 3-D datasets. The segmentation capabilities for research include multiple algorithms for segmentation, interactive change of parameters, monitoring the active contours during iterative procedures, manual segmentation, and an XML database for retrieval and comparisons.

An ultrasound model [39] is integrated with this system to allow HIFU dose and temperature estimations for patient-specific target anatomies. Additional models discussed in the previous section are being studied for implementation and the software architecture allows such modular implementation.

13.3.8 Validation and Clinical Use of the HIFU Therapy Planning System

The utility of any therapy planning and simulation tools must be validated with HIFU experiments initially on simple layered tissue structures, later with more complex shapes, and finally with in vivo animal models. For successful clinical application, the sources of uncertainties in treatment planning must be evaluated. These include patient localization, patient and organ motion during delivery, positioning and immobilization, reproducibility, HIFU effects on imaging, localization of target structures, quality assurance of data transfers from planning system to delivery system, and quality assurance of delivery equipment.

13.4 Challenges and Work Ahead

There has been good progress in developing tools for HIFU therapy planning. Further work must include in vivo experimental validation of computational models, understanding of the HIFU mechanism with a wider range of transducer design, frequency, power and duty cycle, validation of the use of an imaging modality for planning and guidance, incorporation of therapy optimization, control and feedback in simulation, assessment and use of blood flow and tissue perfusion, integration of planning and delivery, robotic control, and training and qualification of the experts.

Validated computational tools will need to be implemented as software tools to demonstrate the feasibility of pretreatment planning and a dosimetry approach for optimizing HIFU delivery to the target tissues for various clinical problems. Finally, user-friendly software implementation of validated computational models must be made available for use by the larger research community. Results from computational models and in vitro and in vivo experiments will provide a strong basis for HIFU dosimetry and treatment planning. These developments along with robust image guidance will help optimize HIFU treatment outcomes, for example, by making the procedure faster or better controlled, and will potentially allow for treatments in targets that are currently difficult, such as moving organs and targets behind bone.

Acknowledgments

The author would like to acknowledge the efforts of graduate students and colleagues who have helped through discussions or have worked on different aspects of HIFU research, some of which is presented in this chapter: Liangshou Wu, Tao Long, Ronald Roberts, R. Bruce Thompson, Matt Hagge, Mark Bryden, Timothy Ryken, Scott McClure, and Shahram Vaezy. The author would also like to acknowledge the financial support in part by a research grant from the Roy J. Carver Charitable Trust and the Institute of Physical Research and Technology at Iowa State University. (*Note:* The content of this chapter represents the opinions and interpretations of the author and does not necessarily reflect the position of any colleagues or funding source.)

References

[1] Blana, A., et al., "High-Intensity Focused Ultrasound for the Treatment of Localized Prostate Cancer: 5-Year Experience," *Urology,* Vol. 63, 2004, pp. 297–300.

[2] Souchon, R., et al., "Visualization of HIFU Lesions Using Elastography of the Human Prostate In Vivo: Preliminary Results," *Ultrasound Med. Biol.,* Vol. 29, No. 7, 2003, pp. 1007–1015.

[3] Chan, A., et al., "An Image-Guided High Intensity Focused Ultrasound Device for Uterine Fibroids Treatment." *Med. Phys.,* Vol. 29, No. 11, 2002, pp. 2611–2620.

[4] Huber, P., J. Jenne, and R. Rastert, "A New Noninvasive Approach in Breast Cancer Therapy Using Magnetic Resonance Imaging-Guided Focused Ultrasound Surgery," *Cancer Research,* Vol. 61, 2001, pp. 8441–8447.

[5] Kennedy, J., et al., "High Intensity Focused Ultrasound for the Treatment of Liver Tumors," *Ultrasonics,* Vol. 42, 2004, pp. 931–935.

[6] Jun-Qun, Z., et al., "Short-Term Results of 89 Cases of Rectal Carcinoma Treated with High-Intensity Focused Ultrasound and Low-Dose Radiotherapy," *Ultrasound Med. Biol.,* Vol. 30, No. 1, 2004, pp. 57–60.

[7] Wu, F., et al., "Extracorporeal High Intensity Focused Ultrasound Ablation in the Treatment of 1038 Patients with Solid Carcinomas in China: An Overview," *Ultrasonics Sonochemistry,* Vol. 11, 2004, pp. 149–154.

[8] McDonnold, N., et al., "MRI-Guided Focused Ultrasound Surgery in the Brain: Tests in a Primate Model," *Magn. Resonance Med.,* Vol. 49, pp. 1188–1190.

[9] Pernot, M., et al., "Adaptive Focusing for Ultrasonic Transcranial Brain Therapy: First In Vivo Investigation on 22 Sheep," *4th International Symposium on Therapeutic Ultrasound,* Kyoto, Japan, September 18–20, 2004.

[10] Dahl, E., et al., "Towards the Use of HIFU, in Conjunction with Surgery, in the Treatment of Malignant Brain Tumors," *5th International Symposium on Therapeutic Ultrasound,* Boston, MA, October 22–23, 2005.

[11] Amin, V., et al., "HIFU Therapy Planning Using Pre-Treatment Imaging and Simulation," *5th International Symposium on Therapeutic Ultrasound,* Boston, MA, October 22–23, 2005, p. 18.

[12] Clement, G. T., "Perspectives in Clinical Uses of High-Intensity Focused Ultrasound," *Ultrasonics,* Vol. 42, No. 10, 2004, pp. 1087–1093.

[13] Vaezy, S., et al., "Use of High-Intensity Focused Ultrasound to Control Bleeding," *J. Vasc. Surg.,* Vol. 29, 1999, pp. 533–542.

[14] Vaezy, S., R. Martin, and L. Crum, "High Intensity Focused Ultrasound: A Method of Hemostasis," *Echocardiography,* Vol. 18, No. 4, 2001, pp. 309–315.

[15] Noble, M., et al., "Spleen Hemostasis Using High-Intensity Ultrasound: Survival and Healing," *J. Trauma,* Vol. 53, 2002, pp. 1115–1120.

[16] Smith, N. and K. Hynynen, "The Feasibility of Using Focused Ultrasound for Transmyocardial Revascularization," *Ultrasound Med. Biol.,* Vol. 24, No. 7, 1998, pp. 1045–1054.

[17] Muratore, R., et al., "Experimental High-Intensity Focused Ultrasound Lesion Formation in Cardiac Tissue," *J. Acoutical Soc. Am.,* Vol. 115, No. 5, 2001, p. 2448.

[18] Craig, R., et al., "Creating an Atrial Communication Using High Intensity Focused Ultrasound: An In Vitro Dosimetry Study," *J. Am. Soc. Echocardiogr.,* Vol. 17, 2004, p. 503.

[19] Treat, L., N. McDannold, and K. Hynynen, "Transcranial MRI-Guided FUS-Induced BBB Opening in the Rat Brain," *J. Acoustical Soc. Am.,* Vol. 115, No. 5, 2004, pp. 2524–2425.

[20] Pake, B., et al., "Tissue Ablation Using High-Intensity Focused Ultrasound in the Fetal Sheep Model: Potential for Fetal Treatment," *Am. J. Obstet. Gynecol.,* Vol. 189, 2003, pp. 70–705.

[21] Hynynen, K., et al., "MR Imaging-Guided Focused Ultrasound Surgery of Fibroadenomas in the Breast: A Feasibility Study," *Radiology,* Vol. 219, 2001, pp. 176–185.

[22] Quesson, B., et al., "Automatic Control of Hyperthermic Therapy Based on Real-Time Fourier Analysis of MR Temperature Maps," *Magn. Resonance Med.,* Vol. 47, 2002, pp. 1065–1072.

[23] McDannold, N., et al., "MRI-Based Thermometry and Thermal Dosimetry During Focused Ultrasound Thermal Ablation of Uterine Leiomyomas," *IEEE Ultrasonics Symposium,* Montreal, Canada, August 22–27, 2004.

[24] Vaezy, S., et al., "Real-Time Visualization of High-Intensity Focused Ultrasound Treatment Using Ultrasound Imaging," *Ultrasound Med. Biol.,* Vol. 27, No. 1, 2001, pp. 33–42.

[25] Konofagou, E., et al., "The Temperature Dependence of Ultrasound-Stimulated Acoustic Emission," *Ultrasound Med. Biol.,* Vol. 28, No. 3, 2002, pp. 221–228.

[26] Ribault, M., et al., "Differential Attenuation Imaging for the Characterization of High Intensity Focused Ultrasound Lesions," *Ultrasound Imaging,* Vol. 20, No. 3, 1998, pp. 160–177.

[27] King, R. L., et al., "Preliminary Results Using Ultrasound Transmission for Image-Guided Thermal Therapy," *Ultrasound Med. Biol.,* Vol. 29, No. 2, 2003, pp. 293–299.

[28] Clement, G. T., and K. Hynynen, "Ultrasound Phase-Contrast Transmission Imaging of Localized Thermal Variation and the Identification of Fat/Tissue Boundaries," *Phys. Med. Biol.,* Vol. 50, No. 7, 2005, pp. 1585–1600.

[29] Azuma, T., et al., "Radiation Force Imaging for Detection of Irreversible Changes Caused by High Intensity Focused Ultrasound Therapy," *IEEE Ultrasonics Symposium,* Rotterdam, the Netherlands, September 13–21, 2005.

[30] Jenne, J., et al., "CT On-Line Monitoring of HIFU Therapy," *Proc. IEEE Ultrasonics Symp.,* 1997, pp. 1377–1380.

[31] Rongmin, X., et al., "A New-Style Phased Array Transducer for HIFU," *Applied Acoustics,* Vol. 63, 2002, pp. 957–964.

[32] Brentnall, M., et al., "A New High Intensity Focused Ultrasound Applicator for Surgical Applications," *IEEE Trans. on UFFC,* Vol. 48, No. 1, 2001, pp. 53–63.

[33] Umemura, S., et al., "Non-Circular Multi-Sector Split-Focus Transducer for Coagulation Therapy," *IEEE Ultrasonics Symp.,* 2000, pp. 1409–1412.

[34] Zderic, V., et al., "Parabolic Reflection High Intensity Ultrasound Based Device with Mechanically-Variable Focusing," *Proc. IEEE Ultrasonics Symp.,* 2003, pp. 1239–1242.

[35]   Kheng, N., et al., "A HIFU Robot for Transperineal Treatment of Prostate Cancer," *Proc. 7th IEEE International Conf. on Control, Automation, Robotics and Vision,* Vol. 2, 2002, pp. 560–565.

[36]   Pather, S., B. Davies, and R. Hibberd, "The Development of a Robotic System for HIFU Surgery Applied to Liver Tumors," *Proc. 7th IEEE International Conf. on Control, Automation, Robotics and Vision,* Vol. 2, 2002, pp. 572–577.

[37]   Fry, W., et al., "Ultrasonics Lesions in Mammalian Central Nervous System," *Science,* Vol. 122, 1955, pp. 517–518.

[38]   Curra, F., et al., "Numerical Simulations of Heating Patterns and Tissue Temperature Response Due to High-Intensity Focused Ultrasound," *IEEE Trans. on UFFC,* Vol. 47, No. 40, 2000, pp. 1077–1089.

[39]   Amin, V., et al., "A Study of Effects of Tissue Inhomogeneity on HIFU Beam," *5th International Symposium on Therapeutic Ultrasound,* Boston, MA, October 22–23, 2005.

[40]   Bilgen, M., and M. Insana, "Effects of Phase Aberration on Tissue Heat Generation and Temperature Elevation Using Therapeutic Ultrasound," *IEEE Trans. on UFFC,* Vol. 43, No. 6, 1996, pp. 999–1010.

[41]   Chavrier, F., et al., "Modeling of High-Intensity Focused Ultrasound Induced Lesions in the Presence of Cavitation Bubbles," *J. Acoustical Soc. Am.,* Vol. 108, 2000, pp. 432–440.

[42]   Chen, W., et al., "Mechanisms of Lesion Formation in High Intensity Focused Ultrasound Therapy," *IEEE Ultrasonics Symp.,* Montreal, Canada, August 22–27, 2002.

[43]   Sokka, S., et al., "Gas-Bubble Enhanced Heating in Rabbit Thigh In Vivo," *IEEE Ultrasonics Symp.,* Montreal, Canada, August 22–27, 2002.

[44]   Curiel, L., et al., "Experimental Evaluation of Lesion Prediction Modelling in the Presence of Cavitation Bubbles: Intended for High-Intensity Focused Ultrasound Prostate Treatment," *Med. Biol. Eng. Comput.,* Vol. 42, 2004, pp. 44–54.

[45]   Mahoney, K., et al., "Comparison of Modelled and Observed In Vivo Temperature Elevations Induced by Focused Ultrasound: Implications for Treatment Planning," *Phys. Med. Biol.,* Vol. 46, No. 7, 2001, pp. 1785–1798.

[46]   Curiel, L., et al., "Firing Session Optimization for Dynamic Focusing HIFU Treatment," *IEEE Ultrasonics Symp.,* San Juan, Puerto Rico, October 22–25, 2000.

[47]   Chavrier, F., et al., "Modeling of High Intensity Focused Ultrasound Induced Lesions in the Presence of Cavitation Bubbles," *J. Acoust. Soc. Am.,* Vol. 108, No. 1, 2000, pp. 1–9.

[48]   Wang, Z., et al., "Study of a 'Biological Focal Region' of High-Intensity Focused Ultrasound," *Ultrasound Med. Biol.,* Vol. 29, No. 5, 2003, pp. 749–754.

[49]   Varian Medical Systems, http://www.varian.com/orad/prd120.html.

[50]   Seip, R., et al., "Automated HIFU Treatment Planning and Execution Based on 3-D Modeling of the Prostate, Urethra, and Rectal Wall," *International Symposium on Therapeutic Ultrasound,* Kyoto, Japan, September 18–20, 2004.

[51]   Bankman, I., (ed.), *Handbook of Medical Imaging: Processing and Analysis, Academic Press Series in Biomedical Engineering,* San Diego, CA: Academic Press, 2000.

[52]   Suri, J., "Two-Dimensional Fast Magnetic Resonance Brain Segmentation," *IEEE Eng. Med. Biol.,* July/August 2001, pp. 84–95.

[53]   Pennes, H., "Analysis of Tissue and Arterial Blood Temperatures in the Resting Forearm," *J. Appl. Physiol.,* Vol. 2, 1948, pp. 93–122.

[54]   Holt, R. G. and R. A. Roy, "Measurements of Bubble-Enhanced Heating from Focused, MHz-Frequency Ultrasound in a Tissue-Mimicking Material," *Ultrasound Med. Biol.,* Vol. 27, No. 10, 2001, pp. 1399–1412.

[55]   Holt, R. G., et al., "Therapeutic Ultrasound," *5th Int. Symp. on Therapeutic Ultrasound,* Boston, MA, October 22–23, 2005.

[56] Lichtenbelt, B., R. Crane, and S. Naqui, *Introduction to Volume Rendering*, Upper Saddle River, NJ: Prentice-Hall, 1998.

[57] Engel, K., M. Kraus, and T. Ertl, "High-Quality Pre-Integrated Volume Rendering Using Hardware-Accelerated Pixel Shading," *ACM SIGGRAPH/EUROGRAPHICS Workshop on Graphics Hardware*, Los Angeles, CA, 2001, pp. 9–16.

[58] Wu, L., V. R. Amin, and T. Ryken. "Slice Cutting and Partial Exposing Techniques for Three-Dimensional Texture-Based Volume Rendering," *Proc. SPIE Medical Imaging*, San Diego, CA, February 11, 2006.

CHAPTER 14

MR-Guided HIFU for Kidney and Liver Applications

Christakis Damianou

14.1 Introduction to HIFU

High-intensity focused ultrasound (HIFU) is a noninvasive procedure for heating tumors without affecting the normal tissue surrounding the tumor. Therefore, application of HIFU is being investigated as an alternative to standard surgical techniques. Although the idea of using HIFU was proposed in the middle of the twentieth century by Lynn et al. [1], its maximum potential for clinical use was established in the early 1990s due to the development of sophisticated systems [2–5]. HIFU has been proven for nearly 60 years to be an effective and efficient method for ablating soft tissue, but its commercial success is still under trial.

The use of HIFU has been explored in almost every tissue that is accessible by ultrasound. Examples of some applications explored include the eye [6], prostate [7, 8], liver [9], brain [10–12], and kidney [13–15]. Recently, the technology of HIFU systems has been improved because advancements have been made in the use of guided imaging with the ultrasonic therapy. Ultrasonic imaging is a simple and inexpensive method, but it offers poor contrast between soft tissues. In contrast, MR imaging offers superior contrast, but is more expensive. Several studies have been conducted in the area of ultrasonic imaging [16, 17] and in the area of MR imaging [18–20], thus enhancing the potential of HIFU.

The main goal of HIFU is to maintain a temperature between 50°C and 100°C for a few seconds (typically <10 seconds) in order to cause tissue necrosis. Typically, a focal peak intensity between 1,000 and 10,000 W/cm^2 is used with a pulse duration between 1 and 10 seconds and a frequency of 1 to 5 MHz.

This chapter focuses on the following concepts of HIFU:

1. Effect of thermal dose on attenuation.
2. Ablation of kidney without MRI guidance. This was done in order to gain experience in kidney ablation using HIFU.
3. Ablation of kidney with MRI guidance. This was done mainly to determine MRI parameters that improve the contrast between necrotic and normal tissue in kidney.

4. Same concept as in items 2 and 3, but applied to liver tissue.

5. Investigation of the effect of tissue interfaces in HIFU ablation under MRI guidance.

14.2   Ultrasonic System and Various Measurement Systems

14.2.1   Attenuation Measurement

The attenuation coefficient can be measured based on the transmission and reception method described by Kossoff et al. [21]. The system consists of a signal generator (HP 33120A, Hewlett Packard, now Agilent Technologies, Englewood, Colorado), a 12-bit, 50-MHz analog-to-digital (A/D) acquisition card (CS1250, GAGE, Lachine, Canada), and two identical flat/circular transducers made from piezoelectric ceramic PZT4 (Etalon, Lebanon, Indiana), 10 mm in diameter. The transducers operate at 4 MHz. Figure 14.1 shows a block diagram of the system illustrated with actual photos of the instruments. One transducer is connected to the signal generator and functions as the transmitter. The transmitter is attached in a small container, which is filled with degassed water. The tissue under measurement is placed inside the small container. In the other side of the container and opposite the transmitter, the other transducer (receiver) is placed. The output of the receiver is connected to the A/D acquisition card. Circulating thermally regulated water though a heating coil controlled the temperature of the tissue. A 50-μm-diaweter T-type copper-costantan thermocouple (Physitemp) was inserted in the tissue in order to measure tissue temperature. Based on the elapsed time and measured temperature, a thermal dose referenced at 43°C was estimated. The temperature was measured using an HP 7500 series B system and an HP 1326B multimeter.

One useful hint when measuring attenuation is that the lower the value, the more reliable the data. A high attenuation coefficient implies air inclusions or reflections in some interfaces, which should not be included in the ultrasonic path when measuring a specific type of tissue.

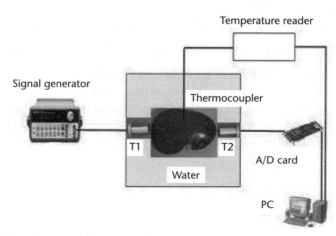

Figure 14.1   System for measuring attenuation.

14.2.2 Ultrasonic System

Figure 14.2 shows a block diagram of the HIFU system with photographs of the actual instruments (final version assembled in May 2006). The system consists of a signal generator (HP 33120A), an RF amplifier (LA 100-CE, Kalmus, Bothell, Washington), a 3-D MR compatible positioning system, and a spherically shaped bowl transducer made from piezoelectric ceramic PZT4. The transducer operates at 4 MHz, has a focal length of 10 cm, and a diameter of 4 cm. The transducer is rigidly mounted on the 3-D positioning system.

The 3-D MR compatible robotic system and the transducer were placed inside an MRI scanner (Signa 1.5 T, by General Electric). The coil used was one that is used commercially for spinal imaging. The signal generator, the RF amplifier, and the PC were placed outside the MRI scanner.

For the kidney ablation the system was utilized outside the MRI scanner. The system in Figure 14.2 also includes the system for monitoring cavitation (see Sections 14.7 and 14.8).

14.2.3 Acoustical Monitoring of Cavitation

For acoustical monitoring of cavitation, we used the acoustical method of Lele [22]. Figure 14.2 shows the arrangement for detecting cavitation acoustically using the actual photographs of the devices. This method includes a receiver, which is placed perpendicular to the beam of the HIFU transducer. Because the HIFU protocol is applied inside the magnet of an MRI scanner, the receiver must be MR compatible. The receiver diameter was 10 mm, its radius of curvature was 10 cm, and it operated with a bandwidth of 11 MHz. The receiver was mechanically coupled to the

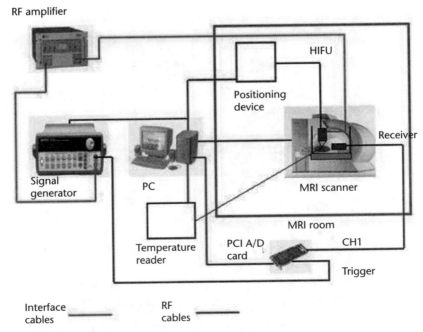

Figure 14.2 Block diagram of the HIFU system.

HIFU transducer. The signal from the receiver was fed to the CS1250 A/D PCI card. The A/D card was synchronized to receive the signal when the HIFU transducer was activated. The received signal was stored in a PC.

The HIFU exposure was controlled so that the temperature never exceeded 100°C (i.e., the temperature causing boiling). Therefore, any cavities that could be created were associated with the occurrence of cavitation and not with boiling, which causes tissue evaporation that results in creation of cavities.

14.2.4 Software

A user-friendly program written in MATLAB was developed to control the system. Different versions of software have been written since 1999. Figure 14.3 shows the May 2006 version of the main window of the software. The software serves six main tasks through various windows or menus: (1) displaying of MR images, (2) transducer movement (the user may move the robotic arm in a specific direction or customize the automatic movement of the robotic arm in any formation by specifying the pattern, the step, and the number of steps), (3) messaging (starting time, treatment time left, and so forth), (4) patient data (age, weight, and so forth), (5) display of motor position, and (6) display of the contents of an MR compatible camera.

14.2.5 Acoustical Field Measurement

The size of the focal region produced by this transducer was obtained by mapping the acoustic pressure field with a needle hydrophone (Specialty Engineering Associates, San Jose, California) having an active element of 1 mm in diameter. The transducer under test was driven by a pulser/receiver (Panametrics 5050R, Waltham, Massachusetts). The hydrophone was connected to the receiver input of the

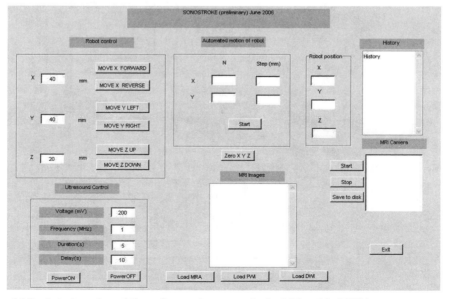

Figure 14.3 Latest version of the software that controls the MRI-guided HIFU system.

pulser/receiver. The output of the pulser/receiver was connected through the CS1250 A/D card to the PC for signal processing. The transducer was moved automatically by the 3-D robotic system. The block diagram of the system that measures the acoustical field of the ultrasonic transducers is shown in Figure 14.4 and is illustrated with actual photographs of the instruments. Raum and O'Brien [23] describe in more detail the principles used.

The full width at half-maximum intensity (D) of the beam was estimated from the measured acoustical field. The spatial average in situ intensity was estimated by the following equation:

$$I_{SAL} = 0.87 P/D^2 \tag{14.1}$$

where P is the acoustic power of the transducer [24]. The total power delivered by the transducer was measured with an ultrasound power meter (UPM-DT-100N, Ohmic Instruments, Easton, Maryland). The estimation of the spatial average intensity is strongly affected by the measurement of D. The maximum error for the intensity measurement was estimated to be 5%.

14.2.6 In Vitro Experiments

The tissue under ablation was placed in the water tank of the system shown in Figure 14.2. Boiling the water to 100°C degassed the water. The tissue was placed on top of an absorbing material (rubber pad) in order to shield adjacent tissue from stray radiation from the bottom of the plastic water tank. The transducer was placed on the 3-D robotic arm and immersed in the water tank, thus providing good acoustical coupling between tissue and transducer. Any bubbles that may have collected under the face of the transducer were removed in order to eliminate any reflections. The sample was gently massaged to remove any trapped air bubbles and

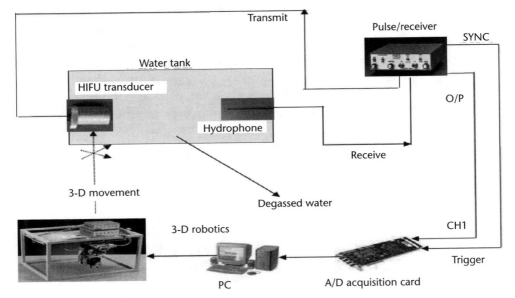

Figure 14.4 Acoustical field measurement system.

the experiment was initiated 10 to 20 minutes after placing the sample in the tank. Although it is difficult to assess whether this procedure was sufficient to remove very small air bubbles, in several samples where this procedure was not followed the presence of air bubbles was easily detected from the resulting lesion appearance.

14.2.7 In Vivo Experiments

The coupling arrangement during animal experiments (Figure 14.5) was slightly different from the method used in the in vitro experiments. An 8 cm × 8 cm × 10-cm rectangular container was made out of plastic. A special holder was designed to hold the container in the laboratory table. The top and the bottom sides of the container were left open so that a low attenuation bag filled with degassed water could be inserted. The transducer was placed inside the water-filled bag (on the top side). The lower side was placed on the animal that lay on a surgical table. The animal was shaved before the experiment. In between the shaved animal and the lower side of the bag, ultrasound gel was placed to ensure that no localized heating at that interface was produced due to reflections.

14.3   Simulation Model

14.3.1   Estimation of Temperature

The temperature versus time history was obtained by numerically solving the bioheat equation by Pennes [25]. The explicit form of this equation is given by:

$$\rho_t c_t \frac{dT}{dt} = k\nabla^2 T - w_b c_b (T - T_a) + q \qquad (14.2)$$

where ρ_t is the density of the tissue, c_t is the specific heat of the tissue, T is the temperature of the tissue, t is the time, k is the thermal conductivity of the tissue, w_b is the blood perfusion rate, c_b is the specific heat of the blood, T_a is the arterial blood temperature, and q is the ultrasonic absorbed power density in the tissue. The first term represents the temperature elevation, the second the conduction, and the third the convection effect due to blood. The blood perfusion is modeled as a uniform heat sink with blood supplied by vessels into the tissue volume at body temperature T_a and exiting at a tissue temperature T. The fourth term represents the absorbed power den-

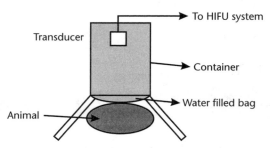

Figure 14.5   Experimental setup for the in vivo experiments.

sity due to the ultrasonic source. The bioheat equation was solved by using finite difference methods with spatial step of 0.1 mm and time step of 0.02 second.

14.3.2 Estimation of Thermal Dose

The effect of hyperthermia depends on the temperature and the duration of the heating. If a constant temperature could be maintained, then the duration of heating would be a reasonable way of expressing thermal dose, with units of time. In reality, however, a constant temperature is not maintained, so it is necessary to find a method of relating a treatment to an equivalent time at a specified reference temperature. A mathematical relation between time and temperature was described by Dewey et al. [26] and given by:

$$t_2 = t_1 R^{(T_2 - T_1)} \qquad (14.3)$$

where T_1, T_2 are temperatures at times t_1 and t_2, respectively, and R is a constant.

The calculation of the thermal dose for changing temperature exposure was done by using the technique suggested by Sapareto and Dewey [27]. The technique uses numerical integration to calculate the time that would give an equivalent thermal dose at a reference temperature under different temperature profiles. The reference temperature of 43°C was chosen because this is the standard temperature used as a reference [27]. For any temperature profile the dose can be found by

$$t_{43} = \sum_0^{t_{final}} R^{(43 - T_t)} \Delta t \qquad (14.4)$$

where t_{43} is the equivalent time at 43°C , T_t is the average temperature during Δt. The default value of R equal to 0.25 was chosen for temperatures smaller than 43°C and a value equal to 0.5 for temperatures higher than 43°C [27]. The temperature after the power was turned off was also considered because during the decay part of the thermal dose is contributed.

14.3.3 Estimation of Lesion Size

The kidney attenuation used in the simulation model depended on the amount of the accumulated thermal dose. It is known that the thermal dose threshold of necrosis referenced at 43°C for kidney is 50 minutes [28]. For a thermal dose below 50 minutes at 43°C, the attenuation at body temperature was used. For a thermal dose higher than 50 minutes at 43°C, the attenuation of necrotic tissue was used (measured using the attenuation method described in this chapter).

14.3.4 Effect of Attenuation

Attenuation α affects the power deposition density Q, which is given by the following equation:

$$Q = 2\alpha I e^{-2\alpha x} \qquad (14.5)$$

where I is the intensity and x is the depth in tissue. For short pulses, the power deposition density is the main factor elevating the tissue temperature.

14.4   Attenuation Measurement as a Function of Thermal Dose

One of the most important issues for HIFU surgery is the understanding and control of the thermal exposure in tissue. When tissue temperature is maintained between 50°C and 100°C for a few seconds (less than 5 seconds), tissue necrosis occurs, and thermal lesions are created. We now have a very good understanding of how thermal lesions vary with the amount of applied acoustical power and the pulse duration for some tissues (brain and liver). However, we do not yet have enough information about the effect of heating on absorption (tissue variable), which also affects the size, shape, and placement of the thermal lesions. Absorption is an acoustical property of tissue with wide variation from tissue to tissue, and it represents the rate at which energy in tissue is converted to heat. For short pulses (less than 5 seconds) it is one of the primary factors contributing to the temperature elevation in tissue because the effect of blood flow is minimized [29]. Attenuation includes absorption, scattering, and reflection, but when minimizing scattering and reflection, attenuation will reflect losses due to absorption for the most part. In this section, the results of attenuation are presented, which are very close to the results of absorption [30].

A number of studies have been conducted on the effect of temperature on tissue attenuation or absorption. Dunn [31] studied the effect of absorption with temperature (up to 30°C) in the spinal cord of mice, where an increase of absorption with temperature was reported. Dunn and Brady [32] studied the absorption as a function of temperature in mammalian central nervous tissue at different frequencies, and it was shown that the absorption decreased for frequencies lower than 1 MHz and increased for frequencies higher than 1 MHz. Gammel et al. [33] studied the effect of attenuation with temperature for excised porcine liver, kidney, backfat, and spleen and for human liver. The frequency was varied from 1.5 to 10 MHz and three different temperatures were studied (4°, 20°, 37°C). It was found that basically, above 5 MHz, the attenuation decreased with temperature for all tissues including the human liver.

Pressure absorption was studied by Fry et al. [34] on rat liver, where for the temperatures studied (30°C, 37°C, 41°C), there were no statistically significant differences in absorption in this temperature range. However, for tissue ablation, attenuation at higher temperatures (50°C to 100°C) is required. Robinson and Lele [35] studied the changes in attenuation (from 30°C to 90°C) in cat brain and showed that the attenuation stays essentially constant up to 50°C, and then increases rapidly. Bamber and Hill [36] studied the effect of temperature on attenuation in the range of 5°C to 65°C on excised liver tissues. Many trends were observed, for example the attenuation of bovine liver decreased and then increased until it reached 65°C, where it stayed constant. In human liver the attenuation dropped with temperature, and for frequencies lower than 2 MHz, the change was insignificant.

Basically in the past years, some studies examined attenuation or absorption as a function of temperature (up to 37°C) and some at higher temperatures. However, no study has examined both attenuation and absorption (at least for one tissue) as a function of temperature (with maximum temperature reached close to temperatures seen in HIFU surgery) or as a function of thermal dose. The paper by Damianou et al. [30] presents a complete study of both attenuation and absorption as a function of thermal dose. This section reports on the effect of a thermal dose on the attenuation of muscle, kidney, and fat.

14.4.1   Results of Attenuation Versus Thermal Dose

Figure 14.6 shows the attenuation of porcine kidney (cortex and medulla), muscle, and fat as a function of thermal dose. Note that the attenuation increases rapidly during necrosis and eventually stabilizes after total necrosis has been produced.

The trend of attenuation with dose shows a trend similar to that seen in the results published by Damianou et al. [30]. The normal kidney tissue included both medulla and renal cortex, for which we found no significant differences in terms of ultrasonic attenuation. The research on muscle and fat is justified, because these two tissues usually surround the kidney. The trend of attenuation with dose in all tissues is similar. There is a rapid increase of attenuation with thermal dose and, when necrosis occurred, attenuation stabilizes. At that point, the attenuation is roughly double the attenuation at room temperature.

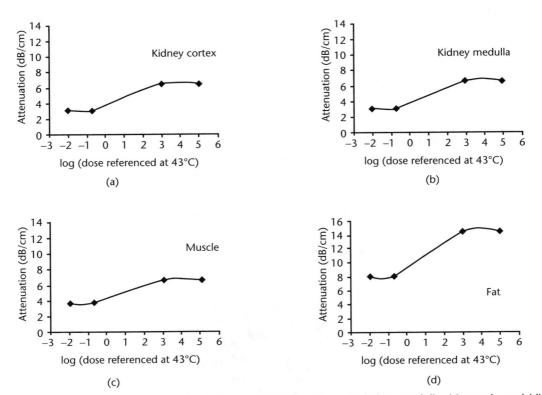

Figure 14.6   Attenuation versus thermal dose for (a) kidney cortex, (b) kidney medulla, (c) muscle, and (d) fat.

14.4.2 Simulation Results Using Variable Attenuation

Table 14.1 shows the simulated length and width using the pure thermal model and using the model that accounts for the variable attenuation. The transducer parameters used are frequency = 4 MHz, radius of curvature = 10 cm, and transducer diameter = 4 cm. The focal in situ spatial average intensity of 2,000 W/cm^2 was applied for 5 seconds at a focal depth of 15 mm. The zero perfusion represents the case of in vitro kidney, whereas the in vivo case is modeled by using the perfusion of 70 kg/m^3-s. Having established a rough estimate of the lesion size using simulation, experiments were carried out in kidney in vitro and then in vivo.

14.5 Kidney Ablation

For applications such as the treatment of benign prostate hyperplasia (BPH), it is sufficient to destroy as much tissue as possible for the purpose of tissue debulking. For applications in oncology, destruction of all viable tumor cells is required and, therefore, protocols in this area must be very accurate and reliable. In this work, methods suggested by Malcolm and ter Haar [24] were used to produce complete, reliable, and consistent ablation of renal tissues (creation of a contiguous array of touching lesions of thermal origin, avoiding boiling and cavitation, and use of cooling to avoid merging of lesions in front of the focal point).

14.5.1 Results

Figure 14.7 shows the length of a lesion created in excised porcine kidney using the in situ spatial average intensity of 2,000 W/cm^2 for 5 seconds. The lesion produced is

Table 14.1 Simulated Length and Width for the Pure Thermal Model and the Model of Variable Attenuation

Quantity	Pure Thermal Model		Variable Attenuation Model	
	Perfusion (kg/m^3-s)		Perfusion (kg/m^3-s)	
	0 (in vitro)	70 (in vivo)	0 (in vitro)	70 (in vivo)
Length (mm)	17.2	16.7	20.6	19.7
Width (mm)	2.6	2.4	3.3	3.1

Note: The transducer parameters used are frequency = 4 MHz, radius of curvature = 10 cm, and transducer diameter = 4 cm. The focal intensity of 2,000 W/cm^2 was applied for 5 seconds at a focal depth of 15 mm.

Figure 14.7 Thermal lesion placed in the cortex demonstrating the excellent repeatability.

placed entirely in the cortex of the kidney. The length is measured along the transducer central axis, whereas the width is measured perpendicularly to the transducer central axis. The lesion length is about 20 mm and the width is about 3 mm. In a certain row of lesions, only one dimension of the lesion (length or width) was measured, because the sample was cut either along the length or the width.

Lesions were created all the way to the medulla, provided that there were no air spaces in the medulla. Air spaces are created because of the absence of blood that normally flows in this region. However, ablation of the cortex tissue is of primary concern, since the renal carcinoma (most important renal cancer) grows inside the cortex and then extends to the peritoneal fat.

Figure 14.8 shows the lesion width with 2,000 W/cm^2 applied for 5 seconds. The lesion measured was between 2.8 and 3.4 mm. The lesions appeared to be repeatable for a given acoustic exposure. Some differences that are observed are attributed to the variation of focal depth arising from the kidney curvature.

Creation of thermal lesions in the cortex is very consistent and repeatable. For the given transducer ($f = 4$ MHz, $R = 10$ cm, and $d = 4$ cm) and the given exposure (in situ spatial average intensity = 2,000 W/cm^2, pulse duration 5 seconds, and 15-mm focal depth), the lesion length is around 20 mm and the lesion width is about 3 mm.

In most of the in vitro experiments, the propagation inside the medulla was difficult due to the air spaces in the medulla. Normally in an in vivo kidney, there is blood inside the medulla, making the medulla a possible target for ultrasound. Figure 14.9 shows lesions that propagated inside the medulla region. Thus, in the case for which there are no bubbles inside the medulla, ablation is feasible.

Figure 14.8   Lesions created in pig kidney in vitro.

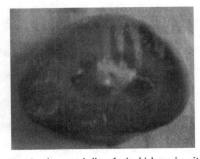

Figure 14.9   Lesions propagating in the medulla of pig kidney in vitro.

Figure 14.10 shows a lesion created by moving the transducer in a patterned movement (square grid of 8×8 with a 3-mm step) in both directions. The intended target was the round shape overlapped to the image. The power used was 2,000 W/cm^2 for 5 seconds. A delay of 10 seconds was used between the pulses in order to eliminate the near-field heating [37]. Based on the width obtained in Figure 14.8, a 3-mm step will produce overlapping lesions. Necrosis with a 3- to 5-mm margin around the target is desired. For the in vitro case, a larger step can be used because ultrasound penetration is excellent and there is no blood flow. Based on the simulation study for the in vivo case, a 3-mm step must be used. Figure 14.11 shows the corresponding depth of the large lesion. The lesion was extended up to the medulla.

The threshold of cavitation was found by detecting the in situ spatial average intensity, which produces a signal in the receiver indicative of cavitation. Thus, single lesions were created for this purpose, by varying the in situ spatial average intensity from 2,000 to 3,500 W/cm^2. For the given transducer and for a pulse duration of 5 seconds, it was observed that cavitation took place for in situ spatial average intensity levels greater than 3,200 W/cm^2 (error was ± 100 W/cm^2). Figure 14.12 shows the lesion length obtained using a 5-second pulse and 4,000 W/cm^2 in situ spatial average intensity in kidney tissue in vitro. Note that, at this acoustic level, a cavitational effect occurred, which was verified by the cavitational receiver. Tissue

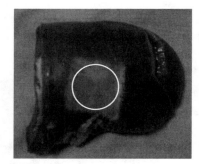

Figure 14.10   Large lesion in pig kidney in vitro. The transducer was moved in a grid pattern ($8 \times$ 8) (top view).

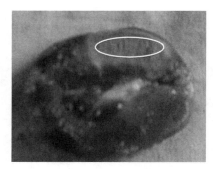

Figure 14.11   Large lesion in pig kidney in vitro. The transducer was moved in a grid pattern ($8 \times$ 8) (side view).

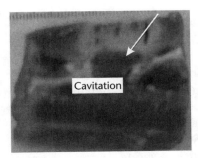

Figure 14.12 Large lesion in pig kidney in vivo showing cavitational activity. The transducer was moved in a grid pattern.

examination revealed gross tissue disruption, verifying the occurrence of an acoustic cavitation event.

With a 4-MHz transducer, 5-second pulses, and in situ spatial average intensity levels below about 3,200 W/cm^2, the lesions created are based solely on thermal effects. Data from Hynynen [38] suggest that the threshold spatial peak intensity at 4 MHz is about 6,000 W/cm^2 (extrapolated because data up to only 2 MHz is provided) or 3,333 W/cm^2 (in situ spatial average intensity). Thus, with the in situ spatial average intensity of 4,000 W/cm^2 used in some experiments, cavitation effects were not avoided.

Figure 14.13 shows the length of an 8×8 matrix of lesions created using 2,000 W/cm^2 for 5 seconds in pig kidney in vivo. The transducer was moved with a step of 3 mm. The lesions were usually firm and white and usually no adjacent hemorrhagic zone was created. Despite the thick fat layer, necrosis of the renal cortex was reliable. The lesion length in the in vivo case appeared to be smaller (about 18 mm compared with 20 mm in the in vitro case). Single lesions produced in the in vivo kidney revealed a lesion width of about 3 mm for 2,000 W/cm^2 at 5 seconds. Thus, the step of 3 mm produced overlapped lesions covering a large target volume.

Lesions were smaller in the in vivo experiments compared with lesions created in dead tissue for the same exposure. The reduced size of the in vivo exposures is attributed to the blood flow and possible reduction of the energy due to attenuation at various interfaces including the skin.

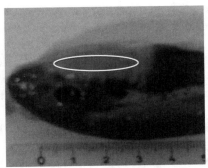

Figure 14.13 Large lesion in pig kidney in vivo (through fat layer) by moving the transducer in a grid pattern.

14.6   Kidney Ablation Under MRI Guidance

The use of MRI to detect changes due to heat was motivated by several studies, which showed that T_1 and T_2 relaxation constants were sensitive to temperature. One example is the study by Le Bihan et al. [39], which showed that the relaxation constants were sensitive to temperature. Other representative examples are the studies by Dickinson et al. [40] and by Matsumoto et al. [41] who demonstrated the temperature sensitivity of T_1 and diffusion.

The use of MRI to guide thermal therapy was fully investigated by groups involved in laser therapy. The most representative examples are the studies by Bleier et al. [42], Matsumoto et al. [41], Mumtaz et al. [43], Kettenbach et al. [44], and Law et al. [45].

The idea of using MRI to guide HIFU was first cited by Fry and Johnson [46]. However, the idea was never explored in the research setting at that time. The combination of ultrasound and MRI was first cited by Jolesz and Jakab [47] who demonstrated that an ultrasonic transducer can be used inside an MRI scanner. The concept of using MRI to monitor the necrosis produced by HIFU was demonstrated in the early 1990s by Hynynen et al. [48] in canine muscle. In the following years similar studies were conducted by Hynynen et al. [19, 20], Cline et al. [49], and Darkazanli et al. [50]. In these studies it was shown that the contrast between necrotic tissue and normal tissue was excellent. This was a great enhancement for HIFU systems because it allowed the therapeutic protocols to be accurately monitored. Therefore, the interest in using MRI as a diagnostic modality to guide HIFU has increased.

In the preceding studies the T_1- and T_2-weighted fast spin echo (FSE) pulse sequences were evaluated extensively. For fast imaging, various gradient pulse sequences were used. It was found that the dimensions of the lesions measured using MRI agreed well with the dimensions measured during the gross examination. Therefore, research efforts were directed toward investigating whether HIFU and MRI can be an effective pair for therapy/diagnosis of tissues accessible by ultrasound.

During the evolution of the HIFU therapy with MRI guidance technology, it was realized that in order to apply this system in the clinic, temperature maps needed to be created. Several studies have been published in this direction [51–56].

Hynynen et al. [15] was the first to guide HIFU surgery in the kidney using MRI. In their study pulse sequences were proposed that are sensitive enough to show temperature elevation during sonication, thereby indicating the location of the beam focus. Fast pulse sequences such as gradient recalled acquisition in the steady state (GRASS) were used with imaging times of a few seconds. The temperature elevation could be clearly located in the gradient echo images.

The effectiveness of MRI in the brain was studied by Hynynen et al. [57], Morocz et al. [58], Chen et al. [59], and Vykhodtseva et al. [60]. In liver the concept has been explored by Rowland et al. [61]. The case of fat was monitored by Hynynen et al. [62]. In the prostate it was explored by Nakamura et al. [63], Smith et al. [64], and Sokka and Hynynen [65]. Therefore, the ability of MRI to guide HIFU has been investigated for almost all tissues accessible by ultrasound.

This technology was also investigated for clinical applications. For example, in the breast it was studied by Pomeroy et al. [5] and Hynynen et al. [66]. HIFU/MRI technology has also proven very useful in measuring some basic quantities of ultrasound. Acoustical streaming was monitored by Starritt et al. [67], ultrasonic fields by Walker et al. [68], and ultrasonic pressure by Sharf et al. [69].

Therefore, the conclusion is that, in just a decade of research, MRI has become an integral part of every aspect of therapeutic ultrasound. Generally, the advantages of using MR for guiding and monitoring HIFU have been demonstrated. These advantages include the following:

1. MR provides good tissue necrosis contrast.
2. The temperature elevation in the focal zone during sonication can be detected with temperature-sensitive techniques.

The MRI images provide feedback data that can be used to monitor dynamically the shape of the HIFU lesion, and therefore HIFU technology can be applied in the clinical setting with confidence.

In this section, the goal was to investigate the effectiveness of MRI to monitor the therapeutic protocols of HIFU use in the kidney. In Section 14.5 the target was to create consistent and reliable necrosis in the cortex of the kidney. In that section, the protocol was applied without any diagnostic guidance. In this section, MR imaging can be used to monitor the therapeutic protocols.

Several MRI pulse sequences were investigated. For high-quality imaging, which can be used at the end of a therapeutic protocol or at some instances of the protocol, FSE techniques (T_1- and T_2-weighted and proton density) were investigated. For fast imaging, the T_1-weighted fast spoiled gradient (FSPGR) pulse sequence was used

14.6.1 HIFU Parameters

In the study in this section moving the transducer in a grid pattern created large lesions. The spacing between successive transducer movements was chosen accordingly so as to create full necrosis coverage of a specific target. The in situ spatial average intensity used was 2,000 W/cm^2 for 5 seconds, which ensures lesioning using thermal mechanisms. This intensity is below the threshold of cavitation in kidney for this particular frequency and pulse duration [70]. The delay between successive ultrasound firings was 10 seconds. Although the study by MacDannold et al. [71] recommends a delay of around 50 to 60 seconds in order to avoid buildup in front of the target, for this application use of a transducer geometry that used a 10-second delay eliminated most of the built-up heating.

14.6.2 MRI Parameters

The various MRI parameters used for the various pulse sequences are listed in Table 14.2.

Table 14.2 Parameters Used for the Various MRI Pulse Sequences

Series	Name	TR (ms)	TE (ms)	Slice Thickness (mm)	Matrix	FOV (cm)	NEX	BW (kHz)	ETL	Flip Angle
1	T_1-weighted FSE	220, 300, 400, 500, 700	9.2	3, 5 (gap 0.3 mm)	256×256	16	1	31.25	8	—
2	T_2-weighted FSE	2500	16,32, 48,64	5 (gap 0.3 mm)	256×256	16	1, 2, 3,4	31.25	8	—
3	T_1-weighted FSPGR	50,100, 150,300	3.7	2 (gap 0.3 mm)	256×256	16	1	62.50	—	75°

Note: TR = time of repetition, TE = time of echo, FOV = field of view, NEX = number of excitations, BW = bandwidth, ETL = echo train length.

14.6.3 Results

T_1-weighted FSE was explored by using different TR (220, 300, 400, 500, and 700 ms). Figures 14.14(a–e) show MR images using T_1-weighted FSE at different TR. Moving the transducer using a 4×4 grid pattern, with a step of 3 mm, created this large lesion volume. This image shows the lesion in a plane perpendicular to the transducer beam. Figure 14.14(f) shows a photograph of the kidney after slicing. All of the parameters used are shown in Table 14.2, row 1.

The contrast between the lesion and normal kidney tissue is excellent with T_1-weighted FSE. The signal intensity of the lesion is homogeneous using this

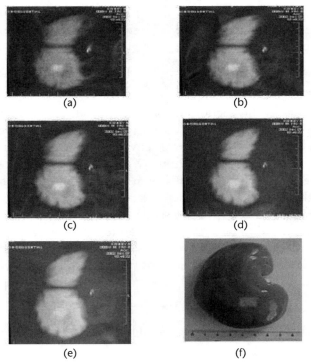

(a)

(b)

(c)

(d)

(e)

(f)

Figure 14.14 Coronal MR images of HIFU ablation using T_1-weighted FSE pulse sequence. (a) TR = 220 ms. (b) TR = 300 ms. (c) TR = 400 ms. (d) TR = 500 ms. (e) TR = 700 ms. (f) Photograph after slicing.

method. The best contrast is observed for TR between 200 and 400 ms. This was proved by plotting the signal intensity versus TR. The difference in the signal of the lesion and kidney cortex is significant in this range. Any air spaces created due to cavitational activity are not visible, because air spaces and lesions have almost the same signal. MRI could detect the lesion immediately after the ablation. In [72], the lesions in the brain appeared in MRI in 15 to 60 minutes using T_2-weighted FSE.

Next, T_2-weighted FSE was explored by using four different TEs. Figures 14.15(a–d) show MR images using T_2-weighted FSE with TE of 16, 32, 48, and 64 ms. These images shows the lesion in a plane parallel to the beam. This large lesion volume was created by moving the transducer using a 4×4 grid pattern with step of 3 mm. Note that the lesion was created entirely in the cortex, leaving the medulla unaffected. Figure 14.15(e) shows the photograph of the kidney after slicing. All the parameters used are shown in Table 14.2, row 2.

A similar evaluation was done for the case of a T_2-weighted FSE pulse sequence. The best contrast can be achieved for TE between 16 and 32 ms. The T_2-weighted FSE was proven to be the best pulse sequence that can detect cavitational activity. This advantage is attributed to the significant difference in signal intensity between the water-filled air spaces and necrotic tissue. Water-filled air spaces appear brighter than thermal lesions. Therefore, for therapeutic protocols created using the cavitational mode, T_2-weighted FSE may be the optimum pulse sequence to be used. For a therapeutic protocol using only thermal mechanisms, T_1-weighted FSE may be suitable, although T_2-weighted FSE can be used as well.

Figure 14.16(a) shows the signal intensity plotted against TR for a single kidney. Figure 14.16(b) shows the graph of signal intensity plotted against TE with the data derived from the images of Figure 14.15. The region of interest (ROI) was cir-

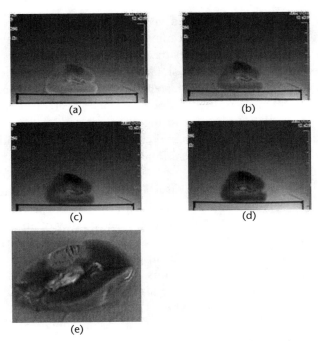

(a) (b) (c) (d) (e)

Figure 14.15 Axial MR images of HIFU ablation using T_2-weighted FSE pulse sequence. (a) TE = 16 ms. (b) TR = 32 ms. (c) TE = 48 ms. (d) TE = 64 ms. (e) Photograph after slicing.

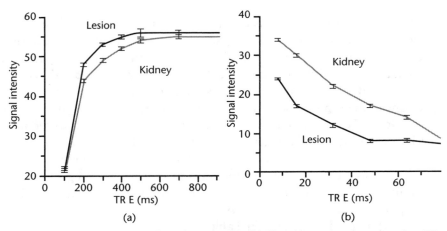

Figure 14.16 (a) Signal intensity plotted against TR. (b) Signal intensity plotted against TE.

cular with diameter of nearly 2 mm. Regarding the T_1-weighted FSE pulse sequence, the best contrast is obtained with TR between 200 and 400 ms. Regarding the T_2-weighted FSE pulse sequence, the best contrast is obtained with TE between 16 and 32 ms.

The structure and water content of necrotic tissue is different compared to kidney tissue and, as a result, the spins diphase more rapidly, resulting in a reduction of T_2 and consequently a reduction of the signal intensity of T_2-weighted images. Thus, necrotic tissue appears darker than kidney tissue. The opposite happens with the other relaxation constant (T_1). The value of T_1 probably increases, and therefore lesions appear brighter than kidney tissue. The increase of T_1 was reported by Dickinson et al. [40].

The FSE methods provided very good contrast and good SNR, however the acquisition time is long (2 to 4 minutes). Therefore, these techniques are useful for monitoring at various instances the development of necrosis during the application of a specific therapeutic protocol. However, to monitor the focal position (initiation of the therapeutic protocol) a faster imaging technique is needed. To find the focal point, a low-intensity pulse that does not causes necrosis is applied for a short period (5 to 10 seconds), and then during that interval an image in the focal plane is acquired. By monitoring the signal changes, it is possible to identify the location of the focal point [15]. Fast techniques are also needed to monitor the growth of necrosis dynamically.

The T_1-weighted FSPGR pulse sequence was investigated for fast imaging. All parameters used are shown in Table 14.2, row 3. Figure 14.17 shows the effect of TR (50, 100, 150, and 300 ms) on the image quality of ablated kidney. Figure 14.18 shows various MR images (the plane is parallel to the beam) of ablation in pig kidney using different ablation exposures. The lesion in Figure 14.18(a) was created based on pure thermal mechanisms. T_2-weighted FSE was chosen to monitor the various HIFU exposures, because air spaces in the necrotic volume are visible, as oppose to T_1-weighted FSE, where the contrast between air and necrotic tissue is poor. The lesion in Figure 14.18(b) was presumably created with stable cavitation.

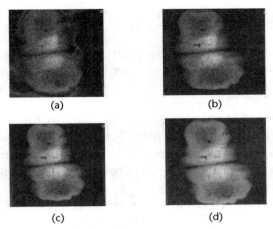

(a)   (b)

(c)   (d)

Figure 14.17   MR images using T_1-weighted FSPGR. (a) TR = 50 ms. (b) TR = 100 ms. (c) TR = 150 ms. (d) TR = 300 ms.

This is indicated by the cavities observed in the image. The possibility that these cavities were the result of air bubbles, trapped in the kidney cortex, is excluded, because the sample was degassed prior to the experiment. We speculate that this is the result of stable cavitation, because there was no unexpected tissue destruction (which can normally result because of transient cavitation). Figure 14.18(c) shows a large lesion volume, which includes one lesion affected by transient cavitation (in situ spatial average intensity was 4,000 W/cm^2 applied for 5 seconds). That lesion includes violent tissue disruption, which is visible in the MR image and in the gross examination. A huge cavity (signal enhancement) in that spot provides additional confirmation of violent cavitation. All the parameters used are shown in Table 14.2, row 2.

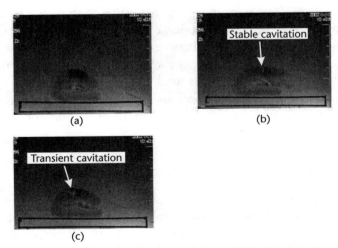

(a)   (b)

(c)

Figure 14.18   MR images of HIFU ablation using T_2-weighted FSE demonstrating the detection of various tissue destruction modes. (a) Thermal. (b) Stable cavitation. (c) Transient cavitation.

14.7   Liver Ablation

Liver metastases and primary liver cancer are one of the main causes of death [73]. The traditional surgical methods may be curative for some cases, but they are not effective for all cases [74]. Therefore, research efforts in the area of HIFU ablation of liver guided by MRI are essential.

A lot of clinical work has been done for the case of liver using laser ablation [75, 76] and for RF ablation [77]. It has been proven that thermal therapies can be utilized for alternative treatments of liver cancer. Therefore, HIFU, which is the least invasive of all thermal therapies, can be a promising technology for thermal ablation of liver cancer.

A lot of work has been done in many directions in the area of liver ablation using HIFU. The threshold of intensity that is needed to cause irreversible damage in liver was suggested by Frizell et al. [78, 79]. This information is very useful, because the intensity needed to create lesions was defined. The thermal effects of HIFU in liver have been well documented [9, 80]. Chen et al. [81, 82] analyzed extensively the effect of HIFU ablation in liver and cancerous liver using histology. The effective delivery of HIFU protocols in real oncological applications of liver was achieved by implanting tumor cells in liver [83–87]. The first attempt to monitor the effect of HIFU using MRI was reported by Rowland et al. [61], who demonstrated that monitoring of thermal lesions in liver is feasible.

In this section the goal was to investigate the effectiveness of MRI to monitor therapeutic protocols of HIFU in liver in vitro. The goal was to create large lesions that included both thermal and cavitational lesions. Therefore, MRI pulse sequences were investigated in order to discriminate between liver and lesion and between thermal and cavitational lesions.

14.7.1   HIFU Parameters

The in situ spatial average intensity used for creating lesions was between 2,000 and 3,500 W/cm^2 for 5 seconds—an exposure that may create either thermal or cavitational lesions. To create large lesions, a grid pattern of 4 × 4 overlapping lesions was used. The spacing between successive transducer movements was 3 mm, which creates overlapping lesions for the intensity and pulse duration used [88]. The delay between successive ultrasound firings was 10 seconds.

If in a grid of, say, 4 × 4 lesions of a specific exposure, one lesion was created under the mechanism of cavitation, then the probability of cavitation (POC) is 1/16 or 6.25%.

14.7.2   MRI Analysis

The MRI parameters used for the various pulse sequences are listed in Table 14.3. The signal intensity was measured by placing a circular ROI of approximately 10 mm in diameter. The contrast-to-noise ratio (CNR) was obtained by dividing the signal intensity difference between the ROI in the lesion and in the ROI of normal liver tissue by the standard deviation of the noise in the ROI of normal liver tissue.

Table 14.3   Parameters Used for the Various MRI Pulse Sequences

Name	TR (ms)	TE (ms)	Slice Thickness (mm)	Matrix	FOV (cm)	NEX	BW (kHz)	ETL	Flip Angle
T_1-weighted FSE	100, 200, 300, 400, 500, 700, 1,000	9.1	3 (gap 0.3 mm)	256 × 256	16	1	31.25	8	—
T_2-weighted FSE	2,500	10, 15, 20, 30, 40, 45, 60	3 (gap 0.3 mm)	256 × 256	16	1	31.25	8	—
FSPGR T_1-weighted	50, 100, 150, 300	2.8	3 (gap 0.3 mm)	256 × 256	16	1	62.50	—	50°

14.7.3   Results

Figure 14.19 shows the POC as a function of in situ spatial average intensity for a 5-second pulse duration in liver. Note that by using an intensity between 2,000 and 3,500 W/cm^2, it is possible to create both thermal and cavitational lesions. The preferred location for cavitation lesions is at the interfaces between liver and the water-filled blood vessels [78, 89]. The threshold of 100% cavitation in liver, which is about 3,500 W/cm^2, is similar to that found for kidney.

This observation is in agreement with Frizell et al. [78] who stated that whatever constitutes the nuclei for the cavitation events may be common to most soft tissues. The POC increases with intensity, and above a certain threshold the POC is 1. Significant POC (above 20%) is caused with an intensity above 2,500 W/cm^2 at 5 seconds. For the case of porcine liver in vitro this threshold of 100% cavitation is about 3,500 W/cm^2 at 5 seconds. The trend of POC as a function of spatial in situ intensity is similar to that found in the study of kidney ablation [88]. Also the length

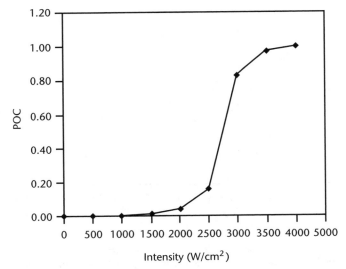

Figure 14.19   POC as a function of in situ spatial average intensity for a 5-second pulse duration in liver.

of a single lesion in liver at an intensity of 2,000 W/cm^2 for 5 seconds, which is around 20 mm, is very close to the lesion measured in kidney [88]. This is not surprising because liver and kidney have similar absorption [29].

Initially, the two traditional FSE techniques (T_1-weighted, and T_2-weighted) were evaluated. The T_1-weighted FSE technique was explored by using different TR (100, 200, 300, 400, 500, 700, and 1,000 ms). Figure 14.20(a) shows the CNR between lesion and liver plotted against TR. All parameters used are shown in Table 14.3, row 1. The T_2-weighted FSE technique was explored by using seven different TE (10, 15, 20 30, 40, 45, and 60 ms). Figure 14.20(b) shows the CNR between lesion and liver plotted against TE. All parameters used are shown in Table 14.3, row 2. Figure 14.20(c) shows the CNR between lesion and liver plotted against TR using T_1-weighted FSPGR. All parameters used are shown in Table 14.3, row 3.

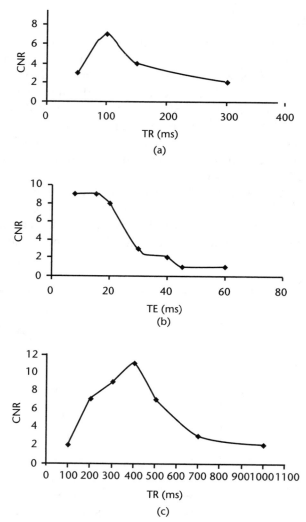

Figure 14.20   (a) CNR between lesion and liver plotted against TR using T_1-weighted FSE. (b) CNR between lesion and liver plotted against TE using T_2-weighted FSE. (c) CNR between lesion and liver plotted against TR using T_1-weighted FSPGR.

The contrast between lesion and normal liver tissue is acceptable when T_1-weighted FSE is used. The best contrast is observed for TR between 200 and 400 ms. This was proved by evaluating the CNR between lesion and liver as a function of TR. The difference in the signal of the lesion and liver is significant in this range. The window that maximizes contrast in liver is much smaller than that for kidney [90]. MRI could detect the lesion 1 to 2 minutes after the ablation.

A similar evaluation was conducted for the T_2-weighted FSE pulse sequence. The best contrast can be achieved for TE up to 20 ms, with better contrast at lower TE, indicating that pulse duration might be the best pulse sequence for the case of liver in vitro. Similar to what was seen for T_1-weighted FSE, the TE window that maximizes contrast is narrow in liver compared to kidney [90]. The T_2-weighted FSE sequence was proven to be the best pulse sequence for detecting cavitational activity. This advantage is attributed to the significant difference in signal intensity between the water-filled air spaces and necrotic tissue. Water-filled air spaces appear brighter than thermal lesions. Therefore, for therapeutic protocols created using the cavitational mode, T_2-weighted FSE maybe the optimum pulse sequence to use.

Mack et al. [91] and Morrison et al. [92] have shown that T_1 and T_2-weighted FSE and gradient echo sequences can effectively monitor lesions created by laser-induced thermotherapy in patients with liver cancer. Therefore, we anticipate that the pulse sequences evaluated in this study will also be successful for treatments in vivo. In the in vivo case, the CNR between the thermal lesion and normal liver tissue will be improved by using contrast-enhanced techniques [93]. Because of the motion artifacts, other techniques should be evaluated, for example, the fast low-angle shot (FLASH) technique. Tsuda et al. [94] used T_1- and T_2-weighted FSE to monitor lesions created in rabbit liver in vivo using RF ablation. They found that the contrast was optimized at TR of 500 ms, which is close to what was found in our study.

Figure 14.21(a) shows the MR image (in a plane perpendicular to the transducer beam) using T_2-weighted FSE with TE = 15 ms. This large lesion was created using a spatial average in situ intensity of 2,500 W/cm^2 for 5 seconds. By using this intensity, 4 out of the 16 lesions (POC = 0.25) were created under the mechanism of cavitation. Figure 14.21(b) shows a cavitation map, which shows a specific grid of single lesions (volume lesion or large lesion) having four lesions, which are created under the mechanism of cavitation. A lesion was classified as cavitational if during the firing of the ultrasonic transducer, the frequency spectrum included subharmonic emissions. Figure 14.21(c) shows a typical frequency spectrum during the creation of a cavitational lesion. The cavitational map was also verified using MRI imaging [Figure 14.21(a)] and using gross examination. With T_2-weighted FSE, the contrast of the cavitational lesion within the large thermal lesion is excellent. The cavitational lesion appears as a scattered bright spot.

The cavitational map was also verified using MRI imaging and using gross examination. This tool can be very useful for evaluating cavitation at different exposures (frequency, intensity, and pulse duration) and at different locations (close to blood vessels, interfaces, tissue type). At the moment we cannot think of any significant use of the cavitational map in the clinical setting, but it may be a very important research tool especially for cavitation studies.

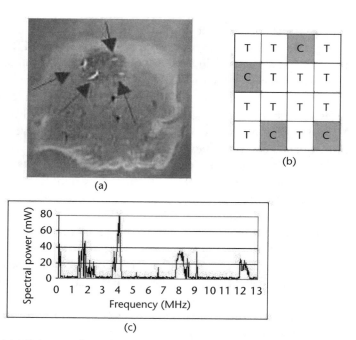

Figure 14.21   (a) MR images (in a plane perpendicular to the transducer beam) of a large lesion using T_2-weighted FSE showing cavitational lesions. (b) Cavitational map of the large lesion. (c) Typical frequency spectrum acquired during the occurrence of a cavitational lesion.

Figure 14.22(a) shows the lesion in a plane parallel to the transducer beam using T_2-weighted FSE. This large lesion was created using a spatial in situ intensity of 2,500 W/cm^2 for 5 seconds. With this pulse sequence, it is possible to identify the lesion created under the mechanism of cavitation. Figure 14.22(b) shows a photograph after slicing. In this photograph, the void indicates a site influenced by the mechanism of cavitation (already confirmed with harmonic emission and MRI).

14.8   Effect of Interfaces in HIFU Ablation

MRI monitoring enhances the effective application of HIFU, but issues remain to be resolved regarding the treatment strategy. This chapter focuses on the effect of interfaces on the effectiveness of HIFU ablation for the case of in vivo kidney. This issue,

Figure 14.22   (a) MR image (in a plane parallel to the transducer beam) of large lesion using T_2-weighted FSE with TE = 15 ms. (b) Photograph after slicing.

which is very critical, was never explored experimentally. Theoretically, it was explored by Fan and Hynynen [95], who developed a computer model that takes into account the refraction and reflection of HIFU beams at tissue interfaces.

To locate the focus point, the MRI pulse sequence T_1-weighted FSPGR was used. This imaging pulse sequence is the most widely accepted pulse sequence [57, 96], because it is acquired in a few seconds. Using FSPGR, the proton resonant frequency is extracted, which is considered to be the most reliable method for estimating tissue temperature. FSPGR gives an indication of heating (decreased signal) and, therefore, it is mainly used to identify the location of the focus using low intensity, which does not cause any irreversible damage to the tissue. It can be used also to estimate the thermal dose during the application of ultrasound [57]. In this study, it was sufficient to study the shape of the temperature changes, which is qualitative information adequately given by FSPGR. The accurate estimation of temperature for quantitative information extracted from the proton resonant frequency shift, although possibly useful for this study, was not considered. The effect of interfaces on temperature is a subject of another study that is under way and will be analyzed using the proton resonant frequency shift.

To create a large lesion, the transducer is scanned in a rectangular or square grid pattern. In this study, two scans were utilized. During the first scan, low-intensity ultrasound was used, which elevates the tissue temperature by 2°C to 5°C. During this diagnostic scan, FSPGR is used to indicate whether ultrasound penetration is acceptable. Then a second scan was performed using HIFU. If, during the first scan, all of the grid points showed a visible decrease of signal (black spot) indicating heating, then, during the second scan, lesions were created at every point in the grid. The first scan, which is used for diagnostic purposes, is a useful tool for evaluating ultrasound penetration. Our intention was to use this diagnostic tool for evaluating ultrasound penetration through an interface. The main goal in this study was to investigate the effect of interfaces on lesion size. MRI adequately estimates the shape of the thermal lesion. Chen et al. [59] confirmed that the lesion estimated by MRI was in good agreement with the lesion measured using histology.

Because the studies by Coleman and Saunders [89] and Frizzell [79] showed that at interfaces there is increased occurrence of cavitation, cavitation was monitored using acoustical methods and using the MRI pulse sequence T_2-weighted FSE, which was proven to be a useful pulse sequence for monitoring cavitation lesions in kidney [90].

Various studies have shown that MRI can enhance the guidance of HIFU [19, 20]. However, issues related to treatment planning remain to be resolved. Thus far, however, many of the issues have been resolved and, therefore, the clinical application of ultrasound has been enhanced. For example, Billard et al. [29] showed that, by using short pulse duration, the effect of perfusion is minimized. Damianou and Hynynen [37] and Malcolm and ter Haar [24] demonstrated that the right spacing of the transducer must be used in order to create complete necrosis. Hynynen [38] suggested that, by using a higher frequency, the threshold of cavitation will be higher and therefore cavitation can be avoided. Hynynen et al. [97] showed that ultrasound can penetrate through the intact skull, by using the appropriate ablating techniques. These are only few examples of studies that focus on treatment planning. We believe that treatment planning is now the major area to be resolved in order to convince clinicians about the effectiveness of HIFU. In this section, particular attention is given to the problem of interfaces. For the case of kidney, two types of interfaces are usually encountered: muscle–kidney and fat–kidney interfaces.

14.8.1 MRI Analysis

Two different pulse sequences were used: T_1-weighted FSPGR for fast imaging and T_2-weighted FSE for high-quality imaging. The following parameters were used for T_1-weighted FSPGR: TR = 50 ms, TE = 2.7 ms, FOV = 16 cm, matrix = 256×256, flip angle = 50°, BW = 62.5 kHz, and NEX = 1. The following parameters were used for T_2-weighted FSE: TR = 2,500 ms, TE = 32 ms, FOV = 16 cm, matrix = 256×256, ETL = 8, BW = 31.25 kHz, and NEX = 1. The evaluation of temperature in fat was monitored using the MRI methods described by Hynynen et al. [62].

14.8.2 Results

Figure 14.23 shows an MR image using a T_2-weighted FSE of lesions in a plane perpendicular to the beam, in rabbit kidney in vivo using different in situ spatial average intensities (1,000, 1,500, 2,500, and 3,500 W/cm^2). Note that, with 3,500 W/cm^2, the lesion was created under the mechanism of cavitation.

To estimate the POC through an interface, multiple lesions were created by moving the transducer in a grid formation (5×4). The spacing between the lesions was 5 mm, so as to create discrete lesions. The pulse duration was 5 seconds. The spatial average intensity was varied for each set of discrete lesions. The intensities were varied between 500 and 4,000 W/cm^2, with a step of 500 W/cm^2. If, out of the 20 lesions, 1 is created under the mechanism of cavitation, then the POC is 5%. The cavitation was confirmed by using four different methods: MRI imaging, detection of harmonic emission, temperature monitoring, and gross examination. Figure 14.24 shows the POC as a function of intensity at 5 seconds, through a muscle kidney interface. In the same figure the POC is plotted for kidney in vitro from a previous study [90] for the same pulse duration (5 seconds) and frequency (4 MHz).

Figure 14.25 shows nine MRI images in a plane perpendicular to the beam acquired by using T_1-weighted FSPGR. The transducer was moved in a 3×3 pattern and the intensity used was 100 W/cm^2 for 10 seconds. This intensity was used in Section 14.5 in vitro and the produced temperature was 2°C to 5°C, which means that

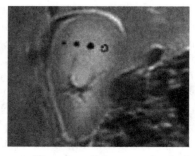

Figure 14.23   Lesions in kidney at different spatial average in situ intensities (1,000, 1,500, 2,500, and 3,500 W/cm^2).

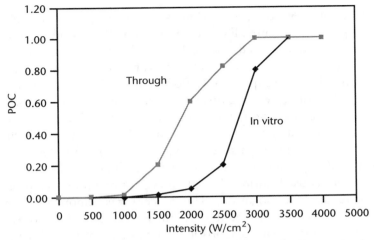

Figure 14.24   POC as a function of in situ spatial average intensity for 5-second pulse duration in kidney in vitro and kidney in vivo through muscle-kidney interface.

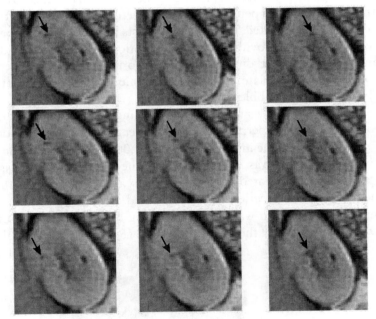

Figure 14.25   MR images using T_1-weighted FSPGR of the focus beam during a 3 × 3 scanning process that used low intensity.

there was no tissue damage with this exposure. These images show that in every one of the nine firings, there is good penetration, because heating is observed (black spot of 1 to 2 mm in diameter). Therefore, this proves that ultrasound penetration through this particular muscle–kidney interface is excellent in all nine locations. Following this diagnostic scan, the intensity was increased to 1,500 W/cm^2 and the nine locations were ablated using a 5-second pulse.

Figure 14.26 shows the MRI image in a plane perpendicular to the beam created by using the T_2-weighted FSE technique. Note that the necrosis coverage is continuous, leaving no untreated spaces. The continuous necrosis coverage was confirmed visually when the kidney was sliced. Next to the large lesion, a single lesion was created using the same exposure (1,500 W/cm^2 for 5 seconds) in order to use it as a reference. Note that, by using this intensity, none of the lesions was created under the mechanism of cavitation, confirming the results of Figure 14.24.

During this diagnostic scan, low intensity is applied and, therefore, there is no need for a delay between successive firings. Therefore, the time needed to perform this type of scan is short. It is up to the HIFU clinicians as to whether or not this diagnostic scan is used during clinical work, but certainly this tool can be very useful during research investigations. This concept was demonstrated for a single shot in in vivo kidney by Hynynen et al. [15], but was never demonstrated in two dimensions (rectangular grid). This diagnostic scan was demonstrated in a plane perpendicular to the beam, but it is possible to demonstrate it in a plane parallel to the beam. When all of the points of the grid show a decrease in temperature (black spot in the T_1-weighted FSPGR), then, during the application of high-intensity ultrasound (therapeutic scan), complete necrosis is observed in the targeted area. If ultrasound goes through an interface that includes bubbles, then the diagnostic scan reveals sites with poor ultrasound penetration and, therefore, later in the therapeutic scan, some spaces are left untreated. This study suggests that an MRI compatible robot should include an axis for providing rotational motion, so that the transducer does not only maneuvers in the xyz dimensions, but also in an angular direction, thus ablating through the appropriate angle. This technique will ensure that interfaces with a major quantity of bubbles are avoided.

Figure 14.27 shows the MRI image using T_2-weighted FSE of a large lesion in the kidney in a plane perpendicular to the beam. This lesion was created by using a 4 × 5 grid of lesions. The intensity used was 2,000 W/cm^2 at 5 seconds and the spacing between the lesions was 3 mm. Note that the necrosis coverage is incomplete, leaving some space untreated. The diagnostic scan of Figure 14.27 has shown that the penetration at the location of the untreated area was poor, because no detectable temperature elevation was observed during the application of the diagnostic scan (low intensity). Out of the 20 locations, only 6 showed the normal 1- to 2-mm-diam-

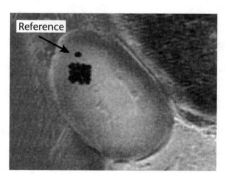

Figure 14.26   MR images (in a plane perpendicular to the transducer beam) of a large lesion (full coverage of the intended target) using T_2-weighted FSE with TE = 32 ms. The spatial average intensity was 1,500 W/cm^2 for 5 seconds.

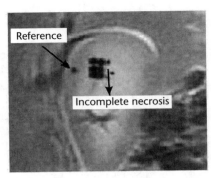

Figure 14.27 MR images (in a plane perpendicular to the transducer beam) of a large lesion (partial coverage of the intended target) using T_2-weighted FSE with TE = 32 ms. The spatial average intensity was 2,000 W/cm^2 for 5 seconds.

eter black spot. It was observed that this muscle–kidney interface contained bubbles (observed by MRI and later by gross examination). During the application of HIFU, at the site where the tissue was not necrosed, no cavitation activity was detected, meaning that reflection was the dominant factor for the incomplete necrosis.

Figure 14.28 is a very interesting case that was encountered. This figure shows the MRI image using T_2-weighted FSE of a large lesion in the kidney in a plane perpendicular to the beam. On top of this site, it was observed that the muscle–kidney interface contained bubbles (observed by MRI and later by gross examination). The large lesion was created using a 5 × 6 grid of lesions. The intensity used was 1,000 W/cm^2 at 5 seconds and the spacing between the lesions was 1.5 mm. The necrosis coverage is incomplete, leaving some space untreated. However, a small area that may look as if it is untreated is, in fact, a lesion created under the mechanism of cavitation (verified by acoustic emission, temperature monitoring, and gross examination). Unfortunately, it was difficult to discriminate using MRI between a cavitational lesion and untreated spaces. Untreated regions in the kidney appear very bright, and cavitational lesions, which include liquid-filled spaces, also appear bright. The diagnostic scan of this figure has shown that the penetration at the location of the untreated area was poor (similar to that observed for the case of Figure 14.27).

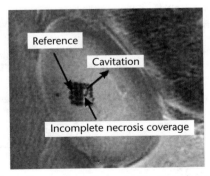

Figure 14.28 MR images (in a plane perpendicular to the transducer beam) of a large lesion (partial coverage of the intended target and occurrence of cavitation) using T_2-weighted FSE with TE = 32 ms. The spatial average intensity was 1,000 W/cm^2 for 5 seconds.

References

[1] Lynn, J. G., et al., "A New Method for the Generation and Use of Focused Ultrasound in Experimental Biology," *J. Gen. Physiol.*, Vol. 26, 1942, pp. 179–193.

[2] Chapelon, J. Y., et al., "A. In Vivo Effects of High–Intensity Ultrasound on Prostatic Adenocarcinoma, Dunning R3327," *Cancer Res.*, Vol. 52, No. 22, 1992, pp. 6353–6357.

[3] Hynynen, K., et al., "The Usefulness of a Contrast Agent and Gradient-Recalled Acquisition in a Steady-State Imaging Sequence for Magnetic Resonance Imaging-Guided Noninvasive Ultrasound Surgery," *Invest. Radiol.*, Vol. 29, No. 10, October 1994, pp. 897–903.

[4] Bihrle, R., et al., "High-Intensity Focused Ultrasound for the Treatment of Benign Prostatic Hyperplasia: Early United States Clinical Experience," *J. Urology*, Vol. 151, No. 5, 1994, pp. 1271–1275.

[5] Pomeroy, O. H., et al., "MR-Imaging Guided Focused Ultrasound Therapy for Breast Fibroadenomas: A Feasibility Study [abstract]," *Radiology*, Vol. 197, 1995, p. 331.

[6] Lizzi, F., et al., "Ultrasonic Hyperthermia for Ophthalmic Therapy," *IEEE Trans. on Sonics Ultrasonics*, Vol. SU-31, No. 5, 1984, pp. 473–481.

[7] Sanghvi, N, et al., "System Design and Considerations for High Intensity Focused Ultrasound Device for the Treatment of Tissue In Vivo," *Med. Biol. Eng. Comp.*, Vol. 29, 1991, p. 748.

[8] Chapelon, J. Y., et al., "Treatment of Localised Prostate Cancer with Transrectal High Intensity Focused Ultrasound," *Eur. J. Ultrasound*, Vol. 9, No. 1, 1999, pp. 31–38.

[9] ter Haar, G., D. Sinnett, and I. Rivens, "High Intensity Focused Ultrasound—A Surgical Technique for the Treatment of Discrete Liver Tumors," *Phys. Med. Biol.*, Vol. 34, No. 11, 1989, pp. 1743–1750.

[10] Fry, W., et al., "Production of Focal Destructive Lesions in the Central Nervous System with Ultrasound," *J. Neurosurg.*, Vol. 11, 1954, pp. 471–478.

[11] Lele, P. P., A Simple Method for Production of Trackless Focal Lesions with Focused Ultrasound," *J. Physiol.*, Vol. 160, 1962, pp. 494–512.

[12] Vykhodtseva, N. I., K. Hynynen, and C. Damianou, "Pulse Duration and Peak Intensity During Focused Ultrasound Surgery: Theoretical and Experimental Effects in Rabbit Brain in Vivo," *J. Ultrasound Med. Biol.*, Vol. 20, No. 9, 1994, pp. 987–1000.

[13] Linke, C., et al., "Localized Tissue Destruction by High-Intensity Focused Ultrasound," *Arch. Surg.*, Vol. 107, 1973, pp. 887–891.

[14] Chapelon, J. Y., et al., "A. Effects of High-Energy Focused Ultrasound on Kidney Tissue in the Rat and the Dog," *J. Eur. Urol.*, Vol. 22, No. 2, 1992, pp. 147–152.

[15] Hynynen, K., et al., "MR Monitoring of Focused Ultrasonic Surgery of Renal Cortex: Experimental and Simulation Studies," *J. Magn. Resonance Imaging*, Vol. 5, No. 3, 1995, pp. 259–266.

[16] Seip, R., and E. Ebbini, "Non-Invasive Estimation of Tissue Temperature Response to Heating Using Diagnostic Ultrasound," *IEEE Trans. on Biomed. Eng.*, Vol. 42, No. 8, 1995, pp. 828–839.

[17] Maass-Moreno, R., C. A. Damianou, and N. T. Sanghvi, "Noninvasive Temperature Estimation in Tissue via Ultrasound Echo Shifts: Part II. In Vitro Study," *J. Acoust. Soc. Am.*, Vol. 100, No. 4, Part 1, 1996, pp. 2522–2530.

[18] Cline, H. E., et al., "MR-Guided Focused Ultrasound Surgery," *J. Comput. Assist. Tomogr.*, Vol. 16, 1992, pp. 956–965.

[19] Hynynen, K., et al., "The Feasibility of Using MRI to Monitor and Guide Noninvasive Ultrasound Surgery [letter]," *J. Ultrasound Med. Biol.*, Vol. 19, No. 1, 1993, pp. 91–92.

[20] Hynynen, K., et al., "MRI-Guided Noninvasive Ultrasound Surgery," *Med. Phys.*, Vol. 20, No. 1, 1993, pp. 107–115.

[21] Kossoff, G., E. Kelly-Fry, and J. Jellins, "Average Velocity of Ultrasound in the Human Female Breast," *J. Acoust. Soc. Am.*, Vol. 53, No. 6, 1973, pp. 1730–1736.

[22] Lele, P., *Effects of Ultrasound on Solid Mammalian Tissues and Tumors In Vivo*, New York: Plenum Publishing, 1987, pp. 275–306.

[23] Raum, K., and D. O'Brien, "Pulse-Echo Field Distribution Measurement Technique for High-Frequency Ultrasound Sources," *IEEE Trans. on Ultrasonics, Ferroelectrics, and Frequency Control*, Vol. 44, No. 4, 1997, pp. 810–815.

[24] Malcolm, A. L., and G. R. ter Haar, "Ablation of Tissue Volumes Using High Intensity Focused Ultrasound," *J. Ultrasound Med. Biol.*, Vol. 22, No. 5, 1996, pp. 659–669.

[25] Pennes, M., "Analysis of Tissue and Arterial Blood Temperature in the Resting Human Forearm," *J. Appl. Physiol.*, Vol. 1, 1948, pp. 93–122.

[26] Dewey, W., et al., "Cellular Responses to Combinations of Hyperthermia and Radiation," *Radiology*, Vol. 123, 1977, pp. 463–474.

[27] Sapareto, S., and W. Dewey, "Thermal Dose Determination in Cancer Therapy," *Int. J. Radiation Oncology Biol. Phys.*, Vol. 10, 1984, pp. 787–800.

[28] Borrelli, M., et al., "Time Temperature Analysis of Cell Killing of BHK Cells Heated at Temperatures in the Range of 43.5 C to 57 C," *J. Radiation Oncology Biol. Phys.*, Vol. 19, 1990, pp. 389–399.

[29] Billard, B. E., K. Hynynen, and R. B. Roemer, "Effects of Physical Parameters on High Temperature Ultrasound Hyperthermia," *J. Ultrasound Med. Biol.*, Vol. 16, 1990, pp. 409–420.

[30] Damianou, C., et al., "Dependence of Ultrasonic Attenuation and Absorption in Dog Soft Tissues on Temperature and Thermal Dose," *J. Acoust. Soc. Am.*, Vol. 102, No. 2, 1997, pp. 628–634.

[31] Dunn, F., "Temperature and Amplitude Dependence of Acoustic Absorption in Tissue," *J. Acoust. Soc. Am.*, Vol. 34, No. 10, 1962, pp. 1545–1547.

[32] Dunn, F., and Brady, J., "Temperature and Frequency Dependence of Ultrasonic Absorption in Tissue," *Proc. 8th Int. Congress on Acoustics, IEEE Ultrasonics Symp.*, 1974, p. 366c.

[33] Gammell, P., D. Le Croissette, and R. Heyser, "Temperature and Frequency Dependence of Ultrasonic Attenuation in Selected Tissues," *J. Ultrasound Med. Biol.*, Vol. 5, 1979, pp. 269–277.

[34] Fry, F., et al., "Absorption in Liver At the Focus of an Ultrasonic Shock Wave Field," *J. Ultrasound Med. Biol.*, Vol. 17, No. 1, 1991, pp. 65–69.

[35] Robinson, T., and P. Lele, "An Analysis of Lesion Development in the Brain and in Plastics by High-Intensity Focused Ultrasound at Low-Megahertz Frequencies," *J. Acoust. Soc. Am.*, Vol. 51, No. 2, 1969, pp. 133–1351.

[36] Bamber, J., and C. Hill, "Ultrasonic Attenuation and Propagation Speed in Mammalian Tissues as a Function of Temperature," *J. Ultrasound Med. Biol.*, Vol. 5, 1979, pp. 149–157.

[37] Damianou, C., and K. Hynynen, "Focal Spacing and Near-Field Heating During Pulsed High Temperature Ultrasound Therapy," *J. Ultrasound Med. Biol.*, Vol. 19, No. 9, 1993, pp. 777–787.

[38] Hynynen, K., "The Threshold for Thermally Significant Caviation in Dog's Thigh Muscle In Vivo," *J. Ultrasound Med. Biol.*, Vol. 17, No. 2, 1991, pp. 157–169.

[39] Le Bihan, D., J. Delannoy, and R. L. Levin, "Temperature Mapping with MR Imaging of Molecular Diffusion: Application of Hyperthermia," *Radiology*, Vol. 171, 1989, pp. 853–857.

[40] Dickinson, R. J., et al., "Measurement of Changes in Tissue Temperature Using MR Imaging," *J. Comput. Assist. Tomogr.*, Vol. 10, 1986, pp. 468–472.

[41] Matsumoto, R., K. Oshio, and F. Jolesz, "Monitoring of Laser and Freezing Induced Ablation in the Liver with T1-Weighted MR Imaging," *J. Magn. Resonance Imaging*, Vol. 2, 1992, pp. 255–562.

[42] Bleier, A. R., F. A. Jolesz, and M. S. Cohen, "Real-Time Magnetic Resonance Imaging of Laser Heat Deposition in Tissue," *J. Magn. Reson. Med.,* Vol. 21, 1991, pp. 132–137.

[43] Mumtaz, H., et al., "Laser Therapy for Breast Cancer: Mr Imaging and Histopathologic Correlation," *Radiology,* 1996, Vol. 200, No. 3, pp. 651–658.

[44] Kettenbach, J., et al., "Monitoring and Visualization Techniques for MR-Guided Laser Ablations in an Open MR System," *J. Magn. Resonance Imaging,* Vol. 8, No. 4, 1998, pp. 933–943.

[45] Law, P., W. M. Gedroyc, and L. Regan, "Magnetic Resonance-Guided Percutaneous Laser Ablation of Uterine Fibroids," *J. Magn. Resonance Imaging,* Vol. 12, No. 4, 2000, pp. 565–570.

[46] Fry, F., and L. K. Johnson, "Tumor Irradiation with Intense Ultrasound," *J. Ultrasound Med. Biol.,* Vol. 4, No. 4, 1978, pp. 337–341.

[47] Jolesz, F. A., and P. D. Jakab, "Acoustic Pressure Wave Generation Within a Magnetic Resonance Imaging System: Potential Medical Applications," *J. Magn. Resonance Imaging,* Vol. 1, 1991, pp. 609–613.

[48] Hynynen, K., et al., "MRI-Guided Ultrasonic Hyperthermia," *1992 RSNA Meeting,* Chicago, IL, December 3, 1992.

[49] Cline, H. E., et al., "Magnetic Resonance Guided Thermal Surgery," *J. Magn. Resonance Imaging,* Vol. 30, No. 1, 1993, pp. 98–106.

[50] Darkazanli, A., et al., "On-Line Monitoring of Ultrasonic Surgery with MR Imaging," *J. Magn. Resonance Imaging,* Vol. 3, No. 3, 1993, pp. 509–514.

[51] Cline, H. E., et al., "MR Temperature Mapping of Focused Ultrasound Surgery," *J. Magn. Resonance Imaging,* Vol. 31, No. 6, 1994, pp. 628–636.

[52] Cline, H. E., et al., "Simultaneous Magnetic Resonance Phase and Magnitude Temperature Maps in Muscle," *J. Magn. Resonance Imaging,* Vol. 35, No. 3, 1996, pp. 309–315.

[53] Kuroda, K., et al., "Temperature Mapping Using the Water Proton Chemical Shift: A Chemical Shift Selective Phase Mapping Method," *J. Magn. Resonance Imaging,* Vol. 38, No. 5, 1997, pp. 845–851.

[54] Kuroda, K., et al., "Calibration of Water Proton Chemical Shift with Temperature for Noninvasive Temperature Imaging During Focused Ultrasound Surgery," *J. Magn. Resonance Imaging,* Vol. 8, No. 1, 1998, pp. 175–181.

[55] de Zwart, J., et al., "Fast Lipid-Suppressed MR Temperature Mapping with Echo-Shifted Gradient-Echo Imaging and Spectral-Spatial Excitation," *J. Magn. Resonance Med.,* Vol. 42, 1999, pp. 53–59.

[56] Kuroda, K., et al., "Temperature Mapping Using the Water Proton Chemical Shift: Self-Referenced Method with Echo-Planar Spectroscopic Imaging," *J. Magn. Resonance Imaging,* Vol. 43, No. 2, 2000, pp. 220–225.

[57] Hynynen, K., et al., "Thermal Effects of Focused Ultrasound on the Brain: Determination with MR Imaging," *Radiology,* Vol. 204, No. 1, 1997, pp. 247–253.

[58] Morocz, I. A., et al., "Brain Edema Development After MRI-Guided Focused Ultrasound Treatment," *J. Magn. Resonance Imaging,* Vol. 8, No. 1, 1998, pp. 136–142.

[59] Chen, L., et al., "Study of Focused Ultrasound Tissue Damage Using MRI and Histology," *J. Magn. Resonance Imaging,* Vol. 10, No. 2, 1999, pp. 146–153.

[60] Vykhodtseva, N., et al., "MRI Detection of the Thermal Effects of Focused Ultrasound on the Brain," *J. Ultrasound Med. Biol.,* Vol. 26, No. 5, 2000, pp. 871–880.

[61] Rowland, I. J., et al., "MRI Study of Hepatic Tumours Following High Intensity Focused Ultrasound Surgery," *Br. J. Radiology,* Vol. 70, 1997, pp. 144–153.

[62] Hynynen, K., et al., "Temperature Monitoring in Fat with MRI," *J. Magn. Resonance Imaging,* Vol. 43, No. 6, 2000, pp. 901–904.

[63] Nakamura, K., et al., "Treatment of Benign Prostatic Hyperplasia with High Intensity Focused Ultrasound, Vol. an Initial Clinical Trial in Japan with Magnetic Resonance Imaging of the Treated Area," *Int. J. Urol.,* Vol. 2, 1995, pp. 176–180.

[64] Smith, N. B., M. T. Buchanan, and K. Hynynen, "Transrectal Ultrasound Applicator for Prostate Heating Monitored Using MRI Thermometry," *Int. J. Radiation Oncology Biol. Phys.*, Vol. 43, No. 1, 1999, pp. 217–225.

[65] Sokka, S. D., and K. Hynynen, "The Feasibility of MRI-Guided Whole Prostate Ablation with a Linear Aperiodic Intracavitary Ultrasound Phased Array," *Phys. Med. Biol.*, Vol. 45, No. 11, 2000, pp. 3373–3383.

[66] Hynynen, K., et al., MR Imaging-Guided Focused Ultrasound Surgery of Fibroadenomas in the Breast: A Feasibility Study," *Radiology*, Vol. 219, No. 1, 2001, pp. 176–185.

[67] Starritt, H. C., et al., "Measurement of Acoustic Streaming Using Magnetic Resonance," *J. Ultrasound Med. Biol.*, Vol. 26, No. 2, 2000, pp. 321–333.

[68] Walker, C. L., F. S. Foster, and D. B. Plewes, "Magnetic Resonance Imaging of Ultrasonic Fields," *J. Ultrasound Med. Biol.*, Vol. 24, No. 1, 1998, pp. 137–142.

[69] Sharf, Y., et al., "Absolute Measurements of Ultrasonic Pressure by Using High Magnetic Fields," *IEEE Trans. on Ultrasonics, Ferroelectrics, and Frequency Control*, Vol. 46, No. 6, 1999, p. 1504.

[70] Damianou, C., "MRI Monitoring of Ultrasonic Cavitation," *Cyprus J. Sci. Technol.*, Vol. Vol. 4, No. 1, 2004, pp. 52–62.

[71] McDannold, N., K. Hynynen, and F. Jolesz, "MRI Monitoring of the Thermal Ablation of Tissue: Effects of Long Exposure Times," *J. Magn. Resonance Imaging*, Vol. 13, 2001, pp. 421–427.

[72] Chen, L., et al., "Study of Focused Ultrasound Tissue Damage Using MRI and Histology," *J. Magn. Resonance Imaging*, Vol. 10, No. 2, August 1999, pp. 146–153.

[73] Di Bisceglie, A., et al., "Hepatocellular Carcinoma," *Ann. Intern. Med.*, Vol. 108, 1988, pp. 390–401.

[74] Bruix, J., et al., "Surgical Resection and Survival in Western Patients with Hepatocellular Carcinoma," *J. Hepatol.*, Vol. 15, 1992, pp. 350–355.

[75] Dick, E. A., et al., "MR-Guided Laser Thermal Ablation of Primary and Secondary Liver Tumours," *Clin. Radiol.*, Vol. 58, No. 2, 2003, pp. 112–120.

[76] Vogl, T. J., et al., "Malignant Liver Tumors Treated with MR Imaging-Guided Laser-Induced Thermotherapy: Experience with Complications in 899 Patients (2,520 Lesions)," *Radiology*, Vol. 225, No. 2, 2002, pp. 367–377.

[77] Elias, D., et al., "Percutaneous Radiofrequency Thermoablation as an Alternative to Surgery for Treatment of Liver Tumour Recurrence After Hepatectomy," *Br. J. Surg.*, Vol. 89, No. 6, 2002, pp. 752–756.

[78] Frizzell, L., et al., "Thresholds for Focal Ultrasound Lesions in Rabbit Kidney, Liver and Testicle," *IEEE Trans. on Biomed. Eng.*, Vol. 24, No. 4, 1987, pp. 393–396.

[79] Frizzell, L., "Threshold Dosages for Damage to Mammalian Liver by High Intensity Focused Ultrasound," *IEEE Trans. on Ultrasonics, Ferroelectrics, and Frequency Control*, Vol. UFFC-35, 1988, pp. 578–581.

[80] Sibelle, A., et al., "Extracorporeal Ablation of Liver Tissue by High-Intensity Focused Ultrasound," *Oncology*, Vol. 50, No. 5, 1993, pp. 375–379.

[81] Chen, L., et al., "Histological Changes in Rat Liver Tumours Treated with High-Intensity Focused Ultrasound," *J. Ultrasound Med. Biol.*, Vol. 19, No. 1, 1993, pp. 67–74.

[82] Chen, L., et al., "Histological Study of Normal and Tumor-Bearing Liver Treated with Focused Ultrasound," *J. Ultrasound Med. Biol.*, Vol. 25, No. 5, 1999, pp. 847–856.

[83] Yang, R., et al., "High-Intensity Focused Ultrasound in the Treatment of Experimental Liver Cancer," *Arch. Surg.*, Vol. 126, No. 8, 1991, pp. 1002–1010.

[84] Sibille, A., et al., "Characterization of Extracorporeal Ablation of Normal and Tumor-Bearing Liver Tissue by High Intensity Focused Ultrasound," *J. Ultrasound Med. Biol.*, Vol. 19, No. 9, 1993, pp. 803–813.

[85] Prat, F., et al., "Extracorporeal High-Intensity Focused Ultrasound for VX2 Liver Tumors in the Rabbit," *Hepatology*, Vol. 21, No. 3, 1995, pp. 832–836.

[86] Cheng, S. Q., et al., "High-Intensity Focused Ultrasound in the Treatment of Experimental Liver Tumour," *J. Cancer Res. Clin. Oncol.* Vol. 123, No. 4, 1997, pp. 219–223.

[87] Chen, L., et al., "Treatment of Implanted Liver Tumors with Focused Ultrasound," *J. Ultrasound Med. Biol.,* Vol. 24, No. 9, 1998, pp. 1475–1488.

[88] Damianou, C., "In Vitro and In Vivo Ablation of Porcine Renal Tissues Using High Intensity Focused Ultrasound," *J. Ultrasound Med. Biol.,* Vol. 29, No. 9, 2003, pp. 1321–1330.

[89] Coleman, A. J., and J. E. Saunders, "A Review of the Physical Properties and Biological Effects of the High Amplitude Acoustic Fields Used in Extracorporeal Lithotripsy," *Ultrasonics,* Vol. 3, 1993, pp. 75–89.

[90] Damianou, C., et al., "High Intensity Focused Ultrasound Ablation of Kidney Guided by ÌRI," *J. Ultrasound Med. Biol.,* Vol. 30, No. 3, 2004, pp. 397–404.

[91] Mack, M. G., et al., "Percutaneous MR Imaging-Guided Laser-Induced Thermotherapy of Hepatic Metastases," *Abdom. Imaging,* Vol. 26, No. 4, 2001, pp. 369–743.

[92] Morrison, P. R., et al., "MRI of Laser-Induced Interstitial Thermal Injury in an In Vivo Animal Liver Model with Histologic Correlation," *J. Magn. Resonance Imaging,* Vol. 8, No. 1, 1998, pp. 57–63.

[93] Aschoff, A. J., et al., "Thermal Lesion Conspicuity Following Interstitial Radiofrequency Thermal Tumor Ablation in Humans: A Comparison of STIR, Turbo Spin-Echo T_2-Weighted, and Contrast-Enhanced T_1-Weighted MR Images at 0.2 T," *J. Magn. Resonance Imaging,* Vol. 12, No. 4, 2000, pp. 584–589.

[94] Tsuda, M., et al., "Time-Related Changes of Radiofrequency Ablation Lesion in the Normal Rabbit Liver: Findings of Magnetic Resonance Imaging and Histopathology," *Invest. Radiol.,* Vol. 38, No. 8, 2003, pp. 525–531.

[95] Fan, X., and K. Hynynen, "The Effect of Wave Reflection and Refraction at Soft Tissue Interfaces During Ultrasound Hyperthermia Treatments," *J. Acoust. Soc. Am.,* Vol. 91, No. 3, March 1992, pp. 1727–1736.

[96] Quesson, B., J. Zwart, and C. Moonen, "Magnetic Resonance Temperature Imaging for Guidance of Thermotherapy," *J. Magn. Resonance Imaging,* Vol. 12, 2000, pp. 525–533.

[97] Hynynen, K., and F. A. Jolesz, "Demonstration of Potential Noninvasive Ultrasound Brain Therapy Through an Intact Skull," *J. Ultrasound Med. Biol.,* Vol. 24, No. 2, 1998, pp. 275–283.

CHAPTER 15

Modalities and Devices for Thermal Ablation

Sunita Chauhan

This chapter includes a brief overview of prevalent thermal ablation modalities. It concentrates on a review of noninvasive surgical techniques (other than radiosurgery) that use thermally induced ablation for treating deep-seated abnormalities. The chapter concludes with devices and systems available for thermal ablation and a representative example of a robotic system, called FUSBOT^{BS}, that was devised at the author's center.

15.1 Introduction

For management of cancers/tumors, surgical resection is considered to be the primary treatment and the gold standard. Surgical resection is effective when cancers are contained in a given organ. For organs situated near lymph nodes, extensive radical resection may be necessary, including the removal of lymph nodes. Normally, a margin area surrounding the cancer is removed to minimize relapse. However, open surgery is associated with several operative and postoperative complications resulting into high morbidity and mortality rates, and many patients are not good candidates for resection. Minimally invasive surgery (MIS) is gaining significance, wherein the affected area is laparoscopically or endoscopically resected. Sometimes, surgery is complicated due to the location of the cancers, which may present severe risks to the patients. For more invasive cancers, it is usually performed along with adjuvant and/or neoadjuvant therapies such as radiotherapy and chemotherapy, which are usually administered postoperatively to reduce the recurrence of the disease.

Therapeutic techniques such as radiation therapy, chemotherapy, and hyperthermia (using modalities such as RF, microwave, or ultrasound) may be effective only in early diagnostic stages. Heat diffusion to sites adjacent to the target is common due to extended periods of exposure and convection by blood perfusion. Achieving temperature control at desired levels is very difficult. Such therapies are given either alone or in appropriate combination. For instance, the cellular membrane permeability (and thus the efficiency) of chemotherapy is increased if hyperthermia is induced prior to, or along with, drug infusion. Changes in tissue geometry and local microcirculation affect tissue response (thermotolerance) to successive intervals of energy deposition.

This chapter provides an overview of prevalent thermal ablation modalities. It concentrates on a review of noninvasive surgical techniques (other than radiosurgery), which use thermally induced ablation for treating deep-seated abnormalities. High temperatures induced in the tissue result in changes in a variety of physical, chemical, and biological mechanisms at the chemical, organelle, cell, and tissue scales. The tissue coagulation threshold is about 60°C. Temperatures exceeding 60°C produced by a short pulse of energy exposure cause irreversible cell damage due to protein denaturation, inhibition of protein synthesis, breaking of chemical bonds in DNA and RNA molecules, and loss of integrity in lipid bilayers. These procedures can be used in either an extracorporeal or minimally invasive manner, as explained in the following sections.

15.2 Thermal Ablation Mechanisms

15.2.1 Hyperthermia and Ablation

Very high temperatures (thermal coagulation) or very low temperatures (cryoablation) cause various changes in cellular composition in biological tissues. These include cytoplasmic and nuclear protein denaturation and DNA and RNA changes in cellular membrane function. At the temperature ranges used in hyperthermia (41°C to 43°C for 30 to 60 minutes), tissue composition is altered and increased cellular membrane permeability, metabolic activity, blood perfusion, pH, and pO_2 levels are observed. This increases radiosensitization and chemosensitization of the cells. The changes, however, are reversible, and long exposures affect the optimum outcomes due to subsistent heat conduction. Further, higher temperatures (ranging from 60°C to 80°C for 2 to 8 seconds), like those used in thermal ablation, result in irreversible changes, for instance, cellular protein denaturation, thermal coagulation, and immediate thermal necrosis within the treatment area, leading to cell death [1–8]. Temperatures beyond 100°C induce coagulation and vaporization of tissue.

In all thermal therapies and thermal ablation procedures, the extent of tissue destruction is decided by the heat dose, which is a product of exposure time and temperature. With mechanical modalities such as ultrasound, tissue damage is a result of both thermal and mechanical stress mechanisms; the contribution of either factor, however, depends on the input intensities and heating rate. In predictive modeling, the bioheat equation is used to determine the spatial distribution of temperature in tissues:

$$\rho c \, \partial T \big/_{\partial t} = \nabla k \nabla T + q - Q_p + Q_m \tag{15.1}$$

where ρ is the mass density (kg/m^3), c is the specific heat (J/kg·K), k is the thermal conductivity (W/m·K), T is the temperature (°C), q is the heat source (W/m^3), Q_p is the perfusion heat loss (W/m^3), and Q_m is the metabolic heat generation (W/m^3). In high perfusion tissues, Q_p is computed as follows:

$$Q_p = \omega_b c_b (T - T_b) \tag{15.2}$$

where ω_b is the blood perfusion per unit volume (kg/m^3·s), c_b is the specific heat of blood (J/kg·K), and T_b is the blood temperature (°C).

For predicting the time required to produce thermal tissue injury, various methods have been suggested. A simple and widely used mathematical model, based on the empirically deduced Arrhenius model (of chemical reaction rates), predicts that the time, t, required to produce tissue death at a constant temperature, T (°C), is given by:

$$t = Ae^{[\Delta E / k (T + 273)]} \tag{15.3}$$

where ΔE is the activation energy, k is Boltzmann's constant (8.617×10^{-5} in eV/K), and A is the preexponential scaling factor, which varies from tissue to tissue. The term E is relatively large for biological tissues, thus time t decreases very rapidly as temperature T increases.

15.2.2   Cryoablation

Cryoablation is an alternative treatment method, for instance, in the palliative management of liver, renal tumors, and colorectal liver metastases at moderate to lower temperatures (–20°C to –40°C for >10 minutes). It involves the extracellular and intracellular formation and growth of ice crystals, resulting in cellular shrinkage, membrane rupture, and cell death during multiple freeze/thaw cycles. Vascular thrombosis resulting from the freezing of blood vessels may cause cell death mediated by ischemia.

The lowest temperatures used and the cooling rate play an important part in optimizing the treatment parameters. The treatment margin is relatively clear during the procedure, and minimally invasive, percutaneous cryoablation can be performed under ultrasonography for guidance. In situations where thermal ablation methods risk causing thermal stricture, such as in the prostate or kidney, cryoablation is preferred because it is not vulnerable to heat sink effects [9, 10]. However, there could be an increased risk of bleeding during the treatment, because the blood vessels are not cauterized as they would be with thermal ablation at high temperatures.

15.2.3   Chemical Ablation

Percutaneous chemical ablation, such as transcatheter arterial embolization and absolute ethanol and acetic acid injection for treating hypervascular lesions, results in several changes in the target tissue. The transport and distribution mechanism in the neoplastic and nonneoplastic tissue is important to deduce the effects. Due to the injection of low-molecular-weight agents such as ethanol, the interstitial fluid pressure increases, resulting in an increase in convective flux of the agent. High-concentration agents produce an increase in diffusive flux. Various effects including cytoplasmic dehydration, denaturation of cellular proteins, and small vessel thrombosis occur as ethanol distributes through tumor interstitium and results in coagulative necrosis [11, 12].

For discussions in this chapter, we concentrate only on the techniques and applications of thermal ablation.

15.3 Therapeutic Modalities for Thermal Ablation

15.3.1 RF Ablation

For tumors that are difficult to resect surgically, percutaneous tumor ablation, for example, in situ radio-frequency ablation (RFA), as a minimally invasive method of destroying tumors presents a preferred alternative. It has been suggested in recent studies that tumor ablation for some tumors may be as effective as resectioning. Commercial RFA systems have been available since 1996. RFA is effective in the control of some heart arrhythmias as well as tumors of the kidney, lung, breast, bone, and adrenal system with significantly increased survival rates. For larger and infiltrative tumors (>3 cm in diameter), however, RFA may provide only palliative and adjuvant treatment in debulking tumors, reducing pain, and enhancing the efficacy of other modalities.

While passing through the biological tissue, the RF electrical currents (in the range of 300- to 1,000-kHz frequencies) achieve a controlled heating of the zone. Basically, two electrodes—an active electrode with a small surface area placed in the target and another electrode with a larger dispersive electrode—are used to close the electrical circuit. In a bipolar ablation system, two active electrodes are used. The factors affecting the necrosed area include electrode length and diameter, tip temperature, and treatment duration. The highest temperatures are achieved at the proximal and distal ends of the electrodes.

For treatment of larger tumors, simultaneous application of RF energy to a multiprobe array system with arrays of two to five electrodes, such as "hooked needle" or "umbrella" electrodes, are commercially available (RITA Medical Systems, Mountain View, California; Radiotherapeutics, Sunnyvale, California). Internally cooled RF electrodes (through which chilled perfusate is circulated), help in preventing tissue boiling and cavitation at the needle tip [7, 8].

The generation of high-frequency ac causes ionic agitation that is converted to heat and induces cellular death due to coagulation necrosis. Internally cooled RF ablation electrodes use dual-channel probes with an electrode at the tip, and cooling fluids such as saline are continuously circulated through one channel and removed through another. Various electrode geometries and a large variation in the electric power in treatment protocols are reported in the literature. Furthermore, the values of electrical and thermal characteristics for whole tissues/organs are reported in the literature; however, the microscopic structures are often overlooked.

Basically, the electrothermal mechanism is explained by assuming the tissues to be purely resistive, because the displacement currents are negligible. With J being the current density (A/m^2), and E the electric field intensity (V/m), the distributed heat source q (Joule loss) in (15.1) is given by:

$$q = J \cdot E \qquad (15.4)$$

To illustrate the thermal-flow coupled problem in order to improve temperature prediction in the circulating blood, the *mass* and *momentum* equations are employed. Fluid dynamics theory can be utilized to compute the velocity field for the saline flowing out of the electrode in irrigated electrodes. The methods used in creating large thermal lesions with cool-tip cluster electrodes include sequential ablation, simultaneous activation of electrodes, and rapid switching of power between electrodes. In the simultaneous method, electrical interference between electrodes may lead to little heating at the center between the electrodes.

15.3.2   Microwave Ablation

Microwave energy has been used for several decades to ablate various types of tissues, for instance, tumors of the uterus, breast, skeletal muscle, liver, prostate, and cardiac tissue. Besides delivering similar results as that of RF energy for tissue ablation, microwaves offer better and more consistent destruction of tumors. The electromagnetic fields can propagate through blood or desiccated or scar tissue and can be deposited directly in the targets at depth. Several factors affect the penetration depth, including the dielectric properties of the tissue, the frequency of the microwave energy, and antenna design. The size of the ablation zone is found to be logarithmically dependent on input power. Undesirable peripheral heating may result by increasing the input power due to heat generated inside the catheters feeding the antenna. Some commercial systems use water cooling of the needle shaft.

Unlike RF ablation, microwave energy heats the tissue by using electromagnetic fields and not by conductive heating, so it does not require a direct tissue contact for creating a lesion. It is faster than RF ablation and creates more uniform heating patterns without the risks of charring. Dielectric radiative heating is produced by enhanced vibration and friction of the molecules as the energy penetrates the tissue and excites the water molecules. After coagulation necrosis, the tissue is left as a nonconductive scar. RF and microwave ablation may be performed in various modes including open, laparoscopic, or percutaneous procedures [13, 14].

In millimeterwave/microwave ablation, localized heating (up to ~150°C) is achieved by electromagnetic energy delivered via catheters to a precise location in the tissue of interest. Desired temperature profiles can be obtained by controlling various parameters such as the delivered power, pulse duration, and frequency. The major components of the apparatus are the microwave source (up to 10W at a 2- to 300-GHz frequency), a catheter/transmission line (to deliver power to the antenna situated at the treatment site), and an antenna at the distal end of the catheter that focuses the radiated beam. Microwave radiation follows the rule of squares, wherein energy density falls off with the square of the distance from the antenna. Depending on the frequency of the incident energy, the power deposition in the target tissue can be more than 2 cm around the antenna.

Several such systems are available. AFx Corporation's (Fremont, California) Flex 2 Lynx device (and the Flex 4 and Flex 10 devices) emits microwave radiation at 2.54 GHz. Systems available for prostate application include Targis and Prostatron (Urologix, Minneapolis, Minnesota), Tmx-2000 (TherMatrx, Evanston, Illinois), and CoreTherm (ACMI, Framingham, Massachusetts). Both single- and multiple-antenna systems (driven simultaneously) are used for localized ablation.

The VivaWave microwave ablation system (Vivant Medical, Mountain View, California) for the treatment of kidney tumors permits the ablation of large lesions and faster therapeutic times by the use of simultaneous irradiation by multiple probes. The ability of a single microwave antenna to ablate a given volume is determined mainly by the input power, antenna efficiency, and application time. Multiple-antenna ablation may be a better alternative that large, single antennas due to better power distribution covering a larger area.

15.3.3   Laser Thermal Ablation

Various modes have been demonstrated for using laser energy in desired target (cancer/tumor) destruction: laser thermal ablation (LTA), interstitial laser thermotherapy (ILT), interstitial laser photocoagulation (ILP), and laser-induced interstitial thermotherapy (LITT) [15–17]. The underlying basic principle for all of these methods remains the same: the use of localized laser energy deposition, resulting in lethal thermal injury to the tissue.

Laser fibers are directly placed in contact with the target site to be treated, and the delivered energy produces conductive and radiant heat leading to cell death. Depending on the input power, the mechanism of cell death could be different. At temperatures above 60°C, local tissue heating results in irreversible damage. Photons from low-intensity laser energy interact with molecular chromophores that are inherent in mammalian cells such as hemoglobin, myoglobin, bilirubin, melanin, and xanthophyll. Tissue composition, vascularity, tissue optical properties, the wavelength of light, and the amount of energy applied are some of the factors that affect the way laser light is scattered, reflected, and absorbed.

Exogenous administration of photosensitizing agents such as photofrin may also accentuate photochemical effects such as in photodynamic therapy (PDT). Higher and rapid energy exposure causes coagulative necrosis, which may reduce optical penetration in the tissue. In LITT, to avoid carbonization and vaporization of tissue at the distal end of the laser fiber, low-energy (3W to 20W) is applied over elongated time (2 to 20 minutes). Temperatures at the commonly used bare-tip quartz fibers (0.5 to 2.5 mm in diameter) may reach 300°C to 1,000°C, producing rapid tissue coagulation and elliptical lesions. The carbonization severely impedes light penetration and heat propagation into the tissue. The use of sapphire-tipped laser fibers significantly reduces the charring effect at the fiber tip. Besides the input energy in the beam, pulse width, pulse rate, and interpulse interval, the shape, size, and material of the fiber tip has a direct influence on the shape and size of the thermal lesions produced:

$$\text{Intensity [W/cm}^2] = \text{Peak power [W]/Focal spot area [cm}^2]$$
$$\text{Peak power [W]} = \text{Pulse energy [J]/Plus duration [sec]}$$

Larger coagulative tissue zones can be achieved by using multiple-fiber systems with beam splitters and simultaneous energy deposition. A short pulse duration maximizes the peak power and minimizes thermal conduction to the surrounding tissue. Initial clinical usage of laser-induced tissue ablation included transuretheral prostate application using a Nd:YAG laser fiber (with a wavelength of 1,064 nm) under TRUS guidance. This, however, resulted in damage to the exposed urethral tissue as well. In ILC, a needle-shaped laser fiber is placed in the prostate adenoma through the urethra under cystoscopic guidance, and a diffusing diode laser beam (830-nm wavelength) is applied interstitially (Indigo laser system, J&J Companies, Cincinnati, Ohio).

Laser technology and computer control software have evolved significantly in the past decade. For photorefractive keratotomy (PRK) of the eye lens, laser sources (such as argon-fluoride excimer or solid-state lasers) are used. PRK is performed with an excimer laser, which uses a cool ultraviolet light beam to precisely ablate very tiny bits of tissue (at the submicron levels) from the surface of the cornea to reshape it for better focusing of light into the eye. Endovenous laser ablation (ELA) is a new technique for treating saphenous vein reflux by eliminating the reflux.

15.3.4   Focused Ultrasound-Based Ablation

Ultrasound waves are mechanical waves that can propagate through biological tissue and can be brought to a tight focus. The physical mechanisms responsible for the therapeutic effects of ultrasound are both thermal and mechanical stress. HIFU is becoming an increasingly attractive modality for tissue ablation because of its unique ability to noninvasively target deep-seated tissue volumes without affecting intervening and surrounding cells [18–25]. However, various factors require attention in order to attain efficacious results from insonification of biological tissues. First of all, the anatomy and functionality of the tissue under consideration should be properly studied in order to better understand the location of the potential targets and to assess the severity of the risks that they may create. The specific location of the tissue/organ in the human body also affects the design of applicator systems. Based on the dimensions, location, and nature of the abnormality, the size and shape of the applicators (and thus the beam shape) can be evaluated by modeling the ultrasonic field in front of the transducers. In certain cases, beam positioning in safe regions can be crucial to the inherent functioning of the biological system, for example, in brain tissue.

When ultrasound energy propagates through a biological medium, a part of it is absorbed and converted into heat. This heat energy is responsible for various reversible or irreversible biological changes depending on the irradiation conditions. For a plane wave of intensity I_0, traveling in a medium with amplitude absorption coefficient, α, the rate at which heat is generated, Q, is given by:

$$Q = 2\alpha I_0 \tag{15.5}$$

If there is no heat loss from the region, the rate of rise of temperature can be calculated by the following relation:

$$\frac{dT}{dt} = \frac{2\alpha I}{\rho c_m} \tag{15.6}$$

where ρ is the density of the medium and c_m is its specific heat per unit mass. This presents a very simplified relation that is applicable to situations where absolutely no thermal conduction occurs and remains valid only some time after the application of a sonic pulse. In practical situations, thermal diffusion occurs at a rate proportional to the temperature gradient and heat transfer to adjoining areas becomes significant very quickly.

Focal ultrasound surgery is a noninvasive means of surgery capable of selective and precisely localized destruction in desired regions. A complete ablation of the affected regions by inducing temperatures on the order of 60°C to 80°C is achieved; the damaged region is called a *lesion*. This technique differs from hyperthermia mainly on the basis of dosage levels (in particular, field intensity and exposure times). The power levels involved here are on the order of 1,000 to 10,000 W/cm². Temperatures thus produced bring rapid, complete, irreversible damage to the exposed tissue. The exposure times are very small (on the order of 1 to 10 seconds) as compared to hyperthermia (10 to 40 minutes) [1–6]. The duration of exposure also has an effect on the time of tissue repair after the treatment. Longer exposure times result in higher damage, unpredictable lesion shapes, and delayed healing. The problems posed by heat diffusion due to blood perfusion are, therefore, less significant.

In the noninvasive approach, the acoustic energy is applied completely externally without any access incisions. The common types of applicators in this category are spherically focused transducers (focused bowls). Electronic phased arrays have been developed for various applications. Either type can be used in various modes: extracorporeal, intracavitary, or directly placed in situ, using minimally invasive techniques. Ultrasound frequencies on the order of 1 to 4 MHz and focal transducers with apertures up to 10 cm in diameter have been reported in the literature. The design of focused ultrasound equipment and the planning of treatment require the ability to model the system to predict in vivo outcomes. The acoustic field of a spherical radiator at a single frequency has already been calculated by many researchers, for example, by O'Niel [26]. A two-dimensional field model based on Huygens' principle and the Fresnel-Kirchhoff diffraction theory has been developed for a large aperture, axially symmetric, focused source of ultrasound. Field variations in the radial as well as axial directions can be studied with this model [27, 28]. The acoustic intensity I at any arbitrary point P in the field at a distance r from the source and time t is given by the following relation:

$$I = \left(A/\rho\text{cm}^2 \right) \left\langle \sum_m \sqrt{\frac{2\lambda}{\pi r}} e^{i(\omega t - kr)} \right\rangle^2 \tag{15.7}$$

where λ is the wavelength of acoustic waves, ω is the angular frequency, k is the wave number, ρ is the density, c is the velocity of acoustic waves in the medium, m represents the number of probes, and A is an arbitrary constant.

Fixed-focus, large-aperture, bowl-shaped transducers incur several limitations during ablation procedures such as reduced flexibility of beam parameters, ineffective coupling for small targets, large treatment durations, and the presence of off-focal hot spots in the intervening tissue while scanning the beam to cover the target region. For minimizing unwanted damage to the normal tissue due to repeated overlapping of the beam energy while scanning a target region in 3-D and improved flexibility of the beam parameters, the use of multiple transducers has been proposed [27–30] by dividing the required power between two or more probes, such that the multiple beams superimpose in the same focal region. The spatial configurations of these probes with respect to each other and with respect to the target tissue were studied by developing 2-D and 3-D numerical models and verified in in vitro and ex vivo tests.

The 3-D model, by analogy with optical field theory, uses Fresnel-Kirchoff diffraction theory to evaluate the pressure amplitude at any location in front of the transducer by taking a double integral over the entire surface of the source as follows:

$$p = \frac{i\rho ck}{2\pi} U_0 \int_0^a R \left(\int_0^{2\pi} \frac{e^{i(\omega t - kr)}}{r} d\phi \right) dR \qquad (15.8)$$

where U_0 is the peak amplitude at the transducer face, a is the radius of the transducer, R is the radial location on the transducer face, ϕ is the angular location on the transducer face, and other symbols have their usual meanings as described earlier.

The computer simulations help in treatment planning for dosage levels and to determine the optimum range of interprobe distances and angles of orientation (defining size and shape and intensity in the superimposed foci). From the simulation studies, it is possible to decide on the range of suitable spatial configurations (location and orientation) of the probes based on the desired size, shape, and level of achievable intensities in specific configurations, as suited to a specific application. Probe positioning mechanisms and scanning of the focal region in the desired target is done using dedicated manipulator systems, called focused ultrasound surgery robots (FUSBOT), a representative example of which is described later in this chapter. Robotic/mechatronic assistance and imaging guidance yield higher accuracy, precision, reliability, and repeatability when manipulating surgical instruments in desired locations.

Phased-array transducers use the same basic acoustic principles as acoustic lenses that focus a sound beam. Variable delays are applied across the transducer aperture. The delays are electronically controlled and can be changed instantaneously to focus the beam in different regions. Phased arrays [25] can be used for electronically steering the beam over the area of interest without mechanically moving the transducer, and thus provide more flexibility in the shape and size of the resulting lesions. With phased arrays, various focal patterns can be planned. A treatment routine can be considered such that the entire tumor region is ablated by creating distant and nonadjacent ablations to minimize cooling time between sonications. However, the inherent disadvantages include increased complexity of scanning electronics (particularly for large arrays), higher cost of transducers and scanners, formation of constructive interference zones that might produce pseudofoci beyond the actual focal regions, and a large number of elements, all of which lead to greater control complexity. With improvement in array technology, both linear and 2-D arrays are now available that can provide high-quality HIFU applicators with increased bandwidth and reduced cross-coupling between the array elements.

There has been good growth of commercial devices and systems utilizing HIFU technology in the past decade. A commercial HIFU device called the Sonablate 200TM (Focal Surgery Inc. of Indianapolis, Milpitas, California) was developed for the treatment of benign prostatic hyperplasia (BPH) and prostate cancers. In the

Sonablate system, a 4-MHz transducer is used for both lesion induction and imaging purposes and can target treatment depths of 2.5 to 4.0 cm.

Another commercial device called the Ablatherm (Technomed International, Lyon, France) utilizes separate transducers: 7.5 MHz for imaging and 2.25 MHz for therapy purposes. This is also targeted for BPH treatment. Another version of this system also called Ablatherm (EDAP TMS S.A., France) is commercially available for treatment of prostate cancers. Insightec Ltd., Israel, commercialized an MRI-guided noninvasive HIFU surgery system called the ExAblate 2000; it has been FDA approved for the treatment of uterine fibroids. It utilizes MRI to visualize treatment planning and monitor the treatment outcomes in real time. A Model JC Haifu focused ultrasound based tumor therapeutic system has been developed by the Chonqing Haifu Technology Company, China, for clinical ablation of solid tumors. China Medical Technologies has developed a HIFU system initiated by the PRC Ministry of Health for treatment of solid tumor/cancers.

15.4 Focal Ultrasound Surgery Robots—FUSBOTs

According to the Robotic Institute of America, a robot is "a reprogrammable, multifunctional manipulator designed to move materials, parts, tools, or other specialized devices through various programmed motions for the performance of a variety of tasks." The essential components of a robotic system, akin to a human body, include the main structure, called a mechanical manipulator; an end-effector to carry, position, and orient working tool(s); internal and external sensors to acquire information about link positions and the environment; a central processing system for planning and controlling tool trajectories in a desired fashion within the robot's workspace; and power source(s) for supplying power to various parts.

The use of robotic principles in surgical assistance, particularly in the area of MIS, provides several advantages. These include higher accuracy, precision, and repeatability in positioning surgical tools and maneuvering controlled trajectories [27, 29, 30–32]. Minimally invasive and noninvasive surgical techniques have been—and remain—prime areas of research in the biomedical arena in the past two decades due to their numerous advantages over conventional surgery methods. Advanced manipulation and computational tools can be used in preplanning, registration, and navigation of surgical devices based on the image data so as to spare the surrounding and intervening healthy tissue. Most of the surgical robotic systems work in interactive semiautonomous or assistive modes under supervisory control of the surgeon.

At the Biomechatronics Research Group's Robotic Research Center, we have devised a range of medical robotic prototypes for focal ultrasound surgery (FUS), named FUSBOTs, such as for breast surgery (FUSBOT^{BS}), urological organ and stomach cancer surgeries using transabdominal and suprapubic routes (FUSBOT^{US}), and neurosurgery (FUSBOT^{NS}) for image-guided FUS applications.

The robotic manipulator design and thus kinematics and dynamics of its mechanical configuration are based on specific applications. Common features include real-time image guidance and interactive and supervisory control by the surgeon. Once the surgical protocol is decided in the preplanning phase, the robots accurately position the HIFU transducer(s) at specified locations coincident with the planned lesion position on a given 2-D image. During the procedure, a real-time lesion positioning and tracking algorithm updates the formation of the lesions in the desired target with a predictive temperature map processed from online imaging data. The dimensions and range of motions of the robotic system correspond to human anthropometry data. The potential of mechatronic/robotic assistance in the operating room and system overview of one representative of these systems, the FUSBOT^{BS}, for noninvasive ablative procedures is described next.

Remote ablation of deep-seated abnormalities by various modalities such as the use of HIFU can provide completely noninvasive procedures if the energy in the beam is carefully targeted. For breast applications, a custom designed five degrees-of-freedom (three for positioning, one for orientation of the end-effector, and one for imaging) robot, FUSBOT^{BS}, for guiding an end-effector through a pre-determined and image-guided trajectory has been developed and tested at our center (Figure 15.1).

The end-effector has a purpose-built jig for mounting the HIFU transducer(s) and it operates in a degassed water tank. HIFU probes are positioned such that the focal zone (of a single probe) or the joint focus (of multiple probes) overlaps within the affected target area. Fragmentation of energy into multiple low-energy beams helps to minimize hot spots in overlying structures [27, 28, 30, 33–35]. For a target tumor area larger in size than the focal zone of the beam, the HIFU probe(s) need to sweep the entire volume of the lesion in 3-D. The probe manipulation modules and robotic work envelope encompass the human torso region and thus are capable of reaching and treating cancers/tumors other than in the breast, such as through acoustic windows using transabdominal and suprapubic routes incorporating a sliding window opening at the top. The end-point compensated accuracy of this system is tested to be within ±0.2 mm. Various laboratory trials in tissue in vitro and ex vivo using the system validate its excellent precision and repeatability.

15.5 Concluding Remarks

The principles of operation for various modalities used in the thermal ablation of remote tissue have been discussed in this chapter. A comparison of methodologies and techniques was made where necessary. All of the ablative modalities presented

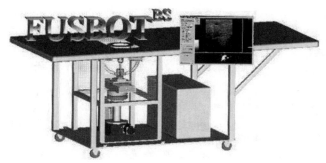

Figure 15.1 The FUSBOT^{BS} system for breast and abdominal route FUS applications.

here have been used in several clinical and preclinical applications; however, it was not the intent here to present a historic review of the clinical results.

Furthermore, online imaging is having a profound impact on ensuring better and safer surgical outcomes. It is an invaluable tool for planning and navigating a complex trajectory of surgical tools into a patient's anatomy in 3-D, through safer routes, and for monitoring the progress of the surgery. Temperature is a prime target parameter in ablative procedures. Monitoring the ablative procedure may help in assessing the tissue response to the thermal dose targeted in the region of interest. MR images show the changes in tissue characteristics induced by temperature elevation during ablation procedures. The use of B- or M-mode diagnostic ultrasound images is a common practice for localizing target lesions in various procedures. But during HIFU ablation sequences, barely any changes can be monitored in such images apart from echogenic regions formed by gaseous cavities/bubbles near cavitation thresholds. Such a situation, however, should be avoided in carefully controlled treatments.

A brief overview of novel image-guided surgical robotic systems, FUSBOTs, dedicated to FUS applications of various parts/organs of the human body, in particular, for breast surgery, was also provided in this chapter. To our knowledge, this is the first ever attempt to devise dedicated biomechatronic systems for FUS worldwide (including the previous study of the author for brain tissue ablation; Ph.D. thesis, Imperial College, London, 1999). The range of benefits reaped in other medical procedures by the use of robotic technology should be extended to noninvasive ablative procedures that share extended problems of image guidance, targeting, and control.

References

[1]   Lele, P. P., "Production of Deep Focal Lesions by Focused Ultrasound—Current Status," *Ultrasonics,* Vol. 5, 1967, pp. 105–112.

[2]   Bamber, J. C., and C. R. Hill, "Ultrasonic Attenuation and Propagation Speed in Mammalian Tissues as a Function of Temperature," *Ultrasound Med. Biol.,* Vol. 5, 1979, pp. 149–157.

[3]   Frizzell, L. A., "Threshold Dosages for Damage to Mammalian Liver by High-Intensity Focused Ultrasound," *IEEE Trans. on Ultrasonics, Ferroelectrics, and Frequency Control,* Vol. 35, 1988, pp. 578–581.

[4]   Fry, F. J., and L. K. Johnson, "Tumor Irradiation with Intense Ultrasound," *Ultrasound Med. Biol.,* Vol. 4, 1978, pp. 337–341.

[5]   Goss, S. A., and F. J. Fry, "The Effects of High-Intensity Ultrasonic Irradiation on Tumour Growth," *IEEE Trans. on Sonics Ultrasonics,* Vol., SU-31, 1984, pp. 491–496.

[6]   ter Haar, G. R., D. Sinnett, and I. H. Rivens, "High-Intensity Focused Ultrasound—A Surgical Technique for the Treatment of Discrete Liver Tumours," *Phys. Med. Biol.,* Vol. 34, 1989, pp. 1743–1750.

[7]   Rossi, S., et al., "Thermal Lesions Induced by 480 KHz Localized Current Field in Guinea Pig and Pig Liver," *Tumori,* Vol. 76, 1990, p. 54.

[8]   Livraghi, T., et al., "Small Hepatocellular Carcinoma: Treatment with Radio-Frequency Ablation Versus Ethanol Injection," *Radiology,* Vol. 210, 1999, pp. 655–661.

[9]   Siperstein, A. E., and E. Berber, "Cryoablation, Percutaneous Alcohol Injection and Radiofrequency Ablation for Treatment of Liver Metastases," *World J. Surg.,* Vol. 25, 2001, pp. 693–696.

[10] Crews, K. A., et al., "Cryosurgical Ablation of Hepatic Tumors," *Am. J. Surg.,* Vol. 174, 1997, p. 614.

[11] O'Berg, K., "The Use of Chemotherapy in the Management of Neuroendocrine Tumors," *Endocrinol. Metab. Clin. North Am.,* Vol. 22, 1993, p. 941.

[12] Giovannini, M., and J. F. Seitz, "Ultrasound-Guided Percutaneous Alcohol Injection of Small Liver Metastases," *Cancer,* Vol. 73, 1994, p. 294.

[13] Climent, V., et al., "Effects of Endocardial Microwave Energy Ablation," *Indian Pacing Electrophysiol. J.,* Vol. 5, No. 3, 2005, pp. 233–243.

[14] Siperstein, A. E., et al., "Laparoscopic Thermal Ablation of Hepatic Neuroendocrine Tumor Metastases," *Surgery,* Vol. 122, 1997, p. 1147.

[15] Heisterkamp J., et al., "Importance of Eliminating Portal Flow to Produce Large Intrahepatic Lesions with Interstitial Laser Coagulation," *Br. J. Surg.,* Vol. 84, 1997, pp. 1245–1248.

[16] de Jode, M. G., et al., "MRI Guidance of Infra-Red Laser Liver Tumour Ablations, Utilising an Open MRI Configuration System: Technique and Early Progress," *J. Hepatol,* Vol. 31, 1999, pp. 347–353.

[17] Morrison, P. R., et al., "MRI of Laser-Induced Interstitial Thermal Injury in an In Vivo Animal Liver Model with Histologic Correlation," *J. Magn. Resonance Imaging,* Vol. 8, 1998, pp. 57–63.

[18] Madersbacher, S., et al., "Tissue Ablation in Benign Prostatic Hyperplasia with High-Intensity Focused Ultrasound," *Eur. Urol.,* Vol. 23, Suppl. 1, 1993, pp. 39–43.

[19] Hynynen, K., et al., "MR Imaging-Guided Focused Ultrasound Surgery of Fibroadenomas in the Breast: A Feasibility Study," *Radiology,* Vol. 219, No. 1, 2001, pp. 176–185.

[20] Koehrmann, K. U., et al., "Technical Characterization of an Ultrasound Source for Non-Invasive Thermoablation by High Intensity Focused Ultrasound," *Br. J. Urol.,* Vol. 90, 2002, pp. 248–252.

[21] Prat, F., et al., "Extracorporeal High-Intensity Focused Ultrasound for VX2 Liver Tumours in the Rabbit," *Hepatology,* Vol. 21, 1995, pp. 832–836.

[22] Wu, F., et al., "Extracorporeal High-Intensity Focused Ultrasound for Treatment of Solid Carcinomas: Four-Year Chinese Clinical Experience," *Proc. 2nd Int. Symp. Therapeutic Ultrasound,* Seattle, WA, July 29–August 1, 2002.

[23] Nakamura, K., et al., "High-Intensity Focused Ultrasound Energy for Benign Prostatic Hyperplasia: Clinical Response at 6 Months to Treatment Using Sonablate 200," *J. Endourol.,* Vol. 11, 1997, pp. 197–201.

[24] Chaussy, C., et al., "HIFU and Prostate Cancer: The European Experience," *Proc. 2nd Int. Symp. Therapeutic Ultrasound,* Seattle, WA, July 29–August 1, 2002.

[25] Clement, G. T., J. White, and K. Hynynen, "Investigation of a Large Area Phased Array for Focused Ultrasound Surgery Through the Skull," *Phys. Med. Biol.,* Vol. 45, 2000, pp. 1071–1083.

[26] O'Niel, H. T., "Theory of Focusing Radiators," *J. Acoust. Soc. Am.,* Vol. 21, 1949, pp. 516–526.

[27] Chauhan, S., M. J. S. Lowe, and B. L. Davies, "A Multiple Focused Probe Approach for HIFU Based Surgery," *J. Ultrasonics Med. Biol.,* Vol. 39, 2001, pp. 33–44.

[28] Haecker, A., et al., "Multiple Focused Probes for High Intensity Focused Ultrasound: An Experimental Investigation," *Proc. 20th World Congress on Endourology and Shockwave,* Genoa, Italy, September 2002, pp. 19–22.

[29] Davies, B. L., S. Chauhan, and M. J. Lowe, "A Robotic Approach to HIFU Based Neurosurgery," in *Lecture Notes in Computer Science 1496,* Berlin: Springer, October 1999, pp. 386–396.

[30] Chauhan, S., *A Mechatronic System for Non-Invasive Treatment of the Breast Tissue, Mechatronics and Machine Vision 2002: Current Practice,* London, U.K., Research Studies Press Ltd., pp. 359–366.

[31] Taylor, R. H., et al., "An Image-Directed Robotic System for Precise Orthopaedic Surgery," *IEEE Trans. on Robotics and Automation,* Vol. 10, No. 3, 1994, pp. 261–275.

[32] Lanfranco, A. R., et al., "Robotic Surgery: A Current Perspective," *Annals of Surgery,* Vol. 233, No. 1, January 2004, pp. 14–21.

[33] Chauhan, S., M. Y. Teo, and W. Teo, "Robotic System for Ablation of Deep-Seated Skull Base Cancers—A Feasibility Study," *Proc. 34th Int. MATADOR Conf.,* Manchester, U.K., July 7–9, 2004.

[34] Chauhan, S., "FUSBOT—An Ultrasound Guided Robot for Focal Ultrasound Surgery," *Proc. Int. Symp. Therapeutic Ultrasound,* Kyoto, Japan, September 18–20, 2004.

[35] Häcker, A., et al., "Multiple HIFU-Probes for Kidney Tissue Ablation," *Journal of Endourology,* Vol. 13, No. 8, 2005, pp. 1036–1040.

About the Editors

Jasjit Suri has spent more than 24 years in imaging sciences, especially in medical imaging modalities and its fusion. He has published more than 200 technical papers (no abstracts) in body imaging, specifically MR, CT, x-ray, ultrasound, PET, SPECT, elastography, and molecular imaging. Dr. Suri has worked at Siemens Research, Philips Research, Fischer Research Divisions, and Eigen's Imaging Divisions as a scientist, senior director of R&D, and chief technology officer. He has submitted more than 30 U.S./European patents on medical imaging modalities. Dr. Suri has also written 17 collaborative books on body imaging. He is a lifetime member of Tau Beta Pi, Eta Kappa Nu, and Sigma Xi and is a member of NY Academy of Sciences, the Engineering in Medicine and Biology Society, the American Association of Physics in Medicine, the SPIE, and the ACM. Dr. Suri is also a Senior Member of the IEEE. He is on the editorial board or a reviewer of *Real Time Imaging*, *Pattern Analysis and Applications*, *Engineering in Medicine and Biology* Magazine, *Radiology*, *Journal of Computer Assisted Tomography*, *IEEE Transactions on Information Technology in Biomedicine*, and *IASTED Imaging Board*. He has also chaired biomedical imaging tracks at several international conferences and has given more than 40 international presentations and seminars. Dr. Suri has been listed in *Who's Who* and was a recipient of the President's Gold Medal in 1980. He is also a Fellow of American Institute of Medical and Biological Engineering and ABI. He is visiting faculty at the University of Exeter, Exeter, United Kingdom, the University of Barcelona, Spain, and Kent State, Ohio, and is the director of the medical imaging division at Jebra Wellness Technologies. Dr. Suri is on board of directors of Biomedical Technologies Inc. and Cardiac Health Inc. He received a B.S. in computer engineering from Maulana Azad College of Technology, Bhopal, India, an M.S. in computer sciences from the University of Illinois, Chicago, a Ph.D. in electrical engineering from University of Washington, Seattle, and an M.B.A. from the Weatherhead School of Management, Case Western Reserve University, Cleveland.

Chirinjeev Kathuria has a B.Sc., with a specialization in U.S. health care policy and administration, and an M.D. from Brown University. He also has an M.B.A. from Stanford University. Dr. Kathuria has had measurable success in building businesses that impact world economies and shift business models. He has cofounded and helped build many businesses, which have generated shareholder wealth and jobs. Dr. Kathuria and his affiliated companies have been featured on many television programs and in media publications. He has extensive experience in the health care industry and has consulted to a broad range of organizations in the United States, Europe, and Asia. Dr. Kathuria also helped develop Arthur D. Little biotechnology and health care policy practice in Europe. He conducted a comparative analysis of the European and U.S. biotechnology industries resulting in a paper entitled "Biotechnology in the Uncommon Market," which was published in Biotechnology magazine in December 1992 which helped change at that time the current thinking of biotechnology development. Dr. Kathuria's coauthored papers include "Selectivity Heat Sensitivity of Cancer Cells," "Avascular Cartilage as an Inhibitor to Tumor Invasion," and "Segmentation of Aneurysms Via Connectivity from MRA Brain Data," the latter of which was published in the Proceedings of the International Society for Optical Engineering in 1993.

Ruey-Feng Chang received a B.S. in electrical engineering from National Cheng Kung University, Tainan, Taiwan, ROC, in 1984, and an M.S. in computer and decision sciences and a Ph.D. in computer science from National Tsing Hua University, Hsinchu, Taiwan, ROC, in 1988 and 1992, respectively. Since 2006, he has been a professor in the Department of Computer Science and Information Engineering, National Taiwan University, Taipei, Taiwan, ROC. He was a professor from 1992 to 2006 and served as the chair from 2003 to 2006 of the Department of Computer Science and Information Engineering, National Chung Cheng University, Chiayi, Taiwan, ROC. His research interests include medical computer-aided diagnosis systems, image/video processing and retrieval, and multimedia systems and communication. Dr. Chang is a member of the IEEE, ACM, SPIE, and Phi Tau Phi. He received the Distinguished Research Award from the National Science Council of Taiwan in 2004.

Filippo Molinari received the Italian Laurea and a Ph.D. in electrical engineering from the Politecnico di Torino, Torino, Italy, in 1997 and 2000, respectively. Since 2002, he has been an assistant professor on the faculty of the Dipartimento di Elettronica, Politecnico di Torino, where he teaches biomedical signal processing, biomedical image processing, and instrumentation for medical imaging. His main research interests include the analysis of strongly nonstationary biomedical signals and the medical imaging applied to the computer-aided diagnosis. Dr. Molinari developed several signal and image processing algorithms, especially in the field of neurosciences and in the functional assessment of disabled subjects.

Aaron Fenster is a scientist and the founding director of the Imaging Research Laboratories at the Robarts Research Institute, the chair of the Imaging Sciences Division of the Department of Radiology & Nuclear Medicine, the associate director of the Graduate Program in Biomedical Engineering at The University of Western Ontario, and the director of the Imaging Pipeline Platform of the Ontario Institute for Cancer Research. Currently, he holds a Canada Research Chair-Tier 1 in biomedical engineering. He was the recipient of the 2007 Premier's Award for Innovative Leadership. Dr. Fenster's group has focused on the development of 3-D ultrasound imaging with diagnostic and surgical/therapeutic applications. His team developed the world's first 3-D ultrasound imaging of the prostate and 3-D ultrasound guided prostate cryosurgery and brachytherapy. This technology has been licensed to commercial companies that are distributing it worldwide for cancer therapy and research. Dr. Fenster's research has resulted in 190 peer-reviewed publications, 32 patents, and the formation of two companies, for which he has been a founding scientist.

List of Contributors

Viren Amin
Center for Nondestructive Evaluation
283 Applied Sciences Complex II
1915 Scholl Road
Ames, IA 50011-3042
e-mail: vamin@cnde.iastate.edu

Elsa D. Angelini
Institut Telecom
Telecom ParisTech
Department of Signal and Image Processing (TSI)
46 rue Barrault
73013 Paris
France
e-mail: elsa.angelini@telecom-paristech.fr

Sergio Badalamenti
Neurology Service
Ospedale Gradenigo
Corso Regina Margherita 10
10153, Torino
Italy
e-mail: sergio.badalamenti@h-gradenigo.it

Ruey-Feng Chang
Department of Computer Science and Information
Engineering and Graduate
Institute of Biomedical Electronics and
 Bioinformatics

Sunita Chauhan
Division of Mechatronics and Design
School of Mechanical and Aerospace Engineering
Nanyang Technological University, North Spine
(N3) Level 2
50 Nanyang Avenue
Singapore 639798
e-mail: mcsunita@ntu.edu.sg

Chii-Jen Chen
Department of Computer Science and Information
Engineering

Dar-Ren Chen
Department of Surgery
Changhua Christian Hospital
Changhua,
Taiwan 500
R.O.C.

Kau-Yih Chiou
Department of Computer Science and Information
Engineering

Bernard Chiu
Robarts Research Institute
100 Perth Drive
London, ON, N6A 5K8
Canada
e-mail: bchiu@imaging.robarts.ca

Kevin D. Costa
Department of Biomedical Engineering
Columbia University ET351
1210 Amsterdam Avenue
New York, NY 10027
United States
e-mail: kdc17@columbia.edu

Christakis Damianou
Frederick University Cyprus
7 Yianni Frederickou
1036, Nicosia
Cyprus
e-mail: cdamianou@cytanet.com.cy

Silvia Delsanto
im3D S.p.A.
Medical Imaging Lab
Via Lessolo 3
10153, Torino
Italy
e-mail: silvia.delsanto@i-m3d.com

Donal Downey
Medical Imaging Department
Royal Inland Hospital
311 Columbia Street, Kamloops
BC, V2C 2T1
Canada
e-mail: ddowney@imaging.robarts.ca

Qi Duan
Department of Biomedical Engineering
Columbia University ET351
1210 Amsterdam Avenue
New York, NY 10027
United States
e-mail: duanqi@gmail.com

Michaela Egger
Robarts Research Institute
100 Perth Drive
London, ON, N6A 5K8
Canada

Marcelo D. Faria
National Laboratory for Scientific Computing (LNCC)
Rio de Janeiro, Brazil

Aaron Fenster
Robarts Research Institute
100 Perth Drive
London, ON, N6A 5K8
Canada
e-mail: afenster@imaging.robarts.ca

Olivier Gerard
Philips France
51 rue Carnot
BP 301
Suresnes, 92156

France
e-mail: olivier.gerard@philips.com

Gilson A. Giraldi
National Laboratory for Scientific Computing (LNCC)
Rio de Janeiro, Brazil

Pierangela Giustetto
Neurology Service
Ospedale Gradenigo
Corso Regina Margherita 10
10153, Torino
Italy
e-mail: pgiustetto@gmail.com

Jeffrey W. Holmes
Robert M. Berne Cardiovascular Research Center
University of Virginia
Box 800759, Health System
Charlottesville, VA 22908
e-mail: holmes@virginia.edu

Shunichi Homma
Department of Medicine, Columbia University
Cardiology Division
Ph. 9 East 111
622 W. 168 Street
New York, NY 10032
United States
e-mail: sh23@columbia.edu

Manivannan Jayapalan
c/o Professor M. Ramasubba Reddy
Biomedical Eng. Division
Department of Applied Mechanics
Indian Institute of Technology Madras
Chennai 600036
India
e-mail: jmvannan@yahoo.com,
  manivannanj@smail.iitm.ac.in

Wei-Ren Lai
Department of Computer Science and Information
Engineering
National Chung Cheng University
Chiayi
Taiwan 621, R.O.C.

Andrew F. Laine
Department of Biomedical Engineering
Columbia University ET351
1210 Amsterdam Avenue
New York, NY 10027
United States
e-mail: laine@columbia.edu

Anthony Landry
Robarts Research Institute
100 Perth Drive
London, ON, N6A 5K8
Canada

William Liboni
Neurology Service
Ospedale Gradenigo
Corso Regina Margherita 10
10153, Torino
Italy
e-mail: william.liboni@h-gradenigo.it

C. P. Loizou
Department of Computer Science and Engineering

Intercollege
P.O. Box 51604
3507 Limassol
Cyprus
e-mail: loizou.c@lim.intercollege.ac.cy,
  panloicy@logos.net.cy

Filippo Molinari
Biolab
Department of Electronics
Politecnico of Torino
Corso Duca degli Abruzzi 24
10129 Torino
Italy
e-mail: filippo.molinari@polito.it

Woo Kyung Moon
Department of Radiology and Clinical Research
Institute
Seoul National University Hospital
Seoul
Korea
e-mail: moonwk@radcom.snu.ac.kr

Grace Parraga
Robarts Research Institute
100 Perth Drive
London, ON, N6A 5K8
Canada
e-mail: gep@imaging.robarts.ca

Marcia Provenzano
National Laboratory for Scientific Computing (LNCC)
Rio de Janeiro, Brazil

Paulo S. Rodrigues
National Laboratory for Scientific Computing (LNCC)
Rio de Janeiro, Brazil

J. David Spence
Siebens-Drake Building
1400 Western Road
London, ON, N6G 2V2
Canada
e-mail: dspence@robarts.ca

Jasjit S. Suri
Eigen LLC
Grass Valley, CA
Idaho Biomedical Research Institute (IBRI)
Idaho State University
Pocatello, ID 83209
United States
e-mail: jsuri@comcast.net

Zhouping Wei
PET Research & Clinical Sciences
Philips Medical Systems
595 Miner Road
Cleveland, OH 44143
United States
e-mail: zhouping.wei@philips.com

Fuxing Yang
Verathon (formerly Diagnostic Ultrasound Corporation)
20001 North Creek Parkway
Bothell, WA 98011
United States
e-mail: fuxing.yang@gmail.com

Index

Related Titles from Artech House

Biomolecular Computation for Bionanotechnology, Jian-Qin Liu and Katsunori Shimohara

Electrotherapeutic Devices: Principles, Design, and Applications, George D. O'Clock

Fundamentals and Applications of Microfluidics, Second Edition, Nam-Trung and Steven T. Wereley

Matching Pursuit and Unification in EEG Analysis, Piotr Durka

Micro and Nano Manipulations for Biomedical Applications, Tachung C. Yih and Ilie Talpasanu, editors

Microfluidics for Biotechnology, Jean Berthier and Pascal Silberzan

For further information on these and other Artech House titles, including previously considered out-of-print books now available through our In-Print-Forever® (IPF®) program, contact:

Artech House
685 Canton Street
Norwood, MA 02062
Phone: 781-769-9750
Fax: 781-769-6334
e-mail: artech@artechhouse.com

Artech House
46 Gillingham Street
London SW1V 1AH UK
Phone: +44 (0)20 7596-8750
Fax: +44 (0)20 7630-0166
e-mail: artech-uk@artechhouse.com

Find us on the World Wide Web at: www.artechhouse.com